ACTIVITY-BASED COST AND ENVIRONMENTAL MANAGEMENT

A Different Approach to ISO 14000 Compliance

ACTIVITY-BASED COST AND ENVIRONMENTAL MANAGEMENT
A Different Approach to ISO 14000 Compliance

by

Jan Emblemsvåg
Considium Consulting Group AS, Norway

and

Bert Bras
Georgia Institute of Technology, U.S.A.

KLUWER ACADEMIC PUBLISHERS
Boston / Dordrecht / London

Distributors for North, Central and South America:
Kluwer Academic Publishers
101 Philip Drive
Assinippi Park
Norwell, Massachusetts 02061 USA
Telephone (781) 871-6600
Fax (781) 681-9045
E-Mail <kluwer@wkap.com>

Distributors for all other countries:
Kluwer Academic Publishers Group
Distribution Centre
Post Office Box 322
3300 AH Dordrecht, THE NETHERLANDS
Telephone 31 78 6392 392
Fax 31 78 6546 474
E-Mail <services@wkap.nl>

Electronic Services <http://www.wkap.nl>

Library of Congress Cataloging-in-Publication Data

Activity based cost and environmental management: a different approach to ISO 14000 compliance / edited by Jan Emblemsvåg and Bert Bras.
p. cm.
Includes bibliographical references
ISBN 0-7923-7247-6 (alk. Paper)
1. Industrial management – Environmental aspects. 2. Environmental management.
3. Cost accounting. 4. Decision making. I. Emblemsvåg, Jan II. Bras, Bert

HD30.255 .E477 2000
658.4'.08--dc21

00-049331

Printed on acid-free paper. Printed in the United States of America

DEDICATION

To those that have a vision; a desire to plan ahead,
towards a better future where we share the bread.
To those that have a drive to counter the illusions,
of societies so devoid from Nature's contributions.

To those that see beyond the numbers that we make,
and want to hear the sages and ask for advice to take.
To those that see the need for a better paradigm,
that stretches out for more than every little dime.

To those that want to see a change in current way
away from fixing problems on a path that lead astray.
To those that have discovered the futility of control
in trying to fix the course after the ball is let to roll.

Yes, many are those that think, read and study well,
but few are those that stand and support you if you fell.
To these we dedicate the essence of our labors,
'cause they are the only true, everlasting neighbors.

To our closest families whose support have been so keen,
without them we often wonder; where would we have been?
To the greatest one we know, the One that truly is,
whose force has kept us going by a steady light of bliss.

CONTENTS

It's easier to resist at the beginning than at the end.

Leonardo Da Vinci

PREFACE

Concerns for man and his fate must form the chief interest of all technical endeavors. Never forget this in the midst of your diagrams and equations.

Albert Einstein

Anything dealing with the environment is difficult. It almost immediately arouses emotional debates and arguments from everybody under the sun. Although we are not even close to understanding the full complexity of the interwoven web of what constitutes what we call 'our environment', one thing is clear: *our environment is changing*, and not for the better. And many agree; redirecting this course requires not some initiatives of the few, but full efforts of the many, and a change in our behaviors and attitudes towards the environment.

Many books have been written suggesting all kinds of approaches to bring about this change, but none has really induced wide-ranging self-motivated and self-sustaining efforts in industry. In fact, the field of environmental management seems to be riddled with jargon and unfamiliar terms that result in high learning curves, frustration, and little action. In the last decade of the 20th century, the International Standards Organization (ISO) has thrown its hat into the environmental ring by producing the ISO 14000 environmental management standards. The aim is to provide clarity and uniformity, but many agree that it still leaves a lot to be desired.

We believe that if environmental management is ever to really induce wide-ranging, self-motivated and self-sustaining efforts in industry, it must connect to the basic motivators for business: *increased competitiveness and profitability*. In this book, we show you *how* to do this, and that is what makes this book unique. Rather than creating a new, unfamiliar, almost mystical environmental assessment and management framework, this book starts directly from familiar principles: *cost accounting and cost management*, specifically Activity-Based Costing and Management.

This book is about demonstrating how you can combine the well-established Activity-Based Costing and Management principles with non-monetary environmental assessment and performance measurement dimensions, and create a single framework that gives you the tools and ability to do *both* environmental management and cost management. We call this approach *Activity-Based Cost and Environmental Management* (ABCEM). It is a simple idea, but you will be surprised how powerful the results are. As the Chinese say - 'simplicity of character is the result of profound thought'.

Our goal with this is simple: We want to give you the ability to include the environment in your decision-making. We want you to realize that competitiveness can be increased *and* environmental impact reduced at the same time. And we want to show *how* you can do it. Hence, this book is *not* about 'gloom and doom'. This book is *not* full of environmental rhetoric.

This book is about how you can empower yourself to make a difference and bring about a change for the better for both your organization and the environment.

We are engineers from background and you may wonder why engineers are writing about Activity-Based Cost and Environmental Management. Well, as engineers, we always think in (physical) processes (activities) and resources, and we like integration, consistency, simplicity and robustness, among others. And, we like to measure, not just 'count'. One day in 1993, while we were going over the 'Future Work' section of a draft of a paper called Activity-Based Costing in Design for Product Retirement, we started wondering about the whole Life-Cycle Assessment mess and the total lack of accepted consistent principles. Accounting is so much better in that respect... Wait a minute... What if we approached environmental assessments from monetary accounting perspectives and principles? *Voila*, the concept of integrated activity-based cost and environmental assessment was born. In this book, we are going one step further and bring in the 'M' word: Management. And stay tuned because there is more to come. For example, what do you think about real-time activity-based economic and environmental performance monitoring of whole production facilities...?

But let us not jump ahead. The primary aim of *this* book is two-fold: 1) we want to educate you about the basic concepts and 2) we want to educate you on how to make your own integrated Activity-Based Cost and Environmental Management implementations. Although it may seem daunting at times, it really is not. And if you do not like the environmental part, at least you will have a good costing system...

A fair warning and invitation to you, readers. Although each of us has written many publications, including dissertations, this is our first 'real' book. If you see or miss something that you would like to give us feedback about, please do. Learning is a life-long experience for all of us.

And as you read this book, you will see quotes from various famous and less famous people from all over the world. Some quotes are meant to humor you, but all of them are to make you think, dream and go beyond, and so is this book.

Jan and Bert

ACKNOWLEDGEMENTS

Happy people are always successful; successful people are not always happy.

Farrokh Mistree

First of all, we would like to thank the Ph.D. committee of Jan Emblemsvåg - consisting of Professors Farrokh Mistree, Steven Liang, C.S. Kiang, Arnold Schneider and Bert Bras (Chair) of the Georgia Institute of Technology and Ray C. Anderson of Interface, Inc. - for recommending that Jan's work be presented in a further publication. As a result, we would like to thank Kluwer Academic Publishers and its editors for their support in publishing this book.

We would like to thank all our colleagues - locally, nationally, and internationally - who have been so supportive of our work. There are too many to name, but we definitely thank those who favorably reviewed our book proposal late 1999 (you know who you are).

At Georgia Tech, past and present students and faculty of the Systems Realization Laboratory, where most of this work was done, deserve some special recognition, in particular Farrokh Mistree who has always been more for us than 'just' a great mentor.

We are also very grateful for the help from 'Herr Director' Tom Graver in proofreading our manuscript and the valuable feedback he provided. He also holds the distinguished record of being the first person who actually used this book already before it was in print to develop an Activity-Based Cost, Energy and Waste model for a local Atlanta company.

From the many other people at Georgia Tech, we like to give some special recognition to Carol Carmichael and Jean-Lou Chameau for their early vision, motivation, and continuous interest and efforts in promoting sustainability at Georgia Tech and beyond. You make a difference.

Many thanks need to go to the companies that we have been so fortunate to work with. Without their support, cooperation, openness and willingness to take a risk, we would have accomplished much less. Some special thanks need to go to some key people of these companies.

At Stokke Gruppen we need to thank CEO Kjell Storeide and Member of the Board Nils Høegh-Krohn for reading our work and recommending it further into the Stokke system in the first place, which ultimately resulted in case studies with Stokke Fabrikker and Westnofa Industrier. At these companies, the support from CFOs Harald Brathaug and Geir Løseth at Stokke Fabrikker while at Westnofa Industrier the support from particularly CFO Jørn Nes (now CEO) and CEO Steinar Loe was indispensable.

We also gratefully acknowledge the invitation from Annik Magerholm Fet at Møre Research to participate in the project with Farstad Shipping. Once that project was started, the help from Technical Managers Jan Henry Farstad and Bjarne Nygaarden was invaluable.

At Interface Flooring Systems, we owe many thanks to CEO and Chairman of the Board Ray C. Anderson, President of Research Mike Bertolucci, and Vice-President of Engineering Dave Gustashaw. Their continued push towards Sustainable Development is unrivaled and motivates us all. Dave Gustashaw deserves some special recognition, not just for his true Southern hospitality, but also for his absolutely unbridled enthusiasm and interest in our mutual endeavors to make sustainability not just an academic theory, but a common practice.

And then there are those that may not have participated in a case study, but provided the early incentives for us to keep moving on. Among those we recognize the always enjoyable interactions with the folks at the Chrysler Corporation, now DaimlerChrysler, specifically Susan Yester, Monica Prokopyshen, and Gerald Winslow. Bill Hoffman at Motorola also deserves recognition for making us realize very early that we were on to something with our activity-based approach.

Also, the financial support from the U.S. - Norway Fulbright Foundation for Educational Exchange, the Research Council of Norway, the National Science Foundation, and Georgia Institute of Technology has kept us fed over the years and made it possible for this project to come this far.

Finally, we would like to thank our families and friends for patiently supporting us during the long hours before completion; we know that it looked like we had disappeared for a while. But now we are back - let's celebrate!

GLOSSARY

When we try to pick out anything by itself, we find it hitched to everything else in the universe.
John Muir

This book presents a multi-disciplinary work and terminology can vary in between disciplines. We use terminology as defined in this glossary. We have tried to use terms consistent with modern management practices like Activity-Based Management, Strategic Cost Management, Total Quality Management, and the ISO 14000 Environmental Management Standards.

Absorption: The extent to which a model traces the relevant resources to the assessment objects. Full absorption models trace *all* relevant resources.

Accountable: In our context, this is the ability to keep accounting information for a given assessment dimension.

Accounting: A statement of debits and credits (Webster 1983).

Action: Every little operation in an activity. For example, for a sewing activity some of the actions may be; getting fabric, measure, sew, stitch, turn around, sew, etc.

Action Chart: A chart of detailed actions associated with an assessment object, aggregated into a single performance measure, e.g., a chart with specific assembly actions capturing assembly time.

Activity: A group of logically connected actions form an activity. E.g., the actions associated with overhauling an engine component can be treated as an activity. Activities can be aggregated into hierarchies or simply into one big hierarchy depending on what is useful.

Activity-Based Costing (ABC): A methodology that measures the cost and performance of activities, resources and cost objects. Resources are assigned to activities, then activities are assigned to cost objects based on their use. Activity-Based Costing recognizes the casual relationships of cost drivers to activities (Raffish and Turney 1991). ABC adopts an attention focusing, long-term, resource consumption orientation (Cooper 1990b).

Activity-Based Management (ABM): A discipline that focuses on the management of activities as the route to improving the value received by the customer and the profit achieved by providing this value. This discipline includes cost driver analysis, activity analysis and performance measurement. ABM draws on Activity-Based Costing as its major source of information (Brinker 1997).

Activity Cost: Sum of all (cost) drivers associated with the activity multiplied by their corresponding consumption intensities.

Activity Energy Consumption: Sum of all energy drivers associated with the activity multiplied by their corresponding energy consumption intensities.

Activity Waste Generation: Sum of all waste drivers associated with the activity multiplied by their corresponding waste generation intensities.

Activity Driver: A measure of the consumption of an activity by a) another activity or b) an assessment object. Activity drivers that measure the consumption by an assessment object are also referred to as 'final' activity drivers, whereas activity drivers that measure consumption of activities by other activities are also called 'intermediate' activity drivers. Examples of activity drivers are the amount of labor, the weight of a product, the number of products, etc.

Allocation: Definitions are based on (Brinker 1997). See also 'Tracing'.

1. An apportionment or a distribution.
2. A process of assigning cost to an activity or a cost object when a direct measure does not exist. For example, assigning the cost of power to a machine activity by means of machine hours is an allocation, because 'machine hours' is an indirect measure of power consumption. Allocations can often be converted to tracing by additional measurements. Instead of using machine hours to allocate power consumption, a company can place a power meter on machines to measure actual power consumption.

There is considerable confusion about this topic due to the early descriptions of an ABC system by Cooper and Kaplan as a system to get 'more accurate fully-absorbed unit costs' when it in essence is a 'contribution margin approach'. The reason for this confusion is that allocation used in volume-based costing systems reflects an arbitrary assignment, see (Kaplan 1992), and is therefore irrelevant for decision making, whereas in ABC it reflects an *estimation*.

Allocation Base: Unit-level product characteristics. The term is used in volume-based costing systems.

Assessment Dimension: The fundamental quantities of interest in an assessment. In this book, the principal assessment dimensions are costs, which can be measured as USD [$], energy consumption [MJ] and waste generation, which can be measured in several ways. Other dimensions such as quality, which can also be measured in several ways and time measured in e.g., hours can be added.

Assessment Object: Any customer, service, organization, project, or product for which separate cost/revenue and/or energy consumption and/or waste generation assessments are needed.

Assessment Object Cost: Sum of all activity drivers associated with the assessment object multiplied by their corresponding cost consumption intensities.

Assessment Object Energy Consumption: Sum of all activity drivers associated with the assessment object multiplied by their corresponding energy consumption intensities.

Assessment Object Waste Generation: Sum of all activity drivers associated with the assessment object multiplied by their corresponding waste generation intensities.

Assumption Cell: A cell in a spreadsheet in which an input is given with an associated assumed uncertainty distribution, representing a source variable.

Consumption Intensity: Unit-'price' of a driver (Cooper 1990a), e.g., dollars per direct labor hour, mega-joules per square yard.

Cost Driver: Any factor that causes a change in the cost of an activity. Raw material quality, number of vendors, employee training, complexity of assembly are all examples of cost drivers. Cost drivers are used in the process view of Activity-Based Costing/Management to identify the root cause of the work and cost of an activity. See also Energy Driver and Waste Driver

Cost Object: Any customer, service, organization, project, flux or product for which separate cost/revenue assessments are needed. See also assessment object.

Critical Assumption Planning (CAP): The process of planning for the critical success factors (see 'Critical Success Factor') to improve performance while reducing uncertainty and risk to a minimum.

Critical Success Factor: The factors that have the greatest impact on the chosen performance measures.

Cross-Consumption: The consumption of an activity by an assessment object causes lower activity consumption by another assessment object. This effect typically occurs when some assessment objects have dominating consumption levels.

Cycle Time: The time that a product is in the production process (Dodd 1997).

Deterministic: The values of the variables are known with 100% certainty.

Driver: Any factor that causes a change in the cost, energy consumption and/or waste generation of an activity. It is in other words a generalization of the term cost driver (see cost driver).

Economic Energy Efficiency (EEE): The EEE is defined as the total energy consumption, see Energy Content, of an assessment object divided by the total costs for the same object; inventory effects are eliminated. The EEE should be as low as possible. The EEE is therefore a relative measure of how well an organization utilizes energy in relation to the resource usage.

Economic Waste Efficiency (EWE): The EWE is defined as the total waste generation of an assessment object divided by the total costs for the same object where inventory effects are eliminated. The EWE is a relative measure of how much waste an organization generates in relation to the resource usage.

Effectiveness: A measure of quality of a decision (correctness, completeness, comprehensiveness) that is made by a designer (Mistree, Smith et al. 1990).

Efficiency: A measure of the swiftness with which information, that can be used by a designer to make decisions, is generated (Mistree, Smith et al. 1990).

Energy: The actual energy released when using/consuming a resource. For example, the heat derived from burning fuel, the kinetic energy from a moving object.

Energy Content or Embodied Energy: The sum of all the energy expenditures in the value chain up to the current point for a resource, e.g., the sum of the energy needed to drill, refine, and transport gas to a gas station for use in a car. A natural phenomenon (e.g. solar heat) has no energy content by definition.

Energy Driver: Any factor that causes a change in the energy consumption of an activity. See also Cost Driver and Waste Driver.

Environment: Surroundings in which an organization operates, including air, water, land, natural resources, flora, fauna, humans, and their interrelation (ISO 14001).

Environmental Aspect: Element of an organization's activities, products or services that can interact with the environment (ISO 14001).

Environmental Impact: Any change to the environment, whether adverse or beneficial, wholly or partially resulting from an organization's activities, products or services (ISO 14001).

Environmental Management System: The part of the overall management system that includes organizational structure, planning activities, responsibilities, practices, procedures, processes and resources for developing, implementing and maintaining the environmental policy (ISO 14001).

Environmental Objective: Overall environmental goal, arising from the environmental policy, that an organization sets itself to achieve, and which is quantifiable where practicable (ISO 14001).

Environmental Performance: Measurable result of the environmental management system, related to an organization's control of its environmental aspects, based on its environmental policy, objectives and targets (ISO 14001).

Environmental Policy: Statement by the organization of its intentions and principles in relation to its overall environmental performance which provides a framework for action and for the setting of its environmental objectives and targets (ISO 14050).

Environmental Target: Detailed performance requirement, quantified where practicable, applicable to the organization or parts thereof, that arises from the environmental objectives and that needs to be set and met in order to achieve those objectives (ISO 14001).

Fixed Cost: There are two distinct definitions depending on whether the term is applied in either activity-based systems or in volume-based systems.

1. Fixed costs are costs that do not vary with the amount of output (Baltz and Baltz 1970). In (Fallon 1983) this is called non-variable costs. Fixed costs divide into two categories (Kerin and Peterson 1998): programmed costs and committed costs. Programmed costs result from attempts to generate sales volume, while committed costs are those required to maintain the organization.
2. A cost element of an activity that does not vary with changes in the volume of cost drivers or activity drivers (Edwards 1998).

Forecast Cell: A cell in a spreadsheet in which an output (result) with associated resulting uncertainty distribution is given, representing a response variable.

Functional Unit: Measurement of the functional outputs of the product system. The primary purpose is to provide reference to which the input and output are normalized (ISO 14041).

Gross Margin: The difference between total sales revenue and total costs of goods sold (Kerin and Peterson 1998).

Impact Analysis: The assessment of the environmental consequences of energy and natural resource consumption and waste releases associated with an actual or proposed action (EPA 1993).

Improvement Analysis: The components of a Life - Cycle Assessment, see Life-Cycle Assessment, that is concerned with the evaluation of opportunities to effect reductions in environmental releases and resource use (EPA 1993a).

Indirect Cost: Cost that is allocated (as opposed to being traced) to an activity or cost object, e.g, the costs of supervision or heat may be allocated to an activity on the basis of direct labor hours (Brinker 1997).

Life-Cycle: Consecutive and interlinked stages of a product system, from raw material acquisition of generation of natural resources to the final disposal (ISO 14040). This must not be confused with the product life-cycle concept used in Life-Cycle Management, see e.g. (Shields and Young 1991), which concerns the four main stages of 'Introduction', 'Growth', 'Maturity' and 'Decline' (Allvine 1996).

Life-Cycle Assessment (Analysis) (LCA): Compilation and evaluation of the inputs and outputs and the potential environmental impacts of a product system throughout its life-cycle (ISO 14040).

Life-Cycle Costing (LCC): As with the term 'life-cycle' there are different possible interpretations. The oldest and most well defined term is related to Life-Cycle Management, see e.g. (Shields and Young 1991). In this book, the LCC definition is adopted from (Sollenberger and Schneider 1996); 'a concept ... which tracks and accumulates costs and revenues of the entire life-span (life-cycle) of a product'.

Life-Cycle Inventory (LCI): The identification and quantification of energy, resource usage, and environmental emissions for a particular product, process, or activity (EPA 1993).

Non-Value Added: An aspect of a process or product that can be eliminated without reducing the value for the external customers. Internal customers, such as other activities and departments, are viewed as undesirable in the sense that they should be eliminated if possible. This view ensures that organizations are kept as slim as possible. Needless to say, that is not always possible.

Overhead Costs: Costs that are traced back to support activities, not e.g. production activities.

Overhead Energy Consumption: Energy consumed by support activities, and not for example production activities. An example can be the electricity in the facility, which cannot be directly traced to any part of the production, thus it is overhead energy consumption.

Overhead Waste Generation: Waste generated by support activities, not for example production activities. An example is the emissions from a gas stove used to heat the facility. The waste generated cannot be directly traced to any part of the production, thus it is overhead waste generation.

Profitability: Assessed revenues minus assessed costs associated with the creation of the revenues. Profit, on the other hand, is actual revenues minus actual costs.

Profitability Resource Efficiency (PRE): The PRE is defined as the profitability of an assessment object divided by the costs - where the costs are determined in a full absorption cost model. Cost of inventories is not included, and financial returns are not included. The PRE should be as high as possible as it measures relatively how effectively the assessment object generates profit.

Resource: An economic, energy related, or waste/mass related element that is consumed by the performance of activities. Resources, like activities, can be aggregated into hierarchies. In special cases, such as waste, resources may be generated by activities instead of consumed.

Resource Driver: A measure of the quantity of resources consumed by an activity. An example of a resource driver is the percentage of total square feet of space occupied by an activity (Brinker 1997).

Risk: Applies to situations for which the outcomes are not known with certainty but about which we have good probability information (Park and Sharp-Bette 1990).

Total Quality Management (TQM): A set of activities whose purpose is continuous process improvement, whose objective is total customer satisfaction and whose core concepts include standardization, efficient use of materials, the critical role of management, design specifications control, reduction of defect rates, statistical quality control and effective use of human resources (Brinker 1997).

Tracing: Also known as direct tracing or direct charge. The assignment of cost, energy consumption or waste generation to an activity or an assessment object using an observable measure of the consumption of resources or generation of waste by the activity or assessment object. Tracing is generally preferred to allocation if the data exist or can be obtained at a reasonable cost. This definition is based on the cost tracing definition in (Brinker 1997). See also 'allocation'.

Trigger: The occurrence of an event that starts as an activity (Brinker 1997).

Uncertainty: Applies to situations about which we do not even have good probability information (see also 'Risk') (Park and Sharp-Bette 1990).

Validity: In logic, validity is most commonly attributed to (Mates 1972):

1. Deductive arguments, which are such that if the premises are true the conclusion must be true.
2. Propositions that are semantically valid, i.e., are true under any alternative interpretation of the non-logical words.

Value-Added: Aspect of a process or a product that adds value to the customer and if eliminated would reduce customer satisfaction.

Value Chain: The linked set of value-creating activities (Porter 1985).

Variable Cost: There are two distinct definitions depending on whether the term is applied in volume-based systems or in activity-based systems.

1. Variable costs are costs that vary with the amount of output (Baltz and Baltz 1970). Like the fixed costs, variable costs are also divided into two categories (Kerin and Peterson 1998): a) cost of goods sold which covers materials, labor and factory overhead applied directly to production, and b) costs that are not directly tied up in production but nevertheless vary directly with volume, e.g., sales commissions, discounts and delivery expense.
2. A cost element of an activity varying with changes in the volume of cost drivers or activity drivers (Edwards 1998).

Volume-Based Costing: An umbrella term for all costing methods that rely upon the distinction of variable and fixed costs to determine the product costs. And because variable costs vary with the amount of output (Baltz and Baltz 1970) and only one single allocation base, it follows that the product costs strongly correlate with the production volume. Contribution Margin Costing and Standard Costing are two well-known volume-based costing methods.

Waste: All unwanted material generated by consumption of activities. The material may be organic and inorganic solids, liquids, and gasses.

Waste Content or Embodied Waste: The sum of all the generated waste in the value chain up to the assessment point of that particular assessment object.

Waste Driver: Any factor that causes a change in the waste generation of an activity. See also Cost Driver and Energy Driver.

ACRONYMS

If a man neglects education, he walks lame to the end of his life.

Plato

Below is a list of the main acronyms used in this book (see also Glossary). Chemical compounds (like CO, SO_2, NO_x) are not included.

Acronym	Explanation
ABC	Activity-Based Costing
ABCM	Activity-Based Cost Management
ABCEM	Activity-Based Cost and Environmental Management
ABCEW	Activity-Based Cost, Energy, Waste (models)
ABM	Activity-Based Management
ACU	Activity-Based Costing with Uncertainty
CAP	Critical Assumption Planning
CI	Consumption Intensity
DFE	Design for Environment
DFX	Design for X
EC	Energy Consumption
ECDM	Environmentally Conscious Design and Manufacturing
EEE	Economic Energy Efficiency
EMAS	Eco-Management and Auditing Scheme
EWE	Economic Waste Efficiency
GAAP	Generally Accepted Accounting Principles
GWP	Global Warming Potential
HFO	Heavy Fuel Oil
IPCC	Intergovernmental Panel on Climate Change
ISO	International Standards Organization
kg	kilogram
LCA	Life-Cycle Assessment
LCC	Life-Cycle Costing
LCI	Life-Cycle Inventory
LCIA	Life-Cycle Impact Assessment
m	meter
MC	Mass Consumption
MGO	Marine Gas Oil
MJ	Mega-Joule
POP	Persistent Organic Pollutant
PRE	Profit Resource Efficiency
PSV	Platform Supply Vessel
pWU	picoWasteUnit
SETAC	Society of Environmental Toxicology and Chemistry
TBT	Tributyltin
TQM	Total Quality Management
VOC	Volatile Organic Compounds
WG	Waste Generation
WI	Waste Index
WU	Waste Unit
WIP	Work In Progress
yd	yard

Chapter 1

INTRODUCTION

Where there is no vision, the people perish.

King Salomon
Proverbs, 29:18

This book is about using environmental metrics with well-established principles from economics and cost accounting. The result is an ability to perform integrated economic and environmental assessment and management; an approach we call Activity-Based Cost and Environmental Management (ABCEM). The concept is simple, the results powerful. In this chapter, we explain why we build upon cost management principles and add new dimensions to develop an approach that is based on accounting monetary cost, energy, and waste. We also provide you with an outline of what you can expect in this book. But first, we start with the motivation for our work.

1. THE ENVIRONMENT AND MANKIND - WHAT'S THE PROBLEM?

People who do not think far enough ahead inevitably have worries near at hand.

Confucius
Analects 15:12

The issues surrounding the environment are sources of great discussions for both experts and laymen. For ages humans have squandered around the earth without any need for noticing what we did to our home. It was not until 1962 when the landslide book *The Silent Spring* (Carson 1962) came out that some of us really started to see that Society and Nature were on a collision course. Many still do not believe that something is wrong, but nonetheless, a majority does now agree that humanity is putting a too heavy burden on Nature. There are many ways to explain why the environmental impact of humans is too high for Nature to sustain and even more explanations are available for remedying the situation.

We do not want to "bore" you too much by citing environmental problems. We all know about the alarming rates with which rainforests are decreasing, we hear about global warming and melting ice-caps, but "so what?" you may think because it does not affect you directly - at least for the time being. But the problem is closer than one might think.

Consider one of the most basic resources on Earth; water (H_2O). We all need it, and without it, we die. But did you know that of all the water on earth, less than 3 percent is fresh, and all but three-thousandths of that is locked up in glaciers and icecaps or is too deep in the earth to retrieve? The freshwater available in rivers, lakes, and accessible groundwater is increasingly polluted and being used at an alarming rate. For example, under North America's High Plains (extending from north Texas to the Dakotas) lies the Ogallala Aquifer, a deposit of Pleistocene groundwater spanning an area larger than California. By 1990, it was being depleted at a rate of 3 to 10 feet a year to provide 30% of America's groundwater-based irrigation. But, it is being recharged at an annual rate of less than a half inch! Half to two-thirds of the economically recoverable Texas portion was already drained by 1980 (Hawken, Lovins *et al.* 1999).

Clearly, these trends are alarming and in 1992, a group of over 1,600 hundred senior scientists (Union of Concerned Scientists 1992), including a majority of the living Nobel laureates in sciences, warned that:

> *A great change in our stewardship of the earth and the life on it is required, if vast human misery is to be avoided and our global home on this plant is not to be irretrievably mutilate.*

Unfortunately, we have no reason to believe that our 'stewardship' has resulted in improvements.

2. THE ECONOMY AND THE ENVIRONMENT - WHAT'S THE PROBLEM?

One basic weakness in a conservation system based wholly on economic motives is that most members of the land community have no economic value.

Aldo Leopold
In "The Land Ethic"

Many believe that we the solution of our environmental problems should be left up to 'the market' and the consumers. However, there are some basic flaws in thinking that the traditional economic supply and demand will fix the problem. One of the easiest ways to explain what causes our environmental impact is to use Paul Ehrlich's equation:

$$I = P \times A \times T$$

It is definitely not very detailed, but this simple equation states that the environmental impact I is equal to the product of the population P with affluence A, and the technological inefficiency T. That population is a factor is obvious. But population size has increased only slightly in the Western or

First World, but our environmental impact has grown exponentially. The real challenge, at least for the highly industrialized and 'developed' world, is our affluence combined with the inefficiency of current technologies.

Affluence is another word for wealth, and the environmental problems we face are related to our systematic and worldwide quest for wealth in many ways. Wealth can be measured in many ways, e.g., the amount of land you have, number of camels or other life-stock, etc., but in most parts of the world wealth is measured in terms of money. As Voltaire said; *"when it comes to money, everybody is of the same religion"*.

That the pursuit of monetary wealth and the protection of the environment can be in conflict is clear. However, we often seem to forget that money is not wealth in itself. Worse, our current economic and financial accounting systems do not seem to account for 'real' wealth in terms of natural resources. For example, at a national level, we commonly measure the wealth of a country in terms of Gross Domestic Product (GDP), and every country's goal seems to be to increase GDP. However, the GDP has some serious flaws. For example, the 1989 Exxon Valdez accident in Alaska was considered to be an enormous disaster. However, the accident *increased* GDP despite the fact that wildlife and Nature in general was seriously damaged. Clearly something is wrong with how we measure wealth.

This de-linking of the financial world from real resources is increasing, as is the terrible shortsightedness of financial institutions, see (Gates 1998). This leads to a situation where the economic system in its shortsighted search for increased GDP and financial returns is blind for the impact on the real resources such as minerals, fish, water and forests. Another way to look at it is that the 'true' price of a resource is unknown. A classic example in a different domain is the price of smoking. It turned out that the price of a cigarette does not cover the expense of health-care for the lung-cancer it causes.

In economics (especially environmental economics), these 'hidden costs' are known as 'externalities'. A negative externality occurs when the social cost of a good exceeds the cost paid by the firm that produces and sells it (i.e., the private cost). The societal cost of a good is its cost to everyone in the society, including people who do not produce or consume it. When it comes to the environment and its resources, there are currently many externalities. A solution to this problem is to 'internalize' the externalities, which means expanding what the private costs capture so that ideally they equal societal costs. However, this is easier said than done, because often we do not know the 'true' societal cost until it is too late, that is, after decisions are made. Consider the smoking example again: the current costs for lung cancer treatment was unknown in the 1950s when many of today's patients bought their first cigarettes. Thus, there is a missing link between the real, natural resources and our ec[illegible] system. Consequently, we believe that there is at least one important [illegible] for the increasing environmental problems; the economic system is i[illegible]e of directing fair attention to the *real* price of resources.

3. THE ENVIRONMENT - WHY SHOULD BUSINESS CARE?

As we shall see, apparent differences between people arise almost entirely from the action of the system they work in, not from people themselves.

W. Edwards Deming

The environment is something executives, managers and engineers cannot afford to ignore for at least two reasons:

1. There are genuine societal concerns about the state of the environment. Many governments have realized that the market-driven economic system does not account for environmental impact and are placing emphasis on environmental impact reduction through legislation and agreements, such as the 1997 Kyoto climate treaty. Unfortunately, in the US, big business (organized in the umbrella organization Global Climate Coalition) used television campaigns and intense lobbying to sink the Kyoto treaty and even trash the very notion of climate change. Thankfully, the picture has changed dramatically the last two years. Giants like United Technologies, Intel, American Electric Power, DuPont and BP have "shifted from being climate skeptics to climate activists" according to Frank Loy (The Economist 2000b).
2. The environmental performance of an organization will impact either directly or indirectly its competitiveness. Quoting President and Chief Executive Officer of ABB, Percy Barnevik: "If you think that today's environmental requirements seem like a light breeze, you should get ready for the storm of tomorrow".

One can argue that the 'good old days' where a product was designed, manufactured and sold to the customer with little or no subsequent concern are over. In the seventies, with the emergence of life-cycle engineering and concurrent engineering in the United States (Winner, Pennell *et al.* 1988), companies became more aware of the need to include serviceability and maintenance issues in their design processes.

Companies are now becoming concerned with the environment because more and more people have realized that there is a cost to society that results from environmental impact. Whether we like it or not, all products and processes affect in some way our environment during their life-span. In , a schematic representation of a system's life-cycle is given. Materials are mined from the earth, air and sea, processed into products, and distributed to consumers for usage, as represented by the flow from left to right in the top half of .

In general, a company's environmental impact comes from (excessive or wasteful) consumption of natural resources and emissions of pollutants to air, water, and land. Recognition of the negative effects of air emissions has led, among others, to the Clean Air Act and Corporate Average Fuel Economy legislation in the United States. Recently, end-of life and disposal issues have received more attention. The emergence of product take-back legislation in Europe has forced manufacturers to think about how to dispose their products appropriately through recycling and reuse (as represented in the flow from right to left and up in the lower half of). The term 'demanufacture' is often used to characterize the process opposite to manufacturing necessary for recycling materials and products. Material demanufacture refers to the process of, e.g., breaking down long polymers into smaller polymers which are then used for stock in new materials.

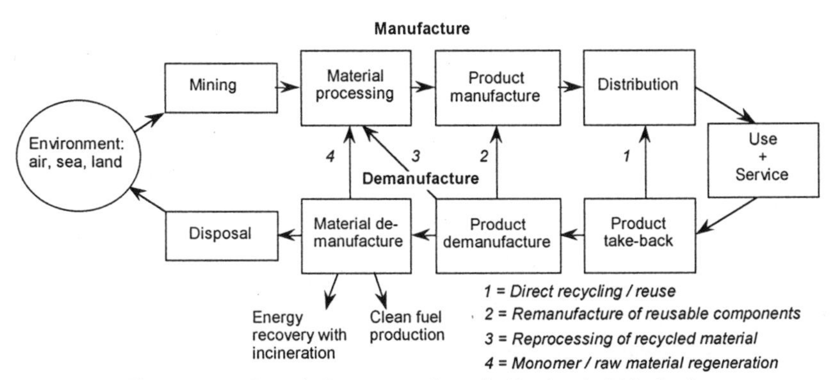

Figure 1 - A Generic Representation of a Product's Life-Cycle.

The most notable drivers for a company or organization to become more environmentally responsible are:

- Legislation: Always a 'popular' way for governments to force companies into becoming more environmentally friendly. The US Clean Air Act has limited the use of a number of materials and European take-back legislation is forcing companies to recycle their products. The US also requires pollution prevention for many companies by law.
- Customer demand: Environmental awareness is increasing among customers. Some customers will even pay more for a 'green' product, and industrial customers (e.g., Original Equipment Manufacturers (OEMs)) do not want (future) environmental liability for a supplier's product.

- Eco-Labeling programs: Eco-labels are analogous to food labeling programs, but instead of listing how many calories a product has, eco-labels try to convey to consumers a sense of how 'green' a product is by stating, e.g., how much CO_2 was released when producing the product. Having an eco-label can be a competitive advantage.
- ISO 14000: The ISO 14000 environmental management standards certification is increasingly becoming an important element in doing business, like the ISO 9000 quality management standards. Some OEMs already require their suppliers to be ISO 14000 certified.
- Economic benefits: Increased efficiency in terms of resource use always makes good business sense. Furthermore, some European countries (e.g., The Netherlands) are shifting their tax system from an income-based system to a resource-based system, meaning that the more resources that you use, the more tax you pay.

Especially ISO 14000 has many companies, large and small, now thinking about the environment like they have not done before. ISO 14000 makes companies think beyond merely complying with the law, but becoming proactive rather than reactive to environmental concerns.

In addition, many have noted that environmentally conscious business practices make good business sense and have many other positive effects. For example, the reduction of material diversity leads to less diverse inventory, volume purchasing, and the opportunity to focus on a reduced number of (core) manufacturing processes. Life extension practices place renewed emphasis on design for serviceability which typically pleases customers. Environmental concerns are also stimuli for finding new creative solutions and products. However, this does not imply that one can always expect a direct financial reward from becoming environmentally responsible.

A good overview of the global state of attitude, activities, and achievements of companies, governmental institutions, and universities towards environmental issues in manufacturing (and beyond) can be found in (WTEC 2000).

4. WHAT'S THE GOAL?

Perfection of means and confusion of goals seem - in my opinion - to characterize our age.

Albert Einstein
In "Out of My Later Years"

Sustainable development is what many see as the goal to achieve, that is, 'development that meets the needs of the present generation without compromising the needs of future generations' as defined by the famous United Nations Brundtland report *Our Common Future* (UN 'Brundtland'

Commission 1987). Sustainable development is focused on the interplay between the environmental, economic, and social dimensions. Sustainability in the social sense means maximum of equality of opportunity, social justice and freedom (Ullring 1996), but this topic is beyond the scope of this book.

According to the US Environmental Protection Agency, achieving this goal involves an interrelationship between sustainable ecosystems and a viable economy with pollution prevention, resource conservation, environmental equity, as illustrated in Figure 2. The goal of sustainable developments sounds very simple, but why haven't we achieved it yet?

Many books have been written about how to achieve sustainable development. But, as the saying goes, 'the devil is in the details' and companies, managers, and engineers alike often run into major problems when attempting to practically implement sustainability.

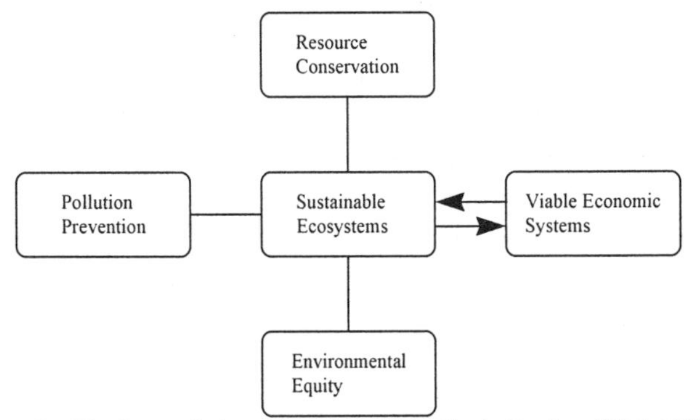

Figure 2 - The Interrelationships of the Life-Cycle Design (EPA 1993a).

5. HOW DO WE GET THERE?

Even a long journey begins with a single step.

Chinese Proverb

Achieving sustainable development from a practical point of view, or 'operationalizing sustainability' (as the latest buzzword seems to be), is not so easy as it often sounds. Sustainable development is a highly complicated issue. It is a multi-dimensional problem involving the environment, economy, and society -- past, present and future. In Chapter 2, we show some of the stages where a company has to go through from merely pollution prevention activities to really becoming sustainable. No matter at what stage a company (or even society) is in its environmental thinking, there will always be the following 'classical' questions:

- *What is our environmental impact?*
- *Where does it occur (the most)?*

- *What should we do about it?*
- *What is it going to cost us?*

Answering those questions is crucial in moving towards sustainable development. Even more crucial is providing companies the ability to answer those questions themselves. Only then will we get self-motivated and self-sustaining efforts to reduce environmental impact.

So, what is needed?

First, we must undoubtedly be able to *assess* both the economic and the environmental impact of our products and processes, because otherwise we simply cannot know what is a viable economic solution or a sustainable solution to a specific problem, product or process.

Secondly, we need to know how to *improve* our product, processes, and organizations. We believe that in order to *truly* reduce the overall environmental impact by an organization, resource use must be investigated from three perspectives; 1) the product perspective, 2) the process perspective and 3) the organizational perspective. The importance of efficient and effective design of *both* products and processes has been widely noted, see e.g. (Devon 1993), (Dutton and Marx 1997) and (Winner, Pennell *et al.* 1988). Of equal importance is to manage companies holistically, that is, companies must think beyond their four walls, consider the total cost and their impact on their environment, for example. Major research efforts that will enable organizations to become more holistic regarding cost management, environmental management and product and process management and design have therefore been initiated and continued internationally.

Thirdly, we need to be able to identify economic and environmental *win-win* situations. 'Money talks' is a frequently heard proverb, especially in the US. Another frequently heard saying is that environmental management and impact reduction 'costs too much', and the question *"How much is it going to cost us?"* is, in our opinion, simply the wrong question. We believe that companies should rather ask:

> *"What is the Return on Investment?"* or
> *"How much value is it going to give us?"*

Thinking in terms of <u>*value*</u> rather than cost may open a company's eyes to new opportunities. For example, Interface Flooring Systems in LaGrange, Georgia, has a photovoltaic solar array standing next to one of their plants. From a cost perspective, the investment does not make sense; the payback time is in the order of decades because electricity from the net is cheap. However, from a value perspective, it is a whole other story. Having the solar array has made it possible for Interface to market carpet made with clean, renewable energy. Some customers are willing to pay extra for this particular product 'attribute', allowing Interface to set a higher price for the same carpet. Suddenly, the payback time is in years, rather than decades.

We can go on and list many other issues that need to be resolved, but the preceding ones are in our opinion crucial from both a strategic and tactical point of view. More importantly, we need to be able to give companies and their engineers, managers, etc. the tools to address the above issued. A reliance on outside experts may not help companies internalize the issues and take ownership of their efforts. What is crucial is to obtain self-motivating and self-sustaining efforts at all levels.

Many environmental assessment and management approaches have been developed to help companies with environmental issues, and in Chapter 2, we review some of these. What we feel is missing from the current state-of-the-art is the following:

- Some basic properties, such as comparability, are missing from many of environmental performance measurements
- The integration environmental assessment and management with other engineering and management tools and practices is still lacking.
- Learning curves are too high.

Comparability and the link between environmental and economic assessment and management are crucial. We must be able to assess and compare *both* the economic and the environmental impact of our products, processes and organizations: otherwise we simply cannot know what is a viable economic solution or a sustainable solution to a specific problem, product or process.

6. RE-LINKING THE ENVIRONMENT, MANKIND, AND THE ECONOMY - OUR APPROACH

Yet to solve a problem which has long resisted the skill and persistence of others is an irresistible magnet in every sphere of human activity. It was this urge to which Mallory alluded when he gave the ingenious reply to this same question - "Because it's there".

Sir John Hunt
In "The Ascent of Everest"

We mentioned in Section 2 that many environmental problems are caused by the fact that they are externalities to the current economic system. There are three distinct possible ways to 'internalize' externalities:

1. Define the cost of the externalities in terms of money.
2. Create a separate system that deals with environmental issues only or
3. Augment the economic system to include environmental issues.

The first approach is difficult, as already mentioned, because we often do not know the true societal cost until it is much later (or even too late).

The second approach, i.e., create a separate system, is the one chosen by the conventional ISO and environmental management community. This is evident from the fact that in the ISO 14000 standards (see also Chapter 2), e.g., ISO 14040, there is no or very little discussion on integrating environmental performance measures with economic measures. We believe that this approach can have both negative management consequences and undesirable behavioral effects.

We have therefore taken the third approach. Yes, many if not most environmental problems are caused by the fact that the economic system does not capture the true costs of the environment and its resources well, but *not* because the economic system does not work *per se*. In fact, we have every reason to believe that the economic system works well given its scope, that is create economic wealth, just take a look at all the wealth in the world. We believe that the *measures* of the economic system must be augmented. Measurements drive behavior, see e.g. (Brown 1995) and by including environmental measures into our existing economic systems, we may get the best of both the environmental and economic worlds. In the following sections, we outline the principles of our approach - Activity-Based Cost and Environmental Management.

6.1 Start from Cost Management

Rather than starting from an environmental assessment or management approach and trying to tie this with traditional business practices, we approach the problem from the opposite side. We propose to build upon modern cost management principles and to extend those into environmental management. This is a bit in contrast to the conventional environmental assessment and management approaches (e.g., like ISO 14000).

Why start from cost management? Here are some reasons:

- First of all, we can leverage the over 200 years of knowledge, experience, and development in the field of cost management.
- Perhaps most importantly, we believe that environmental assessments must be presented in a similar and parallel perspective to costs to get the needed attention from decision-makers.
- Furthermore, building on cost management allows us to put both economic and environmental accounting and management in the same framework so that we can quickly identify economic and environmental win-win situations and tradeoffs. Having separate systems means extra work, increased learning curve, inconsistency and probably lost opportunities.

- Cost management has the General Accepted Accounting Principles (GAAP) in the US and similar principles elsewhere that form a uniform standard. We can build upon these to gain analogous General Accepted *Environmental* Accounting Principles (GAEAP).

In particular, we base our approach on *Activity-Based Costing* (ABC) and *Activity-Based Management* (ABM) which are arguably the state-of-the-art in cost accounting and management. More importantly, the ABC and ABM principles transfer very well to environmental assessments and management. Just look at Figure 3 that represents the fundamental principle of ABC: resources (e.g., materials) are consumed by activities (e.g., manufacturing), which themselves are consumed by objects (e.g., products). This also sounds very logical for environmental assessments where we also talk about resources, but in a broader sense than mere economic resources.

But, as stated earlier, the current economic system is incapable of assigning values to resources that reflect the *true* value (or cost) so an extension and further development of cost management practices is required in order to tackle the environmental issues. Basically, we need to address the questions of *what to measure* and *how to measure it*.

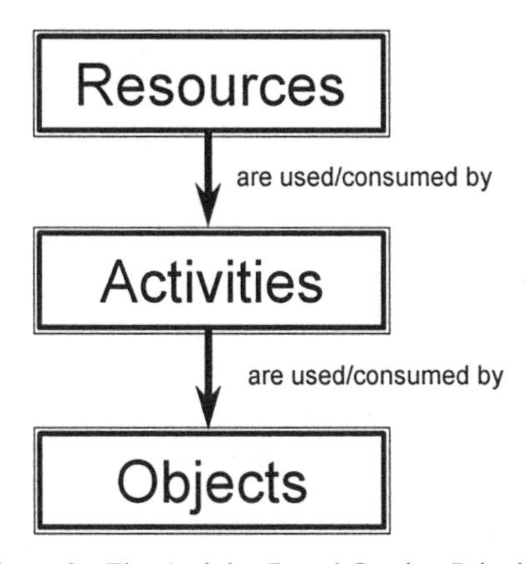

Figure 3 - The Activity-Based Costing Principle.

6.2 Account for Energy Consumption and Waste Generation, as well as Monetary Costs

There are many measures of environmental impact. So many, that it may make your head spin. We already gave you one in Section 2. We will give you some more examples in Chapter 2. ISO 14031, a standard on environmental performance evaluation (see Chapter 2, Section 5) has even more indices that you can use. All of them have their good sides, and bad sides. The basic problem with this enormous variety of measures, however, is that assessments may end up with totally incomparable results.

We believe that to get environmental management started in a company, we have to start with basic, simple, accountable and comparable measures.

For that reason, we state that environmental accounting and management should start (as a minimum) with *energy consumption* and *waste generation* as two basic environmental dimensions to account for, in addition to monetary costs.

Why energy consumption and waste generation? There are a number of reasons for starting with these measures to augment cost management.

- They are fundamental. For every activity (operation) humans undertake, there are inputs and outputs as illustrated in Figure 4. Inputs are material (natural resources) and energy, while the outputs are either useful (products, co-products, reusables and recyclables) or waste.

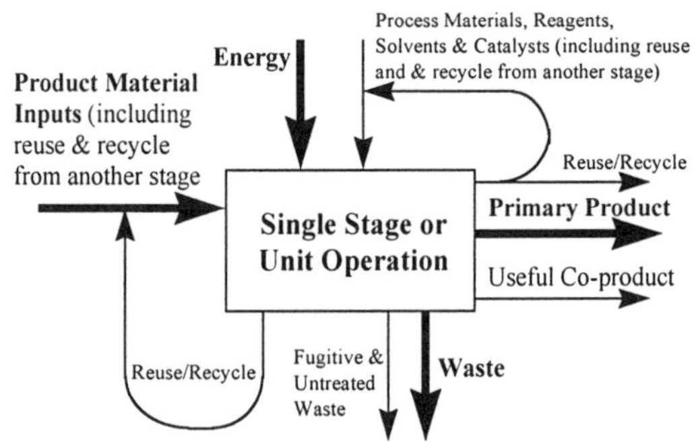

Figure 4 - Environmental Aspects Related to Product Realization (EPA 1993a).

- They are simple. Energy and waste are not strange concepts that require much learning. All of us have experience with energy and waste. Plus, most of us, if not all, know the fundamental laws in physics that deal with conservation of energy and mass.
- They can be easily measured and accounted for (just like money is today). Measurements are key in making the whole system work. Without good measurements, we have no basis for comparison or continuous improvement. Measurements should be as easy as possible to reduce errors and costs. Basic energy and waste measurements can very easily be performed. Care should be taken, however, that the measurements are also valid and accurate. More on this will be given in a later section.
- They are 'objective' and not open to influences on the practitioner level. Objective means that two or more qualified observers should arrive at the same result. Avoiding influences on practitioner level basically are crucial because it ruins comparability and leads to nothing but (endless)

discussions. By sticking to well known fundamental measures, the chances of mistakes are decreased.

- They are comparable. It is easy to compare energy consumption and waste generation from company to company and product to product. Comparability is important because it enables ranking, prioritizations and ultimately decision-making. Also, industry representatives demand comparability, see (Jensen, Elkington *et al.* 1997).
- They complement money. We do not want to duplicate what regular cost accounting can measure, but seek to augment and complement money with measures that can capture the externalities and true amount of resource use. But, the measures should be as intuitive as money, and energy and waste have these qualities. In fact, the only difference to money is what they measure.
- They are useful. We believe, as also argued by (Barlas and Carpenter 1990; Barlas 1996), that we need to investigate useful solutions to problems and not 'academic' solutions. This goes also for environmental management, in our opinion. To understand what is useful we need to go back to the basics of a company and what is possible to account for as seen from the company perspective.
- They are valid. That is, they measure what they are supposed to measure; environmental impact. However, these dimensions are *relative* measures that do not represent *actual* environmental problems (e.g., number of fish killed). We think that this is an advantage because when it comes to measuring actual environmental impact, we tend to get stuck in endless debates. In that sense, the dimensions that we propose are valid because:
    - Costs/revenues represent wealth (revenues) and general resource conversion efficiencies (costs). The economic system captures quite well many aspects of the general resource consumption, but irrelevant and bad managerial accounting practices, see e.g. (Johnson and Kaplan 1987), short-term thinking and disconnection in the financial markets, see e.g. (Gates 1998), protectionism and outrageous subsidies, see e.g. (Brown, Flavin *et al.* 1999), make the economic system partly dysfunctional.
    - Energy represents energy conversion efficiencies. Energy is one of the drivers in socioeconomic development (Olsson 1994) and the strong correlation between energy consumption and CO_2 emissions is unquestionable, see e.g. (Fowler 1990; Seki and Christ 1995). Then, knowing that the energy demand may double worldwide by 2020 (Holberton 1997), it is evidently important to manage energy well. Furthermore, the potential economic saving of improving our energy

management is enormous. In the U.S. alone, (Lovins and Lovins 1997) estimate that $300 billion can be saved annually.

- Waste represents material conversion efficiencies. Waste management has several aspects to it; 1) to reduce/manage the amount of trash generated or 2) to reduce emissions of chemical compounds to the environment. The need for reducing trash is obvious as the U.S. alone generates almost 10 billion tons annually (Brown, Flavin *et al.* 1999), and that this waste is not only taking an enormous amount of space to store but also represent an enormous misallocation of resources. To cut down on emissions is equally important since these emissions affect the environment in many negative ways, see e.g. (IPCC 1993). Thus, improved waste management (trash and emissions) is crucial for sustainable development.

Often we are asked why, from a materials perspective, we focus on waste generation instead of materials consumption. The short answer is that both are important, but not all materials used/consumed cause an environmental impact. Many consumed materials are put to good use. Waste, however, is by definition something unwanted. It is thrown out into the environment where it is likely to cause harm. Plus commonly accepted philosophies like lean manufacturing also stress reduction of waste. Hence, we prefer waste accounting. However, as shown in Chapter 4, you are definitely not limited to waste; you can take any measure you want. In fact, in the Farstad Shipping case study (Chapter 6), we show how you can use mass consumption as a metric. The results, however, are different then when using waste as a metric.

So, our approach is based on assessing and managing three dimensions - cost, energy, and waste. Note that the three perspectives are complimentary. None of them are sufficient to describe, e.g., a process, but combined they describe measurable aspects that relate a process to its economic or environmental status.

6.3 Perform Environmental Accounting and Management along the Value-Chain

Without thinking in terms of value chains, a concept originated by (Porter 1985), there is little hope of getting any further at all in environmental impact assessments. The value chain is 'the linked set of value-creating activities' (Porter 1985) from raw materials extraction through manufacturing to sales and service (if included in sales agreements). In this definition, the value chain concept does not include disposal or recycling or reuse (yet), but it is easy to see how those life-cycle stages can be included in the value-chain.

The value-chain concept is well known and based on modern cost management principles. We can (and should) build on this concept for environmental management.

With respect to environmental management efforts, thinking in terms of the value chain simply means that we need to know what environmental impact the suppliers have made, what their suppliers' environmental impact is, and so on. Just as in the economic world. Except now, we use environmental measures like energy consumption and waste generation as the 'cost', and not money.

The value chain concept also provides another explanation for why we prefer to focus on waste generation (output) versus material use (input). Consider Figure 5, in which the thick arrows signify trade and the like. The output in one link of the value chain is input in the next link and so on. Thus, in principle, a particular company only needs to assess the impact of its outputs because the impact of its inputs are accounted for earlier in the value chain by other upstream companies. The impacts will follow all the resource elements used in that particular company. This is what is used in economic management, and the same approach is also valid for environmental management.

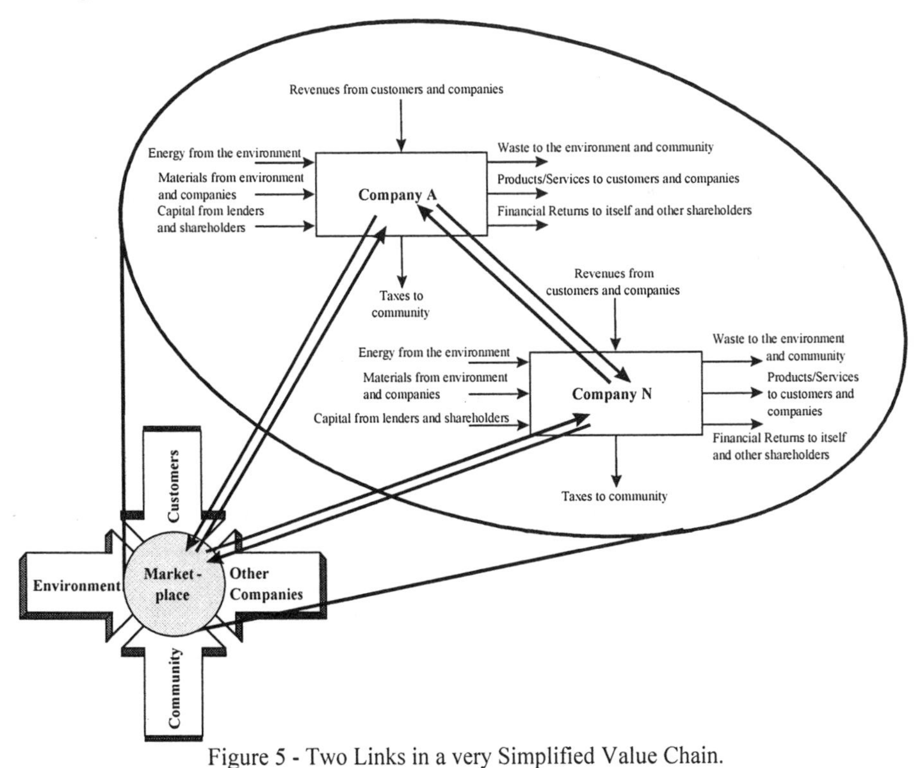

Figure 5 - Two Links in a very Simplified Value Chain.

Analogous to a product or service having a purchase price we also say that it has an 'energy price' and 'waste price', namely its energy content or waste content, respectively. This means that products, components, etc. that are being bought by a company from a supplier not only have a monetary cost, but also an environmental cost in terms of energy consumed and waste generated. Like expense is defined as the money spent on producing, say, a product, the term 'energy content' is used to denote the amount of energy spent on the production of, say, a product. Some refer to this as 'embodied energy' as well, and more on this is given in Chapter 4. Similarly, waste is created in the production of products and services. Thus, these products and services do not only have an energy content, but also a waste content, i.e., the waste generated in creating the product or service.

Thus, simply speaking, a product or service has not one, but (at least) three 'prices' in our approach: monetary cost, energy content, and waste content. We state 'at least' because as we will show in Chapter 4, the sky is the limit: you can add as many 'price' dimensions and values as you want, but the more dimensions, the more information is required, the more it is going to cost you and you may not get more relevant information.

The value chain goes beyond simply describing what flows where. It also incorporates where value is added. In many ways, 'value' can be seen as the opposite of waste, and hence they are related. Many companies are familiar with value-added versus non-value-added activities, and are trying to eliminate the latter. Eliminating non-valued added activities also has, in principle, environmental benefits: if an activity is not required, then it will also not use natural resources (such as energy) and not generate waste.

Thinking in terms of the value chain thus acknowledges the fact that waste and energy can be generated and consumed elsewhere in the value chain, even though the product or service you use at your particular stage in the value chain seems 'clean'.

6.4 Measure Waste using the Waste Index

Wasting materials is a major problem in our society. According to a study referred to in (Anderson 1998), the production of a 10-pound (4.5 kg) laptop personal computer generates 40,000 pounds of waste throughout the entire value chain. This means that only 250 ppm by mass is utilized! Obviously, this cannot go on if we ever are to become sustainable.

However, measuring materials with one index poses a serious challenge: A gram of plutonium is clearly a different problem than a gram of sulfur. Hence, we need to establish an index that can measure such apparently incomparable differences. Measures that are proposed by ISO and SETAC for

Life-Cycle Impact Analyses can easily give incomparable results, as we will discuss in Chapter 2. Plus, measures like acidification, eutrophication, ozone depletion potential, etc. are not intuitive measures for many engineers and managers. Who is capable of trading off (in a scientific manner) global warming with eutrophication, acidification, ozone formation, carciogenity and heavy metals to mention some categories? Although they capture in 'true' environmental impact, we believe that what is needed is a genuinely new approach in order to get more companies motivated to assess their environmental impact.

Our approach to an environmental impact metric is called the Waste Index (WI) and is presented elaborately in Chapter 2, Section 5. We refer to that section, but we would like to point out that the Waste Index has three indispensable features that we consider paramount for further index development:

1. The Waste Index is based on what we refer to as the 'Nature Knows Best' axiom, which states that 'environmental impact can only be measured relatively by benchmarking Nature'. This means essentially that for whatever we release into Nature, we should find out how the release impacts the *balance* of Nature. This philosophy is based on our observation that in general everybody agrees on the following:
   - More releases into Nature are always worse than less (given that we talk about similar substances).
   - The longer a release can be detected in Nature, the worse it is.
   - The larger the area affected by a release, the worse it is.
2. The Waste Index solely relies upon thermodynamics, chemistry and natural balances of the chemical compounds in the environment. What is really important here is that even though the Waste Index is not measuring the true environmental impact, it will be useful because it can be used consistently over time and is beyond debate by practitioners. Then, an ISO like organization could manage information regarding calculation of the Waste Index, and this information could be used by all organizations in the world (in principle) to calculate their generated waste. The result would be a worldwide, comparable and simple waste information system from which managers, engineers and the likes can get the information they truly need to do their job.
3. The Waste Index gives comparable results, which is not the case for aggregated indices according to (Ayres 1995).

We believe that the Waste Index framework can serve as a good basis upon which better indices can be derived or developed. The Waste Index can be accounted for just as energy and costs, which should be the overriding goal of any index development to prevent the lack of data which is currently so prevalent, see e.g. (Jensen, Elkington *et al.* 1997).

7. THE UNIQUE FEATURES AS WE SEE THEM

The problems we have today cannot be solved by thinking the way we thought when we created them.

Albert Einstein

The sum of the previous sections is that our ability to successfully realize products, processes and implement necessary organizational changes in a sustainable manner rests upon our ability to assess and manage the impact of our products and processes in a value-chain and life-cycle perspective. Our approach has four major unique features that we believe will lead to create an environmental assessment and management method that is *truly* useful. These, which are discussed in the four subsequent sections, are:

1. Type of framework; activity-based versus conventional
2. Integration of economic and environmental assessments
3. Different environmental impact indicators
4. Inclusion of uncertainty.

7.1 An Activity-Based versus a Conventional Framework

This first main unique feature is that we believe activity-based frameworks are highly preferable to conventional frameworks. This will be illustrated both in case studies (see Chapters 4 and 5) and argued extensively in Chapter 2 where we discuss conventional environmental management and Chapter 3 where we discuss Activity-Based Costing and Management. The argumentation revolves around two critical issues; 1) accuracy and 2) tracing.

The accuracy of the conventional ISO 14040-43 based assessment methods depends on several constructs (functional units, unit processes and the whole impact assessment). These constructs can result in an inherent method deficiency. Activity-Based and Environmental Management (ABCEM), in contrast, intentionally treats environmental issues with the same framework as costs, i.e. an activity-based framework rooted in the principles of Activity-Based Costing (ABC), and the accuracy of ABC is well documented, see e.g. (Drucker 1995) and Chapter 3.

When it comes to tracing, i.e., the capability to answer the *why*-questions, conventional assessment methods are unclear because practitioners are left to their own devices to do the tracing their own way, if at all. In Activity-Based Cost and Environmental Management, on the other hand, tracing is automatic by sensitivity analyses, as we will show. Furthermore, we will show that

ABCEM has better process-orientation and handling of overhead resources than conventional methods, yielding more accurate and relevant tracing.

7.2 Integration of Economic and Environmental Impact

The second advantage that we have is that we integrate economic and environmental assessments and management. We will both show with case studies and argue that integration has many indispensable advantages that conventional approaches cannot offer. Briefly speaking, the most important advantages are:

- *Implementation* of both cost and environmental management systems becomes easier and therefore cheaper.
- *Tradeoff Capabilities* between cost and environmental performance are secured - a necessity for decision-makers.
- *Consistency* between cost and environmental assessments is secured. This is crucial in order to re-link the 'cost world' with the 'environmental world'.

7.3 Different Environmental Impact Indicators

The choice of environmental impact indicators greatly affects the accuracy and overall usefulness of the method. One of the biggest problems with Life-Cycle Assessment (LCA) as defined in the ISO 14040-43 standards is the impact assessment stage. Most practitioners, engineers, and managers understand Life-Cycle Inventories (LCI) based on mass and energy balances. However, the link between these mass and energy balances and actual environmental impact is usually beyond the grasp of the average engineer and manager. Far too often, we have noticed a blind faith by companies in impact databases, not realizing how over-generalized these are. Impact indicators in conventional LCA rely on several constructs that include subjective weighting, resulting in incomparability. This has also been recognized in the ISO 14042 standard and ISO stresses the importance of good and transparent documentation of the assumptions, but regardless of how much you document the incomparability can occur.

We agree with ISO that a mere inventory based on mass and energy inputs and outputs is not enough, but it is a start. However, we also think that the impact assessments that ISO and SETAC would like to see are still a bit of a 'Holy Grail' with respect to implementation in the business world. We have therefore outlined a new environmental impact indicator called Waste Index that can be regarded as an impact measure beyond cumulative mass/energy data, but which stops short of measuring the actual impact in terms of specific environmental problems, such as, number of frogs killed, global warming, acidification, algae growth, etc. The Waste Index is based on 'benchmarking Nature' resulting in an indicator that relates releases to the efficiency of Nature to degrade these releases. Hence, environmental impact is only measured relatively and not to actual environmental problems - a different approach.

7.4 Inclusion of Uncertainty

Both design and management are disciplines with inherent uncertainty, that is, things are not always known, numbers are uncertain, etc. Handling this uncertainty is therefore crucial in our opinion, and also advantageous, in order to give decision-makers better decision support. We are often baffled by how many researchers deal with environmental issues in a deterministic manner when the amount of uncertainty can be orders of magnitude.

In conventional approaches for economic and environmental management alike, uncertainty is usually spoken of, but never treated as explicitly as needed. In fact, uncertainty is often ignored or treated very simplistically. The result is that we can hardly know what the range of uncertainty is in an assessment, what impacts this range and how can we plan for handling the critical uncertainty factors, i.e., Critical Assumption Planning (CAP). Throughout this book, we will show that Activity-Based Cost and Environmental Management not only handles uncertainty, but also explicitly uses Monte Carlo simulations to improve tracing capabilities, forecasting and Critical Assumption Planning - a clear advantage over other approaches. Whether it is a coincidence or not, we have observed an increase in publications and work related to environmental assessment and management that also is using Monte Carlo simulations as a basis for sensitivity analyses since our initial publication in 1994 (Emblemsvåg and Bras 1994).

8. OUTLINE OF THIS BOOK

He who has begun has half done. Dare to be wise; begin!

Horace (Horatius)

As mentioned in the preface, we have tried to write this book in 'how-to-do' style without the lengthy discussions concerning scientific validation and logic. This the interested reader can find in abundance in (Emblemsvåg 1999). Here, we have taken a 'learn through doing' approach. We have therefore emphasized case studies while at the same time included the necessary theoretical discussions. However, on one hand the concept is simple, on the other hand, 'the devil is in the details' and you may find yourself re-reading some sections of this book. Because some of the terms may be new for you, we have included a glossary, index, and list of abbreviations.

At the start of each chapter, you will find a brief abstract about what you can expect to read and learn about in the chapter. At the end of most chapters, you will also find a section called 'What is Next?' that describes, as it says, what you can expect next. In each chapter, we tried as best as we could to give a summary at the end that should give you insight in what we wanted you to pick up in terms of learning and key points.

In this chapter we have tried to set the stage so that you, the reader, can understand where we come from intellectually, why we think the work presented here is worthwhile reading and what you can expect to read more about later in the book.

In the next chapter, Chapter 2, we give you an overview of the current affairs of environmental management and how we see it. As will be evident, we do not have much confidence in the current approaches being capable of providing businesses with the decision support (and even motivation) they need in order to systematically and thoroughly work towards their sustainability goals. In Chapter 2 we also discuss in detail our new environmental impact indicator called the Waste Index.

In Chapter 3 we present the costing method called Activity-Based Costing, or ABC, that we used as a foundation for our integrated economic and environmental assessment and management approach. Through examples, we will also show you how ABC is superior to conventional costing methods. Many books have been written about it, but we are going to give you the whole flavor in one chapter. We also discuss what is called Activity-Based Management, which is based on ABC. Both ABC and ABM have many features and characteristics that make it an ideal starting point for further development.

Our 'further development' is presented in Chapter 4 - an approach that we call Activity-Based Cost and Environmental Management. Part of this approach is the development of Activity-Based Cost, Energy and Waste models that provide the means for assessing economic and environmental impact, as well as the capability to trace critical factors, and decision support in 'what-if' type scenarios. We should note that in some of our earlier publications the very same approach is referred to as Activity-Based Life-Cycle Assessment (LCA). However, after realizing that our method can do much more than just a LCA we found it more correct to give the method a new name, that is Activity-Based Cost and Environmental Management, and rather view Activity-Based LCA as an instantiation of the more generic Activity-Based Cost and Environmental Management.

In Chapters 5 through 8 we present our case studies, namely WagonHo! (Chapter 5), Farstad Shipping (Chapter 6), Interface Flooring Systems (Chapter 7), and Westnofa Industrier (Chapter 8). With the exception of Farstad Shipping, all deal with manufacturing. Farstad deals with the Activity-Based Cost and Environmental Management of a product in its use phase. The WagonHo! and Farstad case studies have been written in a way that you can actually implement the integrated Activity-Based Cost, Energy, and Waste models yourself in a spreadsheet. The Interface Flooring Systems and Westnofa Industrier case studies are too comprehensive to discuss in detail; we would need another book. However, they illustrate that Activity-Based Cost and Environmental Management can and is applied in real companies.

It is important to note that all the case studies, with the exception of the WagonHo! case study, are real-life with complete sets of process and product data provided by the various companies. Many of the case studies have similar characteristics; that is unavoidable, but that also goes to plainly illustrate that Activity-Based Cost and Environmental Management is reproducible from one case to another, which is essential for comparability. Nonetheless, all case studies illustrate certain special issues (see the chapters). The same case studies can also be found in (Emblemsvåg 1999), but they have been rewritten here to become more 'how to' instead of argumentative and validation-oriented as in (Emblemsvåg 1999).

In the last chapter, Chapter 9, we tie things together again and revisit the most important messages in this book. We also look a little ahead of what the future may look like research-wise when it comes to Activity-Based Cost and Environmental Management and related topics.

You should realize that we articulated the first concepts of Activity-Based Costing and Environmental Management in 1993. The first mention to "*Utilize the ABC method for life-cycle assessments of environmental impact in terms of matter and energy consumption*" already appeared in print in 1994 (Emblemsvåg and Bras 1994). So you are looking at something that has been developed over many years and already 'tried and tested' in many ways. In this book, our aim is to transfer some of our core knowledge and experience gained to you, but there is much more we could tell you about our past, present and future undertakings in this area. We hope you too become excited about the topic, and if you have any questions or comments, or would like some further information, feel free to contact us.

We hope you enjoy reading this book, and we welcome your feedback.

Chapter 2

ENVIRONMENTAL MANAGEMENT AND ASSESSMENT

Whatever wisdom may be,
it is far off and most profound -
who can discover it?

King Salomon
Ecclesiastes 7:24

What is environmental management? This question has many answers that could fill more than just this book. In this chapter, however, we want to give you an overview of a number of environmental management and assessment approaches. We start with showing a classification scheme that can help shed some light in the forest of buzzwords, as it may seem to be for some of you. We will outline the ISO 14000 environmental management standards, as well as some approaches that attempt to link economic and environmental assessments. As you will see, a lot depends on what you define as being 'environmental impact' and how you measure it. In this chapter, we also give an illustrative overview of environmental metrics, including a metric that we developed ourselves: the Waste Index (WI).

1. A CLASSIFICATION OF ENVIRONMENTAL IMPACT REDUCTION APPROACHES

There never was in the world two opinions alike, no more than two hairs or two grains; the most universal quality is diversity.

Michel De Montaigne

In the last decade, the number and variety of people and organizations working on addressing the environmental problems in engineering and industry has grown significantly. The general goal of environmentally conscious approaches to product and process design and management is the reduction of the negative environmental impact of a product throughout its life cycle. Several general approaches to reducing negative environmental impact exist. We have found it useful to distinguish them by differences in scope of organizational and temporal concern and developed a corresponding map shown in Figure 1 (Coulter, Bras et al. 1995; Bras 1997). Ideally, one would like to move from the lower left corner (arguably the current state of practice) to the upper right corner of Figure 1 and achieve sustainable development. In Figure 1, rather than indicate years, we have based the temporal concern gradations on life spans of products, people, and civilizations. The scale is not linear but indicates important distinctions between the approaches. Within a product life cycle we make an additional set of distinctions, indicating manufacturing, use, and disposal as possible lengths of temporal concerns. A product life-cycle could be as short as one or two years for consumer electronics or longer than 30 years for an airplane or ship, and the application of a given approach might change accordingly. Similarly, the scale of organizational concerns was chosen to indicate distinctions. These gradations are fairly self-explanatory; although it is worth noting that 'X products' refers to the negative environmental impact of a group of products. A scope equivalent to 'One Manufacturer' indicates concern about all the activities of a single manufacturing firm whereas 'X Manufacturers' indicates activities among a group of manufacturers.

As we study Figure 1, we note that three classes of approaches can be identified, namely,

- those which are applied *within* a single product life-cycle and focus on *specific* life-cycle stages,
- those that focus on a *complete* product life-cycle and cover *all* life-cycle stages, and
- those that go *beyond* single product life-cycles.

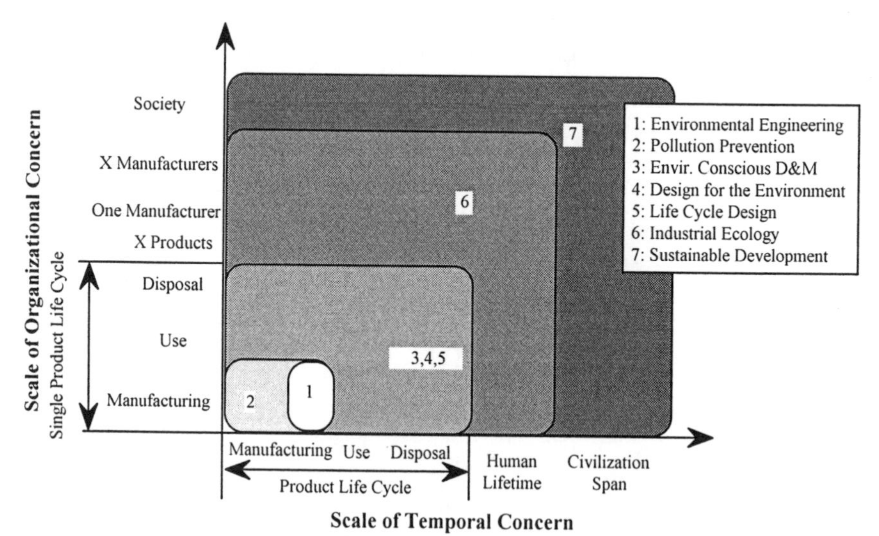

Figure 1 - Environmental and Organizational Scales of Environmental Impact Reduction Approaches (Coulter, Bras *et al.* 1995).

1.1 Approaches Focusing on Specific Life-Cycle Stages

Although its scope is broadening, traditional environmental engineering is primarily concerned with managing the fate, transport, and control of contaminants in water supplies and discharges, air emissions, and solid wastes. In the manufacturing context, the focus of environmental engineering effort is after pollutants have been generated, or at the 'end of the pipe'.

Practiced in industry, pollution prevention usually focuses on elimination of pollutants from existing products and process technologies (Freeman, Harten *et al.* 1992; The National Advisory Council for Environmental Policy and Technology 1993). Pollution prevention efforts are often mandated by legislation. For example, US regulations require pollution prevention programs for companies emitting hazardous substances. Pollution prevention efforts are typically broader than traditional environmental engineering efforts, as indicated in Figure 1. It is sometimes also referred to as waste minimization, green manufacturing or environmentally responsible manufacturing.

With the exception of Design for Environment, environmentally oriented Design for X approaches (such as Design for Recycling, Design for Disassembly, etc.) are also focused on a specific aspect of a product's life-cycle. Although DFX approaches are well known and in many cases well accepted in product design, the danger of focusing too much on specific DFX approaches (or specific aspects of a product life-cycle in general) is that strong concentration on a single environmental aspect may negatively affect other aspects and render the product less environmental friendly as a whole. For example, the shift from thermosets towards better recyclable thermoplastics in car-design may be counter-effective in certain cases, because in order to compensate for the diminished strength of the thermoplastics, their mass has to be increased, leading to potentially higher emissions from the manufacture of the plastics. A heavier product needs more transportation energy, and (thus) an increase in the overall environmental impact of the product, even though recyclability is increased. This understanding has led to the development of approaches that focus on a complete product life-cycle.

1.2 Approaches Focusing on a Complete Life-Cycle

It is generally agreed that environmental considerations cover a product's entire life cycle and that a holistic, systems-based view provides the largest capability for reducing environmental impact of both products and associated processes (Congress 1992; EPA 1993a). In Design for Environment (Navin-Chandra 1991; Ashley 1993; Fiksel 1996), Life-Cycle Design (Alting and Joergenson 1993; EPA 1993a), Environmentally Conscious Design and Manufacturing (Baca 1993; Owen 1993), and Green Design (Congress 1992), the scope of considerations, both in terms of time and the environment, is the life-cycle of one product (see Figure 1). Environmental concerns include all phases of this life-cycle, extending beyond the scope of pollution prevention to include the negative impact resulting from the use and disposal of this product. Similarly, the time scale considered is that of the product life-cycle, from design and manufacturing through use and final disposal or recycling of the materials in the product.

All these approaches have similar goals and encourage a holistic product view. However, it has already been recognized by many that this may not be enough. For example, modern manufacturers often rely on multiple suppliers, have multiple product lines, multiple facilities, often in multiple countries.

1.3 Approaches Going beyond Single Product Life-Cycles

In contrast to the preceding approaches, industrial ecology is not limited to a single product life cycle, but considers the interactions of several product life cycles (of possibly different lengths) over a larger time scale. Industrial ecology provides an integrated systems approach to managing the environmental effects of using energy, materials, and capital in industrial ecosystems analogous to the metabolism (use and transformation) of materials and energy in biological ecosystems (Frosch and Gallopoulos 1989; Ayres, Schlesinger *et al.* 1994; Graedel and Allenby 1995). In industrial ecology, companies, organizations and communities work together to minimize environmental impact and use each others waste in an intelligent manner for creating new products. For example, a carpet industry's production waste can be used by car companies to make sound-deadening materials. On the other hand, polyurethane seat foam taken from recycled cars can be further processed into carpet underlayer. In this example, a carpet manufacturer, car company, and seat recycling company have a symbiotic relationship like found in a biological ecosystems and formed what is termed an industrial ecosystem. Moving towards industrial ecology generally requires cooperation between industries (a vertical move in Figure 1).

The broadest approach, in terms of the scope of environmental and temporal concerns, is sustainable development and technology. The United Nations' World Commission on Environment and Development in its report *Our Common Future*, defines sustainable development as 'development that meets the needs of the present generation without compromising the needs of future generations.' Sustainable development is also the least well defined approach in terms of tools and methods. It is generally agreed that sustainable development requires at least pollution prevention, consideration of life-cycle consequences of production, and an approach that imitates natural or biological processes.

This thinking has caused a few companies to evolve beyond 'business as usual' and to what can be called 'thinking outside the box'. At Interface Flooring Systems, the environmental management approach is centered on *'quantification, qualification, symbiosis'*. This means that once a waste stream's amount has been defined (quantification) and its severity (qualification), an attempt is made not just to reduce it, but to find an outlet that can actually use the waste as a feedstock. These outlets can be other industries and an industrial ecosystem (like promoted by industrial ecology) can be obtained. However, this symbiotic approach can also take place with Nature. DaimlerChrysler is using natural fibers instead of glass fibers in its composites because they realized that the natural fibers can easily be decomposed by Nature at the end of their useful life. Any production scrap is also not a problem anymore. Similarly, Interface Flooring is using natural materials for some of its carpet products, e.g., animal hair and recently corn based fibers. However, both companies noted that entire supply chains had to be set-up. For example, DaimlerChrysler had to ensure a consistent crop quality, which even meant redesigning farming equipment. This does illustrate, however, that certain companies are not just thinking about collaboration with other industries, but collaborating with Nature.

To aid in this process, The Natural Step, a Swedish non-profit organization, has postulated a number of system conditions for sustainability based on the Laws of Thermodynamics.

1.4 The Natural Step

The Natural Step (*Det Naturliga Steget*) organization was founded by the Swedish oncologist Dr. Karl-Henrik Robert in 1989. It was born out of an effort to identify the fundamental scientifically derived principles of sustainability that everybody agreed upon. The Natural Step has phrased four socio-ecological principles that must be fulfilled to create a sustainable society (Robert, Holmberg *et al.* 1994):

1. Substances from the lithosphere (earth's crust and mantle) must not systematically accumulate in the ecosphere.
2. Society-produced substances must not systematically accumulate in the ecosphere.
3. The physical conditions for production and diversity within the ecosphere must not systematically be deteriorated.
4. The use of resources must be effective and just with respect to meeting human needs.

Basically, the first and second principles state that the natural equilibrium of material concentrations should not be disturbed. The third principle states that we should not destroy natural ecosystems. The fourth principle deals with equitable use of natural resources.

Although these principles are too vague to be applicable directly in engineering design and management, they do facilitate a foundation upon which strategy and policy can be built. We therefore view the Natural Step approach as general guidelines for strategy and policy making in both corporations and in society for that matter. In Chapter 7, we present a case study of Interface Flooring Systems, Inc. that employs the Natural Step ideas actively, see (Anderson 1998).

2. ISO 14000 ENVIRONMENTAL MANAGEMENT STANDARDS

When intellectuals appear, the great deceptions intervene. If I were to renounce wisdom and reject knowledge, the people would be hundred times better off.

Lao Tzu
From "Tao Te Ching"

A major factor that has made companies think about their environmental impact is the fact that the International Standard's Organization (ISO) has created a new family of standards for environmental management systems, called ISO 14000. The ISO 14000 family consists of the following standards:

ISO 14001: Environmental Management Systems (EMS) - Specification with guidance for use. This international standard specifies requirements for an environmental management system, to enable an organization to formulate a policy and objectives taking into account legislative requirements and information about significant environmental impacts. It applies to those environmental aspects that the organization can control and over which it can be expected to have an influence. It does not itself state specific environmental performance criteria.

ISO 14004: Environmental Management Systems - General guidelines on principles, systems, and supporting techniques. The general purpose for this standard is to provide assistance to organizations implementing or improving an Environmental Management System.

ISO 14010: Guidelines on Environmental Auditing - General principles on environmental auditing. An environmental audit is defined as a systematic, documented verification process of objectively obtaining and evaluating audit evidence to determine whether specified environmental activities, events, conditions, management systems, or information about these matters conform to audit criteria. This standard is intended to guide organizations, auditors and their clients on the general principles common to the conduct of environmental audits.

ISO 14011: Guidelines on Environmental Auditing - Audit procedures - Auditing of Environmental Management Systems. This standard establishes audit procedures for the planning and conduct of an audit of an EMS to determine conformance with EMS audit criteria.

ISO 14012: Guidelines on Environmental Auditing - Qualification criteria for environmental auditors. This standard provides guidance on qualification criteria for environmental auditors and is applicable to both internal and external auditors. Criteria for the selection of audit teams are not included.

ISO 14020: Environmental labels and declarations - General principles. An environmental label is a claim that indicates the environmental aspects of a product or service. The overall goal of environmental labels is, through communication of verifiable and accurate information (that is not misleading) on environmental aspects of a product or services, to encourage the demand for and supply of those products and services that cause less stress on the environment, thereby stimulating the potential for market-driven continuous environmental improvement.

ISO 14021: Environmental labels and declarations - Self-declared environmental claims (Type II environmental labeling). The objective of this standard is to harmonize the use of self-declared environmental claims.

ISO 14024: Environmental labels and declarations - Type I environmental labeling - Principles and procedures. The standard provides principles and procedures for establishing Type I environmental labeling programs, which are voluntary, multiple-criteria-based third party programs that awards a license which authorizes the use of environmental labels on products indicating overall environmental preference of a product within a particular product category based on life cycle considerations.

ISO 14031: Environmental management - Environmental performance evaluation - Guidelines. According to ISO 14031, Environmental Performance Evaluation (EPE) is an internal management process that uses indicators to provide information comparing an organization's past and present environmental performance wit its environmental performance criteria. It follows a 'Plan-Do-Check-Act' model of management.

ISO 14040: Environmental management - Life-Cycle Assessment (LCA) - Principles and framework. This standard specifies the general framework, principles, and requirements for conducting and reporting LCA studies. The standard does not describe the LCA technique in detail.

ISO 14041: Environmental management - Life-Cycle Assessment - Goal and scope definition and inventory analysis. This standard deals with two phases of LCA: goal and scope definition and Life-Cycle Inventory (LCI) analysis, as defined in ISO 14040. It specifies the requirements and the procedures necessary for the compilation and preparation of the definition of goal and scope for a LCA, and for performing, interpreting and reporting a LCI analysis.

ISO 14042: Environmental management - Life-Cycle Assessment - Life-cycle impact assessment (LCIA). LCIA is the third phase of LCA as described in ISO 14040. The purpose of LCIA is to assess a product and/or service' system's Life-Cycle Inventory analysis results to better understand their environmental significance. The LCIA phase models selected environmental issues, called impact categories, and use category indicators to condense and explain the LCI results.

ISO 14043: Environmental management - Life-Cycle Assessment - Life-cycle interpretation. This standard describes the final phase of the LCA technique, where the results of the Life-Cycle Inventory analysis and (if conducted) the life-cycle impact analysis, or both, are summarized and discussed as a basis for conclusions, recommendations and decision-making in accordance with the goal and scope definition.

ISO 14050: Environmental management - Vocabulary. This standard contains definitions of fundamental concepts related to environmental management, published in the ISO 14000 series of standards.

In a nutshell, ISO 14001 and 14004 deal with the management aspects and provide basic requirements for firms implementing an environmental management system, and ISO 14001 is probably the best known standard in the ISO 14000 family. Many books have been written about ISO 14001, e.g., (Cascio 1996). Concurrently or after the implementation of ISO 14001 and 14004, the ISO 14010 - 14012 standards must be implemented to ensure a correct implementation of the ISO 14001 and ISO 14004 standards, i.e., audits must be performed through internal and/or external auditors. The 14020-24 group of standards basically tell you what you can say and not say when you state, or ask for, an environmental claim and/or improvement. It also ensures that type II labels are not going to be potential trade-barriers used by governments. ISO 14031 stands on itself and focuses on how to quantitatively evaluate one's environmental performance. We discuss ISO 14031 more in Section 5.1. The ISO 14040 through 14043 standards deal with the Life-Cycle Assessment (LCA) procedure and are discussed in Section 4.

Although the actual standard documents are relatively short, interpreting them can be a problem, especially the 14040 - 14043 LCA standards. Other barriers to ISO 14000 registration are listed in (Jump 1995), where the cost effectiveness was also cited as a problem; a study of 34 Fortune 500 companies by (Gloria, Saad *et al.* 1995) shows that just implementing the LCA standards costs between $15,000 to $30,000 *per* product.

Nevertheless, ISO 14000 has been a powerful motivator for companies to consider environmental aspects For example, recently two large US car companies stated that they expect their tier one suppliers to be ISO 14000 certified. In Europe and Japan, ISO 14000 is adopted as a matter of fact. In Japan, ISO 14000 banners can be found in the subway! Besides from competitive pressure, being ISO 14000 certified is perceived as a sign of environmental stewardship.

However, in our humble opinion, ISO 14000 missed a critical opportunity: *It does not emphasize and show enough how environmental management should and can be a natural extension of good management practices.* As a result, many unfortunately see (rightfully or wrongfully) ISO 14000 registration as a specialized standalone issue with a high learning curve; a burden that negates the feeling of opportunity.

3. LIFE-CYCLE ASSESSMENT

Nature always tends to act in the simplest way.

Bernoulli

In 1972 Ian Boustead calculated the total energy consumption of the production of various beverage containers, see (Boustead 1996). This can be viewed as the beginning of LCA. With the advent of ISO 14040, the LCA methodology has started to consolidate. The LCA approach described in ISO 14040 is essentially the same as the one promoted by the Society of

Environmental Toxicology and Chemistry (SETAC). Compare the ISO 14040 framework presented to the SETAC framework presented in (Consoli, Allen *et al.* 1993). The only real differences are the choices of impact categories and the weighting schemes.

A conventional LCA consists of the following steps (Consoli, Allen *et al.* 1993; Jensen, Elkington *et al.* 1997), which is outlined in the ISO 14040 - ISO 14043 standards and in SETAC's 'Code of Practice':

1. Goal definition and scoping - ISO 14040
2. Inventory analysis - ISO 14041
3. Impact assessment - ISO 14042
4. Improvement Assessment (SETAC term)/Interpretation (ISO term) - ISO 14043

We will review each of these steps more in detail, with a critique at the end of each section. In (Ayres 1995) another review of LCA methods in general is given. If you become confused, do not worry. According to (Vigon 1997), LCA terms are 'indecipherable' for non-LCA experts.

3.1 Goal and Scope Definition

During this initial phase of a LCA, one is supposed to define the following:

1. 'Goal', which shall unambiguously state the intended application, including the reasons for carrying out the study and the intended audience, i.e. to whom the results of the study are intended to be communicated.
2. 'Scope', which describes the model of the systems to be studied. The scope should be defined well enough to ensure that the breadth and depth of a study are compatible with and sufficient to address the stated goal.
3. 'Function and functional unit'. The function is the performance characteristics of the system, while the functional unit is selected to measure 'the performance of the functional outputs of the product system'.
4. 'System boundaries', which define the unit processes that will be included in the system to be modeled.
5. 'Data quality', which must be defined by specific characteristics that describe both quantitative and qualitative aspects of data as well as the methods used to collect and integrate those data. There are five data quality indicators; precision, completeness, representativeness, consistency and reproducibility.
6. 'Critical review process' is added in the end as a quality control measure.

Most of these steps are sound and generic for just about any assessment method, except for one step: the definition of functions and functional units. Personally, we feel that these constructs are well meant, but counter-intuitive. Functional units are defined as 'the functional outputs of the product system whose primary purpose is to provide reference to which the input and output are normalized' (ISO 14041). But now consider the following statement from ISO 14041: "If additional functions of any systems are not taken into account in the comparison of functional units, then these omissions shall be documented. For example, systems A and B perform functions x and y which are represented by the selected functional unit, but system A performs function z which is not represented in the functional unit. As an alternative, systems associated with the delivery of function z may be added to the boundary of system B to make the systems more comparable". Are you lost? Assume we understand this for function z, how would we deal with say 20 products with several important functions each, using functional units? We believe that functional units are well meant, but have limited use in practice.

3.2 Inventory Analysis

The purpose of Life-Cycle Inventory (LCI) analysis is simply to quantify inputs and outputs of a product system. It contains the following main issues:

1. 'Data collection'. The data can be site specific or general; in any case, they must be collected for all unit processes within the system boundaries. Even qualitative data are allowed.
2. 'Calculation' is simply a step to manipulate the data to make the amount of data manageable.
3. 'Validation of data' is them employed to ensure the data quality. The purpose is to find areas where the data are insufficient so that better data can be gathered.
4. 'Relating data to unit process'. The purpose is obviously to ensure that the right data are associated with the right unit processes. For each unit process, an appropriate reference flow shall be determined, or functional unit, for normalization purposes.
5. 'Allocation' is employed when more than one product uses the inputs and outputs inside the system boundaries and ISO 14041 describes a procedure for how to allocate. It is a reasonable approach, but we think that principles from modern cost accounting could and should have been used to add consistency (see Chapter 3)

The inventory analysis is relatively straightforward from a mass and energy balance point of view. However, as we will see in Chapter 3, cost accounting terms and principles provide a much clearer approach for allocating mass and energy to a product. Rather than defining new ('indecipherable') terms, why not use what most of the business world is already familiar with?

3.3 Impact Assessment

Consider the impact assessment phase of LCA (ISO 14042). It consists of the following four steps:

1. 'Category definition', where the environmental impact categories are defined to describe the impacts caused by the considered products or product system. Examples of typical impact categories are abiotic resources, biotic resources, land use, global warming, stratospheric ozone depletion, ecotoxicological impacts, acidification and eutrophication.
2. 'Classification', which is a qualitative step based on scientific analysis of relevant environmental processes during which the various inputs and outputs are assigned to the various categories. Since some outputs have to be accounted for in several categories and thereby double and triple accounting may be necessary. The environmental impacts also have to be scaled according to their geographical impact into four groups; local, regional, continental and global. Throughout this process there is an implicit assumption that 'less is better'.
3. 'Characterization', which aims to model categories in terms of indicators. The model should be based on scientific knowledge where possible, but may have simplifying assumption and value-choices.
4. 'Valuation/weighing', which is a step designed to overcome that fact that 'comparison of these categories is not immediately possible'. By assigning weights to the various categories (based on policies, goals, stakeholder opinions, etc.) a number that describes the environmental impact is produced. In (Lindeijer 1996) several approaches for this process are presented.

In our opinion, the impact assessment is an area of potential disaster for the conventional LCA. The problem with impact categorization is that it can lead easily to never-ending discussions and political debate, and ultimately no action, as people disagree about which emission impacts which impact categories, and to what extent, etc. (Jensen, Elkington *et al.* 1997). This is not strange since the impact categories are defined according to environmental problems that are by definition different and, hence, inherently incomparable. During the Kyoto meeting in December 1997, the delegates had enormous problems of agreeing upon just the greenhouse effect, which is just one out of about ten categories. Now consider Table 1, which shows a possible categorization of impacts and the associated weighting scheme (this one is for the Eco-Indicator'95).

After defining the categories, the next step of the impact categorization is classification. Due to the fact that various emissions can contribute to several categories, double, triple counting, etc., is needed. This requires that the practitioners have a good understanding of the effects of every emission, because otherwise they will not be able to classify them correctly. Most engineers and managers will not have the necessary knowledge. They have to rely on third party databases and LCA tools, which they must take for granted and cannot check. This creates a dangerous reliance on black boxes for LCA studies.

Table 1 - The Use of Weighting Factors in the Eco-Indicator (PRé Consultants 1995).

Environmental Effect	Weights	Criterion
Greenhouse effect	2.5	0.1 C° rise every 10 years, 5% ecosystem degradation
Ozone layer depletion	100.0	Probability of 1 fatality per year per million inhabitants
Acidification	10.0	5% ecosystem degradation
Eutrophication	5.0	Rivers and lakes, degradation of an unknown number of aquatic ecosystems (5% degradation)
Summer smog	2.5	Occurrence of smog periods, health complaints, particularly amongst asthma patients and the elderly, prevention of agricultural damage
Winter smog	5.0	Occurrence of smog periods, health complaints, particularly amongst asthma patients and the elderly
Pesticides	25.0	5% ecosystem degradation
Airborne heavy metals	5.0	Lead content in children's blood, reduced life expectancy and learning performance in an unknown number of people
Waterborne heavy metals	5.0	Cadmium content in rivers, ultimately also impacts on people (see Airborne heavy metals)
Carcinogenic substances	10.0	Probability of 1 fatality per year per million people

The next step within the impact categorization is characterization, where one tries to assign the relative contribution of the relevant environmental processes. This is based on scientific knowledge. However, when such knowledge is not available, one simply makes value-choices. The result is that one mixes apples and oranges whose result cannot be comparable, which is a major problem (Ayres 1995). With respect to valuation, when organizations and groups of stakeholders are allowed to choose their own weights, one potential pitfall is they may tend to choose a set of weights that gives them the most favorable end-result for the product being assessed, ruining the credibility of the assessment.

3.4 Interpretation

Interpretation, or 'improvement assessment' as SETAC calls it, concerns three issues to facilitate decision-making:

1. 'Identification of significant environmental issues' in a prudent and justifiable manner which obviously is a necessity if a company should have any use of a LCA.
2. 'Evaluation', where first a qualitative check of the selection of data, processes, etc. is conducted to discuss the possible consequences of leaving out information. Secondly, one needs to do a systematic qualitative or quantitative analysis of any implications of changes in the input data caused by methodological and/or epistemological and/or data uncertainties. The last step (third) is to discuss the variations identified in relation to the goal and scope of the study.
3. 'Conclusions and recommendations', which is similar to any scientific or technical assessment, investigation, etc.

Although ISO 14043 explicitly mentions how to do the sensitivity analyses, in practice these seem to be rather limited. For example, in (Hanssen 1998), which according to (Christiansen, Heijungs *et al.* 1995; Hanssen, Rønning *et al.* 1995) developed 'one of the most comprehensive methods for Environmentally Sound Product Development' (their words, not ours), uncertainty and sensitivity analyses are omitted. More seriously, 'it is not possible to give a general rank of priorities of strategies and options for improvements' (Hanssen 1998). From this we understand that, in this case, the results do not have much value for a company (or for anybody else for that matter).

3.5 Some Remarks

The question whether the environmental management area is converging towards solution or confusion is a highly relevant question. Although ISO 14040-43 has immensely helped to convey to people that there is more to account for in a product's life than manufacture, we do not see LCA is the ultimate tool for business. According to (Ayres 1995), *all* LCA methods suffer from deficiencies, specifically:

- Data deficiency. This problem is due to the fact that environmental impact assessments are a very new field and consequently data are missing.
- Method deficiency. This problem is caused by the use of non-comparable units of measurement.

ISO knows this and deliberately built in a high degree of openness and flexibility and because it wants to see the LCA practice and methodology evolve. A good thing. However, the disadvantage of this approach, and its subsequent rather loose definitions, is that implementation can vary so much, that two studies of the same problem will in principle be incomparable. That can occur if:

- The studies have different system boundaries, goals and/or scope.
- The functional units are chosen differently.
- The impact categorization is different.
- The LCA software used in the two studies has different databases.
- The assumptions made in the process are different.

In our opinion, the probability of having comparable and ultimately useful studies using a conventional LCA approach (such as ISO 14040-43) is slim. This is also well documented in (Jensen, Elkington *et al.* 1997), where both industry and academic representatives are unsatisfied. It is also evident from (Hanssen 1998) who despite having developed the 'most comprehensive LCA method' (their words, not ours) still cannot compare products and thereby produce meaningful priority rankings of improvement options and strategies.

It is interesting to note that SETAC and ISO do not aim for an integrated environmental and cost assessment method. It seems they rather prefer to have two separate methods: one for Life-Cycle Analysis and one for Life-Cycle Costing. This is a very ineffective and inefficient approach (Keegan, Jones *et al.* 1991). It can, in fact, be outright dangerous as it unconsciously prohibits cross-functional and multi-disciplinary teamwork which is so important in today's marketplace, see e.g. (Meyer 1994). So how can economic and environmental aspects be tied? That is discussed next.

4. BRINGING ECONOMIC ASPECTS INTO ENVIRONMENTAL MANAGEMENT

Riches are for spending; and spending for honour and good actions.

Francis Bacon
In "Of Expense"

'Money talks' is a frequently heard proverb, especially in the US. Another frequently heard saying is that environmental management and impact reduction 'costs too much'. It is therefore not surprising that there is a great interest in linking economic aspects to environmental aspects - this book is about just that. In this section, we discuss some concepts that have approached the problem from very different angles, just to give you an idea on what people have thought of. Many of the approaches in the literature, however, in our opinion, do not handle economic and environmental issues to the extent needed because they handle costs too simplistically and/or fail to identify inefficient resource use.

4.1 Environmental Accounting

Environmental accounting, also referred to as life-cycle accounting, total cost accounting, green accounting and full cost accounting (EPA 1993a; Wood 1998), serves primarily to support environmental initiatives and policies by including the costs and benefits that are derived from the effects of the environment on the general ledger. It is therefore essentially a financial reporting and analysis of environmental aspects as they are manifested on the general ledger, see (EPA 1993a) for a thorough discussion.

Consider Table 2 where a number of costs and activities are listed that may be relevant when implementing pollution prevention control and/or environmental accounting. In the 'Costs Traditionally Traced' column we find costs that are typically traced to specific processes. The 'Environmental Compliance Costs' column contains costs that occur as a direct result of environmental compliance efforts, while the in 'Related Oversight Costs' column we find costs that arise as an indirect result of attempting to satisfy various compliance requirements.

The essence of environmental accounting, the way we see it, is to explicitly report and analyze the costs in the two right columns as they appear in the general ledger, in addition to the traditional costs in the left column that are always analyzed. The advantage is that it makes the environmental compliance costs and related oversight costs visible. Normally they are just lumped together in the big sack of general overhead cost, (Hamner and Stinson 1995) making environmental accounting and ultimately environmental management impossible. It is therefore not surprising that Activity-Based Costing is frequently used in environmental accounting.

Table 2 - Business Activities Creating Environmental Compliance Costs, Compliance Oversight Costs and Non-environmental Costs (Hamner and Stinson 1995).

Costs Traditionally Traced	Environmental Compliance Costs	Related Oversight Costs
Depreciable Capital Costs • Engineering • Procurement • Equipment • Materials • Utility connections • Site preparations • Facilities • Installations Operating Expenses • Start-up • Training • Initial raw material • Working capital • Raw materials • Supplies • Direct labor • Utilities • Maintenance • Salvage value	Receiving Area • Spill response equipment • Emergency response plan Raw Material Storage • Storage facilities • Secondary containment • Right-to-know training • Reporting and record-keeping • Safety training • Safety equipment • Container labels Process Area • Safety equipment • Right-to-know training • Waste collection equipment • Emission control equipment • Sampling and testing • Reporting and record-keeping Solid and Hazardous Wastes • Sampling and testing • Containers • Labels and labeling • Storage areas • Transportation fees • Disposal fees Air and Water Emission Controls • Permit preparation • Permit fees • Capital costs • Operating expenses • Recovered materials • Inspection and monitoring • Record-keeping and reporting • Sampling and testing • Emergency planning • Discharge fees	Purchasing • Product/vendor research • Regulatory impact analysis • Inventory control Engineering • Hazard analysis • Sampling and testing Production • Employee training • Emergency planning • Medical monitoring • Rework • Waste collection • Disposal management • Inspections and audits Marketing • Public relations Management • Regulatory research • Legal fees • Information systems • Penalties and fines • Insurance Finance • Credit costs • Tied-up capital Accounting • Accounting system development • Accounting system maintenance

4.2 Exergy Costing

The idea of optimization is a vital part of what is called 'Exergy Costing' which is a part of 'Exergoeconomics' or 'Thermoeconomics' which draw their roots from the fundamental Laws of Thermodynamics. The essence of exergoeconomics is according to (Tsatsaronis, Lin *et al.* 1993) that it 'combines a detailed exergy (second law) analysis with appropriate cost balances to study and optimize the performance of energy systems'. Exergy is the availability to do work (Thabit and Stark 1984). The analyses and balances are usually formulated for single components of an energy system. Exergy costing, one of the basic principles of exergoeconomics, means that exergy, rather than energy or mass, is the basis for costing energy carriers.

Exergy costing sounds very interesting but it has at least two inherent problems that make it not very useful in our opinion:

1. Exergy costing is based on the notion of optimization. But in a deregulated market situation, i.e., prices are set by the market, optimization simply does not make sense. Exergy costing is therefore limited only to regulated markets where demand and price can be easily determined in advance. Such markets are vanishing in the US and Europe, e.g., many countries have now 100% deregulated electricity markets.
2. More seriously, however, is that exergy costing is based on sub-optimization, that is, the analyses are performed 'for single components of an energy system' (Tsatsaronis, Lin *et al.* 1993).

These and other problems made a number of researchers abandon the whole idea. Others, such as (Yang, Wang *et al.* 1992), have come up with an improvement of exergy costing called exergy-anergy costing. Anergy is the energy <u>not</u> available to do work. Thus, exergy + anergy = energy. The very same basic problems still persist, however. Nevertheless, exergy costing has been employed over the life-cycle of an energy system (Thabit and Stark 1984), but the two inherent aforementioned problems were not resolved.

In (Cornelissen 1997), a related approach is presented called Exergetic Life Cycle Assessment (ELCA), built on similar ideas as Exergy Costing. ELCA is a very detailed approach to identify thermodynamic deficiencies, i.e. irreversibilities and it is like ISO 14040-43 LCA; 'the framework of the ELCA and the LCA are similar. The goal definition and scoping of the LCA and ELCA are completely identical. The inventory analysis of the ELCA is more extensive' (Cornelissen 1997). Basically all the indicators used in ISO 14000 LCA are replaced with exergy while the framework is kept the same. The question is then whether exergy analysis better captures environmental performance than many of the conventional indicators. We believe it does to some extent. The greatest advantage is that exergy provides an indicator free from political content. A problem with using exergy is that toxicity cannot be captured by exergy and is thus ignored. Therefore, in short, ELCA has overcome some of the flaws of the ISO 14000 LCA, but in the wake has created some new ones.

4.3 Emergy Analysis

Another very different approach is called 'emergy analysis', pioneered by Howard Odum and presented in (Odum 1996). This approach attempts to assess sustainability directly, without assessing environmental impact and costs explicitly. Emergy analyses are based on the belief that on every scale phenomena are self-organizing to include energy-aligned material cycles, concentrate in centers, and pulse concentrations in time (Odum 1998). Emergy is defined as (Brown and Ulgiati 1998) 'the amount of energy of one form directly or indirectly required to provide a given flow or storage of energy or matter'. Another important definition in this theory is 'transformity', which is defined as 'the ratio of emergy needed to produce a flow or storage to the actual energy of that flow or storage'.

The purpose of the emergy analysis is to evaluate all system components on a common basis (Martin 1998), and provide a quantitative and systematic comparison between economic use activities and the supporting ecosystems to *maximize* total performance (Tilley 1998). The reference point in emergy analyses is the sun, since everything, it is assumed, originated from the sun, and the unit of emergy is therefore often given in terms of solar emergy joules, or solar emjouls, in short 'sej'. In order that the free work of nature can be compared to the paid services of humans, the units of emergy are translated to emergy-dollars (EmDollars or Em$). The translation is simply the amount of emergy divided by the emergy/money ratio of the economy (Odum 1996).

Emergy analyses are undoubtedly an innovative approach that has scientific foundations, but there are several major potential pitfalls from a performance measurement perspective. First of all, natural systems are complex. Many peculiar, but vital, relationships exist in Nature. Due to this complexity, models would have to be grossly simplified to have the necessary information and to be computational possible. This simplification gives room for a large amount of uncertainty. It can also lead to outright wrong decisions if subtle aspects of the ecosystems are either ignored or simply not detected.

Another problem relates to the economy. Large economies will have a low emergy/money ratio. Consequently, in two countries where the emergy is exactly the same but the economies differ significantly, the environmental impact of the *same* project will be *different* in terms of Em$ in the economically weaker country. Hence, the Em$ concept is highly misguiding. Exchange rates between various economies will also lead to fluctuating emergy estimates and countries in dire straits will have devaluated ecosystems. This issue will become more and more important as the globalization of the economy proceeds. A weakened Brazilian Real will, e.g., reduce the value of the Brazilian rain forests and thus promote economic activity in the rain forests, since the Em$ has dropped.

Although interesting, we find the emergy approach dangerous because it *appears* to measure sustainability, but in a distorted way. Measuring sustainability directly may still be beyond our means due to its complexity.

5. MEASURING ENVIRONMENTAL IMPACT

Not everything that counts can be counted, and not everything that can be counted counts.

Albert Einstein

The objective of this section is to step back and look at a fundamental aspect of environmental impact assessments that sometimes seems to be forgotten; i.e., *what* to measure and *how* to measure it. When it comes to indicators, there are many to choose from. We have chosen some typical ones to illustrate the complexity of this field.

5.1 ISO 14031 - Environmental Performance Evaluation

ISO 14031 explicitly deals with Environmental Performance Evaluation (EPE) that is an internal management process that uses indicators to provide information comparing an organization's past and present environmental performance with its environmental performance criteria, according to ISO 14031. The proposed approach follows a Plan-Do-Act sequence:

- Plan: Plan the EPE and select indicators for EPE (these indicators may be selected from existing indicators, or new ones can be developed).
- Do: Collect data relevant to the selected indicators; convert this data into information describing the organization's environmental performance; assess the information describing the environmental performance in comparison with the organizations environmental performance criteria; report and communicate the information describing the organizations environmental performance.
- Check and Act: Review and improve the EPE.

To measure the performance, the standard describes two general categories of performance indicators:

1. Environmental Performance Indicators (EPIs) which consist of two types:
   - Management Performance Indicators (MPIs) that provide information about management's efforts to influence the environmental performance of the organization's operation (e.g., the number of pollution prevention initiatives implemented, number of achieved objectives and targets)
   - Operational Performance Indicators (OPIs) that provide information about the environmental performance of the organization's operations (e.g., quantity of energy used per year, quantity of water per unit product).
2. Environmental Condition Indicators (ECIs) that provide information about the condition of the environment (e.g., concentration of a specific contaminant in ground- or surface water, or ambient air). This information can help an organization to better understand its actual or potential impact.

The standard basically says that many different types of indicators can be used for performance evaluation, as long as they promote understanding, and we could not agree more. But, we also note that having a wide choice of options may result in findings that are incomparable. ISO 14031 provides a large number of MPI, OPI, and ECI examples. In the following, we provide some indicator examples that are widely known/used of our own.

5.2 Sustainability Indicators

Although all indicators can arguably be described as sustainability indicators due to the broadness of the sustainability issue, in this section we present examples of indicators that are especially designed for keeping track of certain aspects of sustainability. Some indicators are designed for other purposes as well, e.g. Gross Domestic Product (GDP) which is used to described wealth creation in an economy.

In (Corson 1994) an overview of many indicators is presented. Here we just present the sustainability indicators briefly in Table 3. The letters G, R, N and L denote global, regional, national and local, respectively, in terms of the geographical impact area that the various indicators apply to. Also, the table is split into two main dimensions: 1) 'natural resources and environment' and 2) 'social environment'. We are not going to discuss all the indicators in Table 3, but just point out a few issues. First, many of these indicators are very specialized and therefore are not easy to use in design and/or management. Secondly, other indicators are too macro to be useful. Thirdly, some of the indicators are outright misguiding. The best known of these must be the GDP, see (Gates 1998) for an excellent discussion. Consider the fact that the GDP is actually increasing for every accident. The Exxon Valdez accident in Alaska increased the GDP because of the money spent on clean-up. Clearly, an increase in GDP can be very misguiding when it comes to measuring progress towards sustainability.

Table 3 - Selected Sustainability Indicators Used by Major Organizations. Based on (Corson 1994).

Dimension	Indicator
Natural Resources and Environment	
Energy	1. Energy Use, Total and per Person (G, N, L) (World Resources Institute and International Institute for Environment and Development 1992). 2. Energy Efficiency Index (N) (World Resources Institute and International Institute for Environment and Development 1992). 3. Percent of Energy from Renewable Sources (G, N, L) (World Resources Institute and International Institute for Environment and Development 1992). 4. Energy Imports as Percent of Consumption (N) (World Resources Institute and International Institute for Environment and Development 1992). 5. Fossil Fuel Reserves (G, N) (World Resources Institute and International Institute for Environment and Development 1992).
Non-Fuel Minerals	1. Aluminum Consumption per Person (N) (United Nations Environment Programme 1992). 2. Percent of Aluminum Recycled (N) (United Nations Environment Programme 1992). 3. Metal Reserves (G) (World Resources Institute and International Institute for Environment and Development 1992). 4. Metal Reserves Index (N) (World Resources Institute and International Institute for Environment and Development 1992).
Solid Waste	1. Municipal Solid Waste, Total and per Person (G, N, L) (World Resources Institute and International Institute for Environment and Development 1992; United Nations Environment Programme 1993). 2. Percent of Glass and Paper Recycled (N, L) (United Nations Environment Programme 1992; United Nations Environment Programme 1993).
Hazardous Waste	1. Hazardous Waste Generated, Total, per Person and per km^2 (N, L) (United Nations Environment Programme 1992; United Nations Environment Programme 1993).
Atmosphere and Climate	1. Greenhouse Gas Emissions, Total and per Person (G, N) (World Resources Institute and International Institute for Environment and Development 1992). 2. Atmospheric Concentration of Carbon Dioxide (G) (World Resources Institute and International Institute for Environment and Development 1992). 3. Average Global Air Temperature (G) (Brown, Flavin *et al.* 1992).
Acidification	1. Emissions of Sulphur and Nitrogen Oxides, Total and per Person (N) (United Nations Environment Programme 1992; World Resources Institute and International Institute for Environment and Development 1992). 2. Acidity of Rainfall, Surface Water and Soil (L) (United Nations Environment Programme 1992).

Air Pollution	1. Emissions of Traditional Air Pollutants (N) (United Nations Environment Programme 1993). 2. Concentrations of Carbon Monoxide, Nitrogen and Sulphur Oxides, Ozone (L) (United Nations Environment Programme 1992).
Ozone Layer Depletion	1. Consumption of Ozone-Depleting Chemicals, Total and per Person (G, N, L) (United Nations Environment Programme 1992). 2. Atmospheric Concentration of Ozone-Depleting Chemicals (G) (World Resources Institute and International Institute for Environment and Development 1992).
Noise	1. Percent of Population Disturbed by Traffic Noise (N, L) (OECD 1993).
Fresh Water Supply	1. Water Withdrawals, Total and per Person (G, N, L) (World Resources Institute and International Institute for Environment and Development 1992; United Nations Environment Programme 1993). 2. Water Withdrawals as Percentage of Water Resources (N, L) (World Resources Institute and International Institute for Environment and Development 1992). 3. Renewable Water Supply per Person (N) (World Resources Institute and International Institute for Environment and Development 1992).
Fresh Water Quality	1. Nitrogen and Phosphorus Concentration in Major Rivers (R) (World Resources Institute and International Institute for Environment and Development 1992; OECD 1993). 2. Concentration of Nitrogen, Phosphorus and Organic Chemicals, Surface Water and Groundwater (L) (United Nations Environment Programme 1992). 3. Biological and Chemical Oxygen Demand (L) (United Nations Environment Programme 1992).
Food and Agriculture	1. Index of Food Production per Person (G, N) (World Resources Institute and International Institute for Environment and Development 1992; United Nations Environment Programme 1993). 2. Food Import Dependency Ratio (N) (World Resources Institute and International Institute for Environment and Development 1992). 3. Pesticide Use (N) (United Nations Environment Programme 1992; World Resources Institute and International Institute for Environment and Development 1992).
Land and Soil	1. Land Degradation as Percentage of Vegetated Land (G, R) (World Resources Institute and International Institute for Environment and Development 1992).
Forests	1. Percent of Land Area in Forest and Woodland (G, N, L) (World Resources Institute and International Institute for Environment and Development 1992). 2. Deforestation Rate (G, N) (World Resources Institute and International Institute for Environment and Development 1992; United Nations Environment Programme 1993). 3. Reforested Area as Percentage of Deforested Area (G, N) (World Resources Institute and International Institute for Environment and Development 1992).
Natural Habitat	1. Percent of Land under Protected Status (G, N) (World Resources Institute and International Institute for Environment and Development 1992). 2. Number and Extent of Protected Areas (G, N, L) (World Resources Institute and International Institute for Environment and Development 1992). 3. Protected Area Index (N) (World Resources Institute and International Institute for Environment and Development 1992).
Wildlife	1. Percent of Wildlife Species at Risk (G, N, L) (World Resources Institute and International Institute for Environment and Development 1992). 2. Species Risk Index (N) (World Resources Institute and International Institute for Environment and Development 1992).
Marine Resources, Fisheries	1. Marine Fish Catch as Percentage of Estimated Sustainable Yield (G, R) (Brown, Flavin *et al.* 1992; World Resources Institute and International Institute for Environment and Development 1992). 2. Coastal Ocean Pollution Index (N) (World Resources Institute and International Institute for Environment and Development 1992). 3. Municipal and Industrial Discharges to Coastal Waters (L) (United Nations Environment Programme 1992). 4. Total Suspended Solids and Biological and Chemical Oxygen Demand in Coastal Waters (L) (United Nations Environment Programme 1992).
Transportation	1. Total Production of Automobiles and Bicycles (G, N) (Brown, Flavin *et al.* 1992). 2. Passenger Cars per 1000 People (G, N, L) (United Nations Environment Programme 1993). 3. Measures of Passengers and Freight Carried by Air, Rail and Road (G, N) (Brown, Flavin *et al.* 1992; World Resources Institute and International Institute for Environment and Development 1992).
Economy	1. Gross Product per Person, World and Domestic (G, N) (United Nations Environment Programme 1993). 2. Unemployment Rate (N, L) (United Nations Environment Programme 1993). 3. Inflation Rate (N) (United Nations Environment Programme 1993).
Socioeconomic Equity	1. Percent of Population Living in Absolute Poverty (G) (United Nations Environment Programme 1993). 2. Income Ratio of Highest 20% of Households to Lowest 20% (G, N) (Brown, Flavin *et al.* 1992; United Nations Environment Programme 1993). 3. GDP per Person for Developing Nations as Percentage of GDP for Industrial Nations (G)

	(Sivard 1993). 4. Years of Schooling, Females as Percentage of Males (N) (United Nations Environment Programme 1993). 5. Percent of Parliament Seats Held by Women (N) (United Nations Environment Programme 1993).
Social Environment	
Human Development	1. Human Development Index (N, L) (United Nations Environment Programme 1993). 2. Life Expectancy at Birth (G, N) (United Nations Environment Programme 1993). 3. Expenditures for Education and Health per Person and as Percent of GNP (G, N, L) (Sivard 1993).
Housing	1. Average Number of Persons per Room in Housing Units (N, L) (World Resources Institute and International Institute for Environment and Development 1992).
Utilities	1. Percent of Households without Electricity (N, L) (World Resources Institute and International Institute for Environment and Development 1992). 2. Telephones per 1000 People (N, L) (United Nations Environment Programme 1993).
Security	1. Intentional Homicides per 100 000 People (N, L) (United Nations Environment Programme 1993). 2. War-related Deaths (G, N) (Sivard 1993). 3. Military Expenditures as Percentage of Combined Expenditures for Education and Health (G, N) (Sivard 1993; United Nations Environment Programme 1993).
Population	1. Annual Rate of Population Increase (G, N, L) (Population Reference Bureau 1993). 2. Birthrate per 1000 People (G, N, L) (Population Reference Bureau 1993). 3. Population Density (G, N, L) (Population Reference Bureau 1993). 4. Access to Birth Control Index (N) (Population Crisis Committee 1992b). 5. Percent of Married Couples Using Birth Control (G, N) (World Resources Institute and International Institute for Environment and Development 1992; United Nations Environment Programme 1993).
Health	1. Life Expectancy at Birth (G, N, L) (World Resources Institute and International Institute for Environment and Development 1992; United Nations Environment Programme 1993). 2. Infant Death Rate and Child Death Rate (G, N, L) (World Resources Institute and International Institute for Environment and Development 1992; United Nations Environment Programme 1993). 3. Calorie Supply and Protein Consumption per Person (N) (World Resources Institute and International Institute for Environment and Development 1992). 4. Access to Safe Drinking Water (N) (World Resources Institute and International Institute for Environment and Development 1992).
Education	1. Literacy Index (N) (United Nations Environment Programme 1993). 2. Schooling Index (N) (United Nations Environment Programme 1993).
Culture	1. Daily Newspaper Circulation per 1000 People (G, N) (United Nations Environment Programme 1993). 2. Radios per 1000 People (G, N) (United Nations Environment Programme 1993). 3. Book Titles Published per 100 000 People (N) (World Resources Institute and International Institute for Environment and Development 1992).
Political Participation and Involvement	1. Percent of Population Voting in Elections (N, L) (Shapiro 1992). 2. Political Freedom Index (N) (Population Crisis Committee 1992a). 3. Civil Rights Index (N) (Population Crisis Committee 1992a).
Governmental Stability and Effectiveness	1. Changes in Government Indicator (N) (Population Crisis Committee 1990). 2. Communal Violence Indicator (N) (Population Crisis Committee 1990). 3. Government Efficiency Index (N) (Shapiro 1992).

5.3 Macro Level Indicators

One of the best-known macro level environmental impact measures is Paul Ehrlich's environmental impact equation (Burkhardt 1998):

$$I = P \times A \times T \qquad (1)$$

In plain words, the environmental impact (I) is equal to the product of the world population (P) with affluence (A) and technological inefficiency (T). Clearly, its level is too high for use in business management. The same lack of specificity exists also in the so called 'Master Equation' (Graedel and

Allenby 1995), which is used by several experts to sum up causes for environmental impact:

$$\text{EI} = \text{Population x } \frac{\text{GDP}}{\text{Person}} \text{ x } \frac{\text{Environmental Resource Use}}{\text{Unit of per Capita GDP}} \tag{2}$$

In this equation, GDP is Gross Domestic Product of a country and EI is Environmental Impact. Obviously, such equations are not useful in business at all because they deal with input variables such as GDP. Clearly, a product has no measurable influence on the GDP. Hence such indicators cannot capture phenomena on a product level.

5.4 Global Warming Potential

In (IPCC 1993) a detailed analysis of climate changes is presented, where a possible enhanced Greenhouse Effect induced by man is investigated using the Global Warming Potential (GWP), which is also used in the ISO 14000 standard. GWP is the time integrated commitment to climate forcing from the instantaneous release of 1 kg of trace gas expressed relative to that from 1 kg of carbon dioxide (Shine, Derwent *et al.* 1993). The formula is:

$$\text{GWP} = \frac{\int_0^n a_i \cdot c_i \, dt}{\int_0^n a_{CO_2} \cdot c_{CO_2} \, dt} \tag{3}$$

where

a_i: The instantaneous radiative forcing due to an unit increase in the trace gas, i, concentration.

c_i: The concentration of trace gas, i, remaining at time, t, after its release.

n: The number of years over which the calculation of trace gas, i, is performed.

Similar analyses have also been performed by (Derwent 1990; Fisher, Hales *et al.* 1990; Lashof and Ahuja 1990). Some problems with evaluating the GWP are:

- The atmospheric lifetimes of gases and future variation (especially CO_2).
- The dependence of the radiative forcing of a gas on its concentration and the concentration of other gases with special overlapping absorption bands.
- The calculation of the indirect effects of the emitted gases and the subsequent radiative effects of these indirect greenhouse gases (ozone poses a particular problem).
- The specification of the most appropriate time period over which to perform the integration.

As can be seen from GWP, this indicator is estimating other gases relative to the trace gas CO_2, which is believed to be of primary importance. The focus on CO_2 is due to the 'greenhouse' properties of CO_2, but there are natural cyclic variations of the climate that do not relate to the CO_2 concentration in the atmosphere. In fact, 'the inherent variability of the climate system appears to be sufficient to obscure any enhanced greenhouse signal to date. Poor quantitative understanding of the low-frequency climate variability (particularly on the 10 - 100 years scale) leaves open the possibility that the observed warming is largely unrelated to the enhanced greenhouse effect' (Wigley and Barnett 1993). A second problem related to the carbon cycle is the identification of sources and sinks (Watson, Rodhe *et al.* 1993):

- The uptake of CO_2 by the oceans is underestimated.
- There are important unidentified processes in terrestrial ecosystems that can sequester CO_2.
- The amount of CO_2 released from tropical deforestation is at the low end of the current estimates.

Given all these problems in understanding the role of CO_2 in our world, we find it risky to base all our work on GWP. The GWP, however, is probably one of the best and most likely it *does* capture what it is supposed to capture, i.e., the potential for various gases to cause global warming. The problem is that the GWP only applies to atmospheric gasses. To remedy that situation, researchers have come up with several different types of potentials such as Acidification Potential (AP), Ozone Depletion Potential (ODP), etc.

5.5 Chemical Indicators

Finding metrics in the fields of atmospheric chemistry, soil chemistry and general chemistry, to mention some, is very easy. There is in fact, an overwhelming number of metrics to choose from. The only problem is that all these metrics are so specialized, even more specialized than the GWP, that they cannot be used by non-chemists for purposes such as engineering design and management. Some examples are:

- Level I Fugacity calculation, which describes how a given amount of chemical partitions is at equilibrium between six media; air, water, soil, bottom sediment, suspended sediment and fish (Mackay, Shiu *et al.* 1992a).
- Level II Fugacity calculation, which simulates a situation in which chemical is continuously discharged into the multimedia environment and achieves a steady-state equilibrium condition at which input and output rates are equal (Mackay, Shiu *et al.* 1992a).
- Level III Fugacity calculation, which is developed to find the intermediate transport rates by the various diffusive and non-diffusive processes described in (Mackay and Paterson 1991). Whereas Level I and Level II calculations assume equilibrium to prevail between all media, this is

recognized as being excessively simplistic and even misleading (Mackay, Shiu *et al.* 1992a). The expressions for the transport rates are found by selecting values for 12 intermediate transport velocity parameters.

In addition to these we have the physical-chemical properties that can be used as indicators for certain purposes, e.g., half-life to keep track of radioactivity. In (Howard, Boethling *et al.* 1991; Mackay, Shiu *et al.* 1992a; Mackay, Shiu *et al.* 1992b; Mackay, Shiu *et al.* 1993; Mackay, Shiu *et al.* 1995; Mackay, Shiu *et al.* 1997) comprehensive lists of chemicals and the values of the various metrics are given.

5.6 Biological Indicators

Biological indicators are based on using control groups of biological entities as a relative indicator of the environment. In (Thingstad 1997), for example, birds called Pied Flycatchers (*Ficedula Hypoleuca*) are used to determine the level of variability in the environment. This approach could conceptually be used in some situations where for example toxicity is an issue. There are also indicators for vegetation, such as the 'Two-Axis Adjusted Vegetation Index', see (Xia 1994). Such indicators are used to quantitatively study of vegetation status such as forest canopies and crop production by employing satellites. Such indicators are clearly valuable and useful for policy-makers. Direct use in business decision-making is a whole other story.

5.7 Mass- and Energy-Balance Indicators

There are many environmental indicators in the literature based on mass- and energy balances. In essence, they measure material and energy efficiencies and most are very simple. For example, in (Cunningham 1994) a very simple waste index is presented, which is simply a ratio between the unwanted material (waste) created in the process and the material spent in the process. This is simply a process efficiency metric, a well-known concept in engineering. It can be surprising, how much can be learned from such simple mass and energy process efficiency metrics and we would recommend all to first look at their efficiencies before plunging into using complicated and unclear metrics.

The advantage of mass and energy balances and basic mass and energy efficiencies is that they are simple. But the problem is how to relate them to actual environmental impact because they fail to capture any qualification, i.e., the difference, say, between a ton of lead or a ton of plutonium. The index in (Cunningham 1994) was multiplied with a 'difficulty factor' in order to relate the efficiency to environmental impact. How that factor is determined was not mentioned. In the following section, we provide our 'middle-of-the-road' approach to measuring environmental impact, called the Waste Index.

6. MEASURING IMPACT USING THE WASTE INDEX

All nature is but art unknown to thee;
All chance, direction which thou canst not see;
All discord, harmony not understood;
All partial evil, universal good;
And, spite of pride, in erring reason's spite,
One truth is clear, Whatever is, is right.

Alexander Pope

6.1 Basic Assumption - Measure Impact Relative to Nature

Given the seemingly endless debate over the correctness environmental impact indicators, we as 'dumb' engineers took a step back and inspired by the Natural Step thought about what everybody agreed on regarding environmental impact. To us, it seemed that everybody agrees in general on the following:

- More releases into Nature are always worse than less (given that we talk about similar substances.
- The longer a release can be detected in Nature, the worse it is.
- The larger the area affected by a release, the worse it is.

This thinking led to an indicator that we call The Waste Index (WI) and which is based on the following basic assumption:

Any substance in a sufficient amount beyond the natural amount of the substance in a control volume (environment) can be considered waste (pollution).

Reversing this basic assumption establishes a more general statement: *Whatever the concentration of a substance that exists naturally in Nature is, that is what it is supposed to be*. Any disturbance to the natural equilibrium of materials is therefore a measure of environmental impact. This leads to what we call our 'Nature Knows Best' axiom:

Environmental impact can only be measured relatively by benchmarking Nature

In keeping with this, 'Waste' is, therefore, defined as all materials (solid, liquids, or gases - toxic, radioactive or not) created by a human or industrial activity that exceed the natural amount in a specific geographical area of the environment called 'Control Volume'. More specifically, the 'Control Volume' is the geographical area that is affected by either the generated waste itself, or by the substances into which the waste is decomposing.

In general, waste does not exist indefinitely, but decompose or are absorbed into natural substances. However, some chemical compounds persist longer than others. Hence, the amount of disturbance of the natural balance, the length of time it persists, as well as the area it affects, all directly correlate to the seriousness of the impact. This represents a paradigm shift; away from trying to assess the actual environmental problems to assessing simply the relative impact on the balance in Nature.

6.2 Calculating the Waste Index

The preceding assumptions and the 'Nature Knows Best' axiom form the basis of our Waste Index - a simple mathematical metric that measures environmental impact by benchmarking nature. The general expression for the Waste Index (WI) is

$$WI = \sum_{i=1}^{M} \frac{V^i}{A^i_N} \cdot \int_0^{T^i_N} d^i(t)\ dt \tag{4}$$

The Waste Index equation consists of the following elements:

- The degradation function, $d^i(t)$, is a mathematical expression for the rate of degradation of a substance *i* in a control volume. It is based on the following first-order differential equation in (Scow and Hutson 1992):

$$-\frac{dC}{dt} = k \cdot C \tag{5}$$

  where C is the concentration of the substance, whose units can be either [kg/m^3] or [g/l], and k is a constant. The degradation function can in principle be of any shape. By assuming a first-order degradation as in Figure 2, and constant release R^i, the WI can be expressed as:

$$WI = \sum_{i=1}^{M} V^i \cdot \frac{R^i}{A^i_N} \cdot \int_0^{T^i_N} (1 - e^{-\frac{(T^i_N - t)}{t}})dt \tag{6}$$

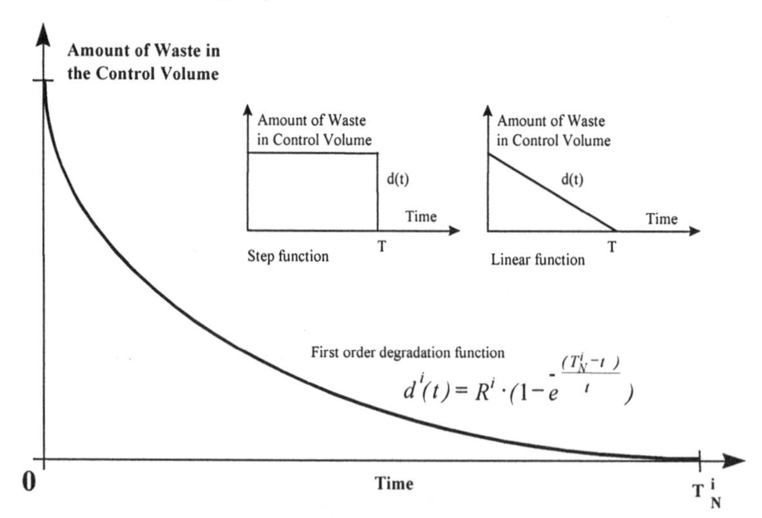

Figure 2 - Degradation of Generated Waste.

In the simplest case, the degradation function is a step function, as in (Emblemsvåg and Bras 1997a), resulting in a very conservative WI.

However, a more appropriate function, yet simple, is a linear degradation function, which yields:

$$WI = \sum_{i=1}^{M} R^i \cdot \left(\frac{V^i \cdot T_N^i}{2 \cdot A_N^i} \right) \tag{7}$$

- Released amount R(t) [kg/h or ton/year] of the released substance, i.e. waste. For simplicity, we often assess R(t) on an annual basis and assume it to be constant, denoted as R.
- Natural amount A_N [kg] is the amount that exists naturally of the released substance, or of the substances into which the released substance would decompose, in the specified control volume.
- Estimated time T_N [year] the released substance, the waste, will affect the control volume. This is the time it takes Nature to remove the disturbance in the control volume and return to the pre-existing natural balance. In order to use simple degradation formulas, it is assumed that all releases occur as a single release, although more sophisticated release and degradation models can surely be used, if desired. Given this, there are three options for measuring this balance time T_N, in order of preference:

1. Direct measurement; the time is measured from the very first emission to the point at which 99.99% of the releases' effects are gone (by convention) - that is, the time the control volume needs to achieve balance again. Direct measurements can be done experimentally in predefined standard environment such as the U.S. Standard Atmosphere, see (COESA 1976; IPCC 1993). Although a standard atmosphere is not the same as the real atmosphere, it will give consistent results over time.
2. Half-times can be used, but then the entire chain of chemical reactions must be investigated and a library of chemical reactions is needed, such as (Jacobson and Hampel 1946 - 1959). Such publications are rare today, arguably due to the shear work of making them. But, by using modern information technology, managing such vast amounts of information could be an important role for organizations like ISO, in our opinion.
3. For toxic substances, biological indicators can be used. For example, in 1989 Tributyltin (TBT), a resistance reducing compound, was banned in coating used on ships smaller than 25 meters (Council Directive 89/677/EEC 1989). Roughly six years later (Evans 1995; Waldock 1995) and (CEFIC 1996) reported that biological recovery was observed in many coastal dogwhelk and oyster populations. Thus, by using dogwhelk as a biological indicator, similarly to what (Thingstad 1997) did with birds, we can assume that the T_N is roughly six years for TBT.

- V [-] is a dimensionless ratio of the largest affected control volume for a specific waste release *i* (V^i_{Waste}) [m^3] divided by the control volume of the entire system (V_{System}) [m^3], as given in Equation 8:

$$V^i = \frac{V^i_{waste}}{V_{System}} \tag{8}$$

Multiple (say M) waste releases will disperse differently, and each will affect a different control volume V^i_{waste}. Equation 8 represents the normalization of each specific affected volume by the largest possible volume considered, or affected, V_{System}.

Finally, it is extremely important to use consistent units so that the WI will be consistent. The unit of the WI is called a 'Waste Unit' (WU). A single Waste Unit is very large; the global annual release of CO_2 from fossil fuel burning represents roughly 1 WU. We therefore scale it to 1 *pico* WU = 10^{-12} WU = 1 pWU.

6.3 Strengths and Weaknesses

From the definition of the WI, we can identify two potential weaknesses:

- The WI is a simple index with one conceptual problem: The current WI cannot handle biological changes (deforestation, bacteria cultures etc.), because this is a very complicated issue due to reproduction, interplay between various species and so forth. However, for most man-made products this is of little or no concern. Where it is, the issues are often so complex that science often has not provided sufficient hard data for objective decision-making. Nevertheless, we believe that WI represents a significant conceptual improvement compared to other indicators, despite this problem.
- From a practical point of view we see that there are several variables that must be determined for every chemical compound. This may seem a daunting task, and to some extent it is if all chemicals are to be investigated. However, the WI is based on thermodynamics and chemistry whose laws are the same wherever you are. Because the WI does not try to assess the *actual* (or absolute) environmental impact, but rather just the *relative* environmental impact, the values for the various chemical compounds only have to be assessed once if researchers would agree (by convention) upon what natural conditions and degradation function to use. With respect to the latter, we suggest a linear function because it is simple. For the ISO 14042 (LCIA) recommended indicators it is not that simple since they try to assess the actual environmental impact. Hence, ISO recommended indicators require much more information.

Besides the aforementioned problems, lack of data is a problem to some extent, particularly for uncommon materials. This problem will be alleviated as use of the Waste Index increases. Another issue is that the Waste Index cannot handle all kinds of environmental problems. But this can be an advantage from a comparability perspective since indicators that attempt to capture too much easily become incredible. The primary positive aspects of the Waste Index definition are:

- The WI does not have to be normalized according to some chosen functional unit, which would jeopardize the capability of comparison between different products and processes. Instead we normalize releases to Nature itself.
- The WI is continuous and can have any value. This is important because we want the WI to handle various materials and chemical compounds under various conditions.
- The WI is consistent in use of units and it is comparable from product to product - system to system. This is a major problem for all the LCA indicators according to (Ayres 1995). The WI is a relative index describing how much better/worse a product is to another product, and it is apolitical and comparable given a set of axioms and conventions.
- The use of a control volume - instead of discrete and subjective scales such as local and regional - facilitates a continuous and unambiguous scaling that always works well.
- It also meets the four socio-ecological principles of The Natural Step organization of a sustainable society (Robert, Holmberg *et al.* 1994) in the following way:
  a) *Substances from the lithosphere (earth's crust and mantle) must not systematically accumulate in the ecosphere.* The WI takes this care of this, as the T_N will increase as the capability of the ecosystem to handle the releases deteriorates. A_N represents the natural state.
  b) *Society-produced substances must not systematically accumulate in the ecosphere.* T_N will again increase and A_N is the natural state.
  c) *The physical conditions for production and diversity within the ecosphere must not systematically be deteriorated.* Again, the T_N and A_N will capture this. T_N is calculated using thermodynamic and chemical models.
  d) *The use of resources must be effective and just with respect to meeting human needs.* This will be ensured as the waste drivers can be traced effectively and thereby allowing proper use of resources. Whether the use of resources is just or not is a political and ethical issue and cannot be captured by any metric.

All good and well, you may say, but how does the Waste Index compare to other indicators? That is the topic of the following sections.

6.4 Comparing the Waste Index with Global Warming Potential

In Table 4, four different types of car fuels are compared using both the WI and GWP. The WI is computed using Equation 7. The NO_x, CO_2 and CH_4 data in Table 5 are from David Hart, Centre for Environmental Technology, Imperial College, London and Günter Hörmandinger, Energy Policy Research, London. These data are combined with the data in Table 5, obtained from (IPCC 1993), yielding the WI and GWP results in Table 4. Please note that the large values of WI and GWP are due to the use of concentration instead of actual substance amount in the calculations. However, since the control volume is the same, the relative difference between the values is maintained and a comparison is therefore valid, even though the absolute numerical results are 'wrong' with an unknown, but fixed, correlation factor.

Table 4 - Comparing the Waste Index and the Global Warming Potential.

Type of Car Fuel	**NO_x** [g/km]	**CO_2** [g/km]	**CH_4** [g/km]	**Waste Index** [WU/km]	**GWP** [GWP/km]
Petrol Car	0.2600	209	0.0420	2.29E+08	2.91E+02
Diesel Car	0.5700	154	0.0300	3.66E+08	3.29E+02
	219.2 %	73.7 %	71.4 %	**159.7 %**	**113.1 %**
Methane Car	0.0400	130	0.0720	7.98E+07	1.47E+02
	15.4 %	62.2 %	171.4 %	**34.8 %**	**50.6 %**
Natural Gas Car	0.0240	83	0.0590	5.03E+07	9.43E+01
	9.2 %	39.7 %	140.5 %	**21.9 %**	**32.4 %**

Table 5 - Input Data for Calculating the WI and GWP (IPCC 1993).

Material Name	**GWP**	**Balance Time [years]**	**Control Volume**	**Concentration**
Carbon Dioxide	1	125.0	Atmosphere	2.80E-04
Methane	68	10.0	Atmosphere	8.00E-07
Nitrous Oxide	303	150.0	Atmosphere	2.88E-07

We see that WI and GWP have similar trends and correlate 96%. The WI is, however, assessing diesel fuel to be 60% more harmful than petrol, while GWP does not distinguish much between the two. Since there is no 'objective' benchmark to measure against, we can only conclude that WI and GWP work similarly, with one very significant difference; *the GWP applies only to gasses*. There is also a conceptual difference; GWP is benchmarking releases to a trace gas (CO_2), while WI is benchmarking a released compound (or what the released compound will decompose into) to the natural amount of that compound in Nature. Hence, the GWP is depending upon the significance of CO_2 with respect to global warming, while the WI is not depending upon the significance of any compound with respect to any specific environmental effect. Thus, *the WI can be viewed as a generalization of the GWP*.

6.5 Comparing the Waste Index with EPS and Eco-Indicator 95

The GWP is often used as an indicator for the Global Warming category in conventional environmental indicators such as the Eco-Indicator. It is therefore more interesting, but verily more difficult, to compare the Eco-Indicator, the EPS-Indicator and the WI. Such a Comparison is difficult because it is hard to know how the former indices were computed, e.g. what was included, what not, etc. Nevertheless, we used the IDEMAT software by (TU Delft 1996) as our source of information along with (COESA 1976; IPCC 1993) and (Mackay, Shiu *et al.* 1992a). Furthermore, we could only calculate the WI for atmospheric gasses due to lack of information. In IDEMAT, environmental analyses of several materials and liquids are given where the Eco-Indicator and the EPS-Indicator are used. Here, we are presenting some of these analyses and compare them to the WI. In Table 6 the balance times [years] and natural amounts [kg] for the released waste are presented.

Table 6 - Balance Times (T_N) and Natural Amounts (A_N') for the Waste Released.

Parameter	CO_2	CO	NO_x	N_2O	SO_2	Methane	n-Pentane
Balance Time	120	0.21	0.075	150	0.10	10	10.136
Natural Amount	2.11 E+15	3.60 E+11	1.58 E+09	1.48 E+12	6.04 E+10	2.20 E+12	2.20 E+12

The information in Table 6 and V^i = 76%[1] entered in Equation 7, i.e. a linear degradation function, yields the $V^i \frac{T_N^i}{2A_N^i}$ ratios for the various waste releases found in Table 7. This information is kept in a separate database for easy access. It is interesting to note that N_2O has the most severe environmental impact, while CO_2 is the least significant. What makes CO_2 so important is the enormous amount of releases. Also note that we have reasons to believe that the SO_2 releases are underestimated in the sense that the whole chain of chemical reactions after acid rain has been formed is not taken into account. For example, acid rain causes releases of heavy metals in soil, which can have devastating long-term effects

The final WI is achieved by multiplying the $V^i \frac{T_N^i}{2A_N^i}$ ratios for the various waste releases with their respective magnitude found in Table 8. The left columns are the materials that are produced and assessed by the three indicators in the three right columns. In the two columns to the right in Table 8 the corresponding values for the Eco-Indicator and the EPS-Indicator are

[1] All these releases are atmospheric, and the atmosphere constitutes 76% of the entire possible control volume on earth, i.e. the volume of atmosphere, freshwater and the oceans. Since we do not have any data for releases to soil, soil is not included in the total control volume. Also, releases to soil are very local and hence insignificant *unless* they are very toxic.

shown. All the columns in the middle are the releases and the amounts generated during production of the materials.

Table 7 - $V^i \dfrac{T_N^i}{2A_N^i}$ for the Various Waste Releases.

	CO_2	CO	NO_x	N_2O	SO_2	Methane	n-Pentane
$V^i \dfrac{T_N^i}{2A_N^i}$ [pWU/kg]	0.0215	0.2210	18.0396	38.4597	0.6280	1.7245	1.7479

Table 8 - Comparison of WI and the Eco-Indicator and EPS-Indicator in Production of Selected Materials and Liquids.

Production of	Generates Releases R^i of [g]							WI	Eco-Indicator	EPS
1 kilogram of	CO_2	CO	NO_x	N_2O	SO_2	Methane	n-Pentane	[pWU]	[mPts]	[mELU]
X10CrNiS 18 9			7.80E+00	5.56E-04	1.29E+01	1.54E+00	1.69E-03	0.1511	21.10	4,070
Fe360			1.16E+00	9.26E-04	3.39E-01	1.53E+00	1.65E-03	0.0237	4.66	560
CuZn30	7.01E+03	4.70E+00	2.73E+01		2.84E+01	8.65E-02	6.47E-02	0.6613	111.00	11,750
E-glass Fibre		8.34E-02	3.02E+00		2.28E+00	1.67E-02		0.0559	0.84	145
Glare 1 3/2-0.3			1.52E+01	2.50E-03		1.12E+01	2.22E-03	0.2932	31.90	7,378
Hylite 2/1-0.2		1.33E-07	1.24E-03	1.74E-01		2.23E+01		0.0452	18.60	1,493
Concrete, Plain			1.04E-01	1.22E-04	6.43E-02	3.52E-02	9.89E-05	0.0020	0.66	20
Crude Oil	1.80E+02	7.00E-02	2.20E+00		9.00E-02	1.70E+00		0.0465	0.32	442
Diesel	2.84E+02	8.00E-02	2.90E+00		1.80E+00	2.90E+00		0.0644	0.62	466
Natural Gas	2.09E+00	5.04E-03	2.01E-04			1.32E-03		0.0001	0.47	449
Petrol	2.70E+02	8.60E-02	3.07E+00		1.10E+00	3.02E+00		0.0670	0.63	458
Leather	2.38E+03	2.47E-01	3.56E+00		1.79E+00	1.21E+02		0.3240	31.70	679
Nitrile Rubber			1.08E+01		9.14E+00	1.50E+01		0.2266	3.67	990
HPDE	9.40E+02	6.00E-01	1.00E+01		6.00E+00	2.10E+01		0.2402	2.78	768
Cotton			7.39E-01	8.15E-01	7.02E-02	1.03E+00		0.0465	5.40	30,329
Polyester	1.63E+04	4.59E+01	5.36E+01		7.06E+01	5.67E+01		1.4678	17.20	1,687
Pitch Pine				3.20E-03	5.98E-02	9.77E-01	8.23E-03	0.0019	1.14	125

As an example, for Fe360, i.e. construction steel, we take the $V^i \dfrac{T_N^i}{2A_N^i}$ value of, e.g., NO_x in Table 7 which is 18.0396 pWU/kg, and multiply it by 1.16g = 0.00116 kg (see Table 8). This yields a WI for NO_x of 0.0209 pWU. This calculation is done for each individual waste release (N_2O, SO_2, methane and n-pentane) associated with production of 1 kg Fe360. Next, all WI results for the individual waste releases are added together, yielding a total WI sum of 0.023686 pWU (see Table 8). Hence, with the information from IDEMAT, NO_x releases constitute 88.3% of the environmental impact of producing 1kg of Fe360 when assessed using the WI.

Obviously, the validity of some analyses in IDEMAT is questionable, because how is it possible to produce, e.g., stainless steel (X10CrNiS 18 9) without emitting CO_2? This clearly suggests that the analyses presented in IDEMAT are flawed by either excluding emissions or by the fact that parts of the value chain up to the assessment point (production) are missing. Hence, the information the comparison is based upon is not 100% reliable.

Nevertheless, in Table 9 the results of a simple correlation analysis are presented. We see that the WI correlates significantly, i.e. over 85% for most releases and about 44% to the Eco-Indicator, while it hardly correlates to the EPS-Indicator. But the EPS Indicator, on the other hand, correlates <u>negatively</u> with methane, which has a GWP of 68.

Table 9 - Correlation Matrix for Releases and the Indices.

	CO_2	CO	NO_x	N_2O	SO_2	Methane	n-Pentane	WI	Eco-Indicator	EPS Indicator
CO_2	100.0 %	94.7 %	98.5 %	N/A	99.1 %	30.8 %	N/A	99.5 %	37.5 %	35.3 %
CO		100.0 %	92.3 %	N/A	95.9 %	29.1 %	N/A	93.4 %	8.3 %	8.4 %
NO_x			100.0 %	-34.8 %	99.0 %	26.6 %	85.5 %	98.0 %	44.3 %	3.6 %
N_2O				100.0 %	-25.7 %	-4.9 %	84.5 %	-17.3 %	-18.9 %	94.1 %
SO_2					100.0 %	27.7 %	87.9 %	97.6 %	38.3 %	0.9 %
Methane						100.0 %	-28.9 %	44.1 %	14.2 %	-15.5 %
n-Pentane							100.0 %	88.1 %	94.2 %	77.7 %
WI								100.0 %	43.7 %	3.9 %
Eco-Indicator									100.0 %	29.3 %
EPS-Indicator										100.0 %

From Table 9 we assert that the EPS indicator handles atmospheric emissions poorly. We can also see that both the Eco-Indicator and the EPS Indicator correlate only slightly with the atmospheric releases, but more interestingly they hardly correlate between themselves. This suggests that even with the same information, these two indices produce different results. So, what do the Eco-Indicator and the EPS-Indicator really measure? It may be that these indicators use large weights on solid and water related releases, compared to air releases. Another possibility is that the numerical values of these indicators are reflecting the *believed* environmental impact. If so, these indicators may measure aspects irrelevant to the true environmental impact.

We can continue this analysis by plotting the various materials (observations) and the indicators, and we get the plots in Figure 3 through Figure 5. Note that the observations are counted from the top of Table 8 so that Material 1 is a material denoted X10CrNiS 18 9, which is chrome vanadium stainless steel. We see that the WI values vary with the various materials. This is intuitive and also logic, because it is extremely unlikely that two materials have the same environmental impact.

If we look at the Eco-Indicator and the EPS-Indicator, on the other hand, there are several areas in the plots where the graphs are virtually flat. This is, in our opinion, another indication of unsoundness. It is also interesting to see how wide the differences can be for two indicators based on conventional LCA philosophy.

From the plot analyses we realize that the WI is giving a realistic distribution of results without flat regions. It is also interesting to see that, according to the WI, the production of polyester is 14,678 times worse than the production of natural gas. This seems to be realistic since polyester is a highly processed product while the natural gas is basically taken straight from the ground. According to the Eco-Indicator the production of CuZn30 (brass) is 347 times more damaging to the environment than the production of crude oil, which is also not unrealistic. But the EPS-Indicator, on the other hand, assesses the production of cotton to be 11,516 times more environmentally damaging than the production of plain concrete. This cannot be true unless very persistent pesticides are used and taken into account.

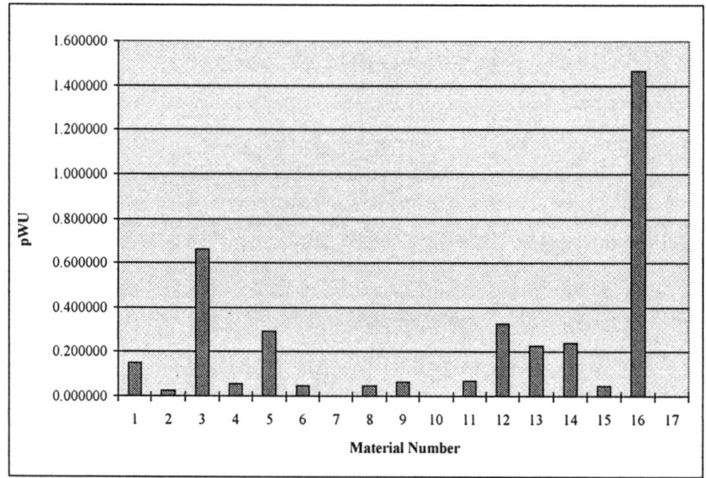

Figure 3 - Waste Index Plot for Production of Selected Materials.

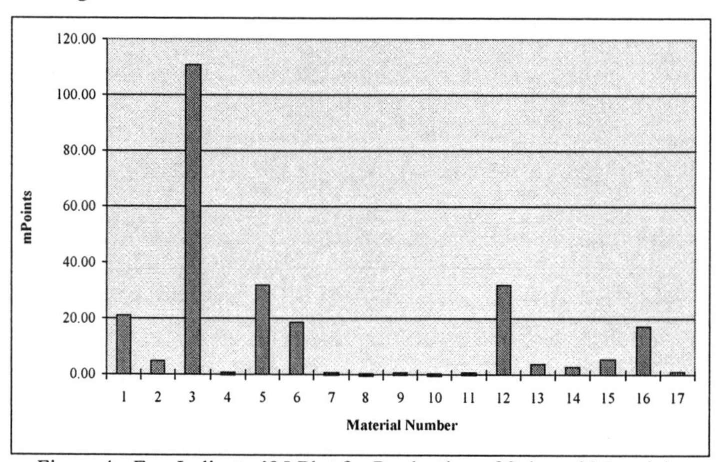

Figure 4 - Eco-Indicator'95 Plot for Production of Selected Materials.

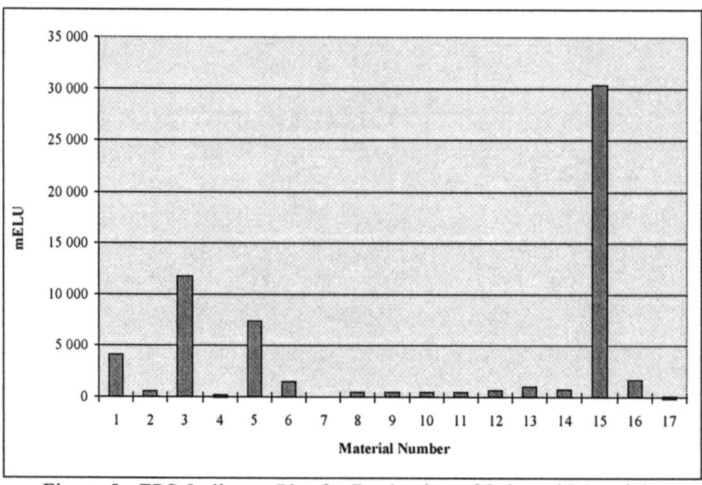

Figure 5 - EPS-Indicator Plot for Production of Selected Materials.

6.6 Biological Diversity Degradation and Human Health Risks Issues

Clearly, the WI is only concerned with the general state of the environment. The problems of biological diversity degradation and human health risks are therefore not considered *per se*. Obviously, it is important to handle the biological diversity, because preserving the biological diversity is important for humans both directly and indirectly via the use of biological entities. Many natural systems sustain cycles of disturbance (fire, hurricanes, disease) and rejuvenation (Ricklefs 1990) and rely on re-colonization. But as habitats become more and more fragmented (e.g., forests by the spread of agriculture and urban development) disturbances can so thoroughly destroy an area that little chances are left for complete recovery, even with substantial help from humans (Janzen 1988). Furthermore, any reduction in biological diversity can upset the balance of a system and alter its function (Wilson 1988). Thus, it seems that preserving a high biological diversity is just as important in the long run as any other measurement of the state of our environment. In fact (Lovelock 1988) proposes a theory - the Gaia Theory - where living systems (biological entities) and nonliving systems co-evolve and in fact *produce* the environment. It is also important to handle human health risks because humans are large mammals that can become extinct even though the general state of the environment has changed relatively little.

In principle, we can include these issues in the definition of our Waste Index by using penalty functions that will come into play when released wastes exceed certain threshold limits and thereby increasing the WI. However, after some initial investigation, we believe that such an approach is not likely to work at this moment, because we only have *partial* assessments and data for 5% (National Academy of Sciences 1996) of the chemicals in the environment. Some data for exposure limits for humans can be found in (American Conference of Governmental Industrial Hygienists 1996). Thus, an attempt to incorporate biological diversity and human health risk issues into the WI is currently not realistic, in our opinion. We believe that currently these two issues should be handled by simply using separate indicators. Those indicators serve as 'flashlights' alerting a practitioner when releases have negative effects related to human health and/or biological degradation. The following expressions can be useful for the biological diversity and human health risk indices for a released waste i, respectively:

$$I^i_{BD} = \begin{cases} 1 & if \quad \dfrac{R^i}{V^i_{BD}} \geq C^i_{BD} \\ 0 & if \quad \dfrac{R^i}{V^i_{BD}} < C^i_{BD} \end{cases} \tag{9}$$

$$I^i_{HH} = \begin{cases} 1 & if \quad \dfrac{R^i}{V^i_{HH}} \geq C^i_{HH} \\ 0 & if \quad \dfrac{R^i}{V^i_{HH}} < C^i_{HH} \end{cases} \tag{10}$$

Where;

- I^i_{BD} is the indicator for potential biological diversity degradation caused by the release of R^i. When it is one (1), it implies that there is a local problem related to biological diversity degradation caused by a specific release *i*.
- V^i_{BD} is the control volume of the released waste R^i the biological diversity degradation is expected to affect.
- C^i_{BD} is the concentration of the released waste R^i at which the biological diversity degradation is expected to be effectuated.

In other words, the indicators are zero as long as the resulting increased concentration of a substance resulting from a release (R^i) into the control volume is smaller than the concentration level for which biological degradation is expected to appear. It works similarly for human health (HH) risks, see Equation 10. These indicators should be used, but independently, with the WI when information is available. We feel that these indicators should not be used to affect the numerical magnitude of the WI because it reduces consistency and ultimately comparability.

6.7 Answers to Frequently Asked Questions about the Waste Index

There are some questions that are frequently asked by people presented the WI. The questions are inevitable due to the different approach the WI has to environmental issues compared to the conventional indicators. In his section we consequently present these questions and our answers, see Table 10.

The results from our case studies seem to indicate that the WI is sound, in our opinion, but as long as we do not understand all the relations in Nature we will never quite know. A good example of this is presented in (Seehusen 1998) where a discussion between two scientist is cited about a claim that catalytic converters in cars have *increased* the formation of ground-level ozone due to the changed ratio between NO_x and VOC in the air, despite initial reductions of NO emissions. If so, the use of catalytic converters in cars has increased the environmental problems it was supposed to reduce! If this is true, it clearly illustrates the need for benchmarking Nature.

Table 10 - Frequently Asked Questions and Their Answers Regarding the WI.

Question	Answer
How does the WI handle releases of particles in the air?	The dispersion of the particles is handled by assessing the affected control volume affected. Depending on the chemical composition of the particles, which must be determined, the degradation times and natural amounts of the various chemicals in the chosen control volume will be used to compute the WI value.
How does the WI handle a solid piece of metal lying on the ground?	A solid piece of metal has no emissions (unless it leeches) thus it is only a control volume issue. Because the control volume is so minute compared to the largest control volume, the effect is negligible. If it leeches, then it is handled as a dispersion of particles (see above).
How does the WI handle releases of toxic chemicals and persistent compounds?	Such releases are handled as any other release because toxic releases and even persistent compounds also eventually will degrade. A critical issue here is to establish the chain of chemical reactions that will occur during the degradation so that the actual times and natural amounts can be found.
How is waste such as scrap in production handled?	Production waste has three components; 1) the cost of such waste, 2) the energy content and 3) the waste content of the waste itself. The waste *content* is measured by the WI. If the scrap is emitted to the environment, it will be assessed based on its emission characteristics.
How does the WI handle dumping trash on a waste site?	As everything else, trash has a waste content that can be assessed based on upstream releases. In addition, as trash decomposes gases and fluids are released, which are assessed straightforward.
How does the WI handle renewable versus nonrenewable energy?	The issue of renewability is of no concern for the WI. What matters is that energy has a waste content based on how much waste has been generated earlier upstream - renewable or not. Only this matters.
How does the WI handle water use and other resource use issues?	It does not handle such issues, unless there are emissions associated with the use. General natural resource use issues are invariably so complex that we do not believe a generic approach is possible at this time.

7. WHAT IS NEXT?

Time is a great teacher, unfortunately it kills all its pupils.

Hector Berlioz

In this chapter, we gave you an idea of what environmental management approaches all entail. Obviously, there is a lot out there and you may not see the forest through the trees anymore. It should also be clear that there is no unified approach. In fact, even the approach that is supposed to bring this unification, ISO 14000, still leaves a lot to be desired and certain aspects of it have a very high learning curve. It is with that in mind, that we choose to build environmental assessment and management upon the known foundations of cost accounting and management, which brings us to the next chapter, where we will explain a cornerstone in our method, Activity-Based Costing (ABC).

Chapter 3

ACTIVITY-BASED COSTING

Which of you, intending to build a tower, sitteth not down first, and counteth the cost?

Jesus
Luke (Lucas) 14:28.

Due to the importance of Activity-Based Costing (ABC) and Activity-Based Management (ABM) in our work we present these topics quite extensively. However, this book is not solely about ABC/ABM and we refer the reader who is interested in even more detail to the widely available literature on ABC and ABM. For example, good in-depth discussions of ABC and ABM can be found in (Cooper 1990a; O'Guin 1990; Turney 1991a; Edwards 1998) and (Cokins, Stratton *et al.* 1992; Cokins 1996). In this chapter, however, we want to give the reader a relatively complete and comprehensive view on *why* we (and others) prefer the activity-based costing and management concepts.

1. THE ABC CONCEPT

Never ask for money spent
Where the spender thinks it went
  Nobody was ever meant
  To remember or invent
What he did with every cent.

Robert Frost
The Hardship of Accounting

To understand the rise of ABC, one must really understand the problems of traditional costing methods, discussed next, because ABC came as a result of these problems.

1.1 The Problem with Traditional Cost Accounting

What is common to the greatest number has the least care bestowed on it.

Aristotle

The most widely used cost accounting approaches are still traditional volume-based costing systems. These are also referred to as ‘conventional costing systems’, or ‘traditional costing systems’. Many varieties exist, but traditional volume-based cost accounting and financial reporting typically classifies costs in different categories (e.g. expense, capital/asset and product cost). Then, those costs that belong to the product cost category are grouped into the various accounts that the company has, such as payroll and materials inventory. There are several essential features of volume-based costing systems (Hardy and Hubbard 1992):

1. For product costing purposes, the firm is separated into functional areas of activity - that is, manufacturing, marketing, financing and administration.
2. The manufacturing costs of direct materials, direct labor, and manufacturing overhead are inventoriable costs, i.e. are accounted for in inventory assessments.
3. Direct material and direct labor costs are considered to be traceable (i.e., chargeable) directly to the product.
4. Manufacturing overheads of both production and manufacturing service departments are treated as indirect costs of the product but are charged to the product by the use of predetermined overhead rates.
5. When a single, plant-wide, predetermined overhead rate is used, overhead is charged indiscriminately to all products without regard to possible differences in resources utilized in the manufacture of one product versus another. As one plant manager expressed; “we spread overhead to all products like peanut butter”.
6. The functional costs of marketing, financing and administration are accumulated in cost pools and are treated as costs of the period in which they are incurred. These costs are not treated as product costs.

There are a lot of advantages related to the volume-based way of performing costing; it is widely used and understood well, simple, and fairly accurate when direct labor has a large portion of product costs.

However, faced with growing competition, managerial accountants started to become aware that the relevance of the numbers they produced for end-users and decision-makers was becoming lower and lower. In their widely acclaimed book *Relevance Lost: The Rise and Fall of Management Accounting*, published in 1987, Johnson and Kaplan explain how the early costing systems were focused towards decision-making, but then became more and more focused towards external financial reporting. The result is that (Johnson and Kaplan 1987):

> *Today's management accounting information, driven by procedures and cycle of the organization's financial reporting system, is too late, too aggregated, and too distorted to be relevant for managers' planning and control decisions.*

They were (and are) not alone in their criticism; very negative phrases have been associated with volume-based costing systems such as; ‘number one enemy of production’ (Goldratt 1983), ‘undermining production’ (Kaplan 1984), and ‘systematically distorting product costs’ (Cooper and Kaplan 1987). The question was even asked whether cost accounting was an asset or a liability (Fox 1986)!

A major problem with the traditional volume-based costing systems is that they typically rely on direct labor as allocation base for the overhead cost allocation, see e.g. (Cooper 1990a). This works well if the business relies heavily on manual labor (e.g., a gardening business, a garage, or even an engine remanufacturer). But many manufacturing companies have introduced some degree of automation, the overhead has grown and production costs are not directly proportional anymore to direct labor. This makes the volume-based costing systems we have today poor tools for understanding true product costs.

Example: Consider two product models that are produced in equal amounts and require the same amount of labor and material, thus having the same direct cost. Product A only needs to be assembled from components. Product B, however, also needs welding before final assembly which is done using, say, a robotic welder. In a volume-based system, the equipment costs of welding and assembly would be aggregated and then allocated to products A and B. Because both products are produced in equal amounts using an equal amount of labor, the unit cost for these two products will be exactly the same, according to a volume-based costing system that uses labor as an allocation base. Do you see the problem?

Intuitively, you will probably immediately say, "Hey, wait a minute, product B requires robotic welding equipment, the other not, thus product B should cost more". Exactly, product B is under-cost and product A is over-cost. Most of our readers may wonder why this is not obvious. Well, surprisingly, it is not in cost accounting for the last 80 years. An ABC system (to be discussed next), on the other hand, would see that there are two different activities for the two products and would trace the cost of the robotic welding equipment only to the product that needs welding (product B). Thus, in an ABC system, the unit cost of product B (which requires welding) would be higher than product A (which does not require welding), as expected.

Although this example is very simple, the effects can be enormous. Think about what would have happened if the company decided to produce more of the welded products in favor of the ones that require assembly alone. Using a traditional costing system, the unit overhead cost is the same, and with the same direct cost, the profit should be the same as well, right? Wrong, as we just saw, and as many companies that switched to ABC suddenly realized, they were producing the wrong (i.e., under-cost) products.

Nevertheless, in a survey presented by (McIllhattan 1987) 94% of the companies reported to use labor hours to allocate overhead costs. This is quite shocking news given the known limitations of using only direct labor as allocation base. Thus, it can be argued, to some extent, that many in industry do not know what their products cost and that many companies therefore have survived despite their cost management systems. Obviously, this is not a desirable situation, and in the recession in the 1980s many companies learned this the hard way, see e.g. (Cooper 1997). Another (numerical) example of the traditional costing systems' inadequacy to handle overhead correctly is given in section 1.3, but first we explain how 'Activity-Based' Costing works.

1.2 Activity-Based Costing

A problem well defined is half solved.

John Dewey

Since the formalized[1] ABC framework first appeared in the late 1970s/early 1980s, it has undergone quite substantial changes. In (Cooper 1990c) an older way of implementing an ABC system is presented, while in (Edwards 1998) the most up-to date discussions (as of 1998) can be found. Many books about the topic are available, such as (Brinker 1994) and (Cokins 1996), to which we refer interested readers wanting to know more about the topic.

In a nutshell, there are two key differences between ABC and traditional costing systems:

- In ABC, it is assumed that a cost object (e.g., product) consumes activities (e.g., welding), and activities consume resources (e.g., electricity), whereas in a conventional system it is assumed that a cost object directly consumes resources.
- ABC uses what are called resource and activity drivers at several levels to trace costs from resources to activities to objects in a causal, directly proportional manner, while a conventional system uses only unit-level allocation bases such as direct labor.

These two differences are explained in more detail in the following, but the bottom line is that because of these differences, ABC can handle and trace overhead costs better than volume-based costing systems.

1.2.1 Activities Consume Resources

ABC is a costing system that is based on the formulations of activities, resources and cost objects, as shown in Figure 1 (and Figure 2), where cost objects consume activities which in turn consume resources. Activities are

[1] Gordon Shillinglaw at Colombia and George Staubus at Berkeley had articulated activity-based concepts by the early 1960s. In the early 1960s General Electric accountants may have been the first to use the term 'activity' to describe work that causes costs. However, the process of codifying the concept to what ABC is best known for today is mainly attributable to Professor Robin Cooper at Harvard Business School (Johnson 1992).

basically anything you do in, say, a manufacturing organization, for example welding, assembly, maintenance, shipping, ordering, etc. A resource is something you need to perform the activities, for example personnel, electricity, water, trucks, machines, buildings, land, etc. A cost object can be anything for which a cost/revenue assessment is needed. A product is a classical cost object, so is a service and an organization/department, as well as a project, and even a customer, etc.

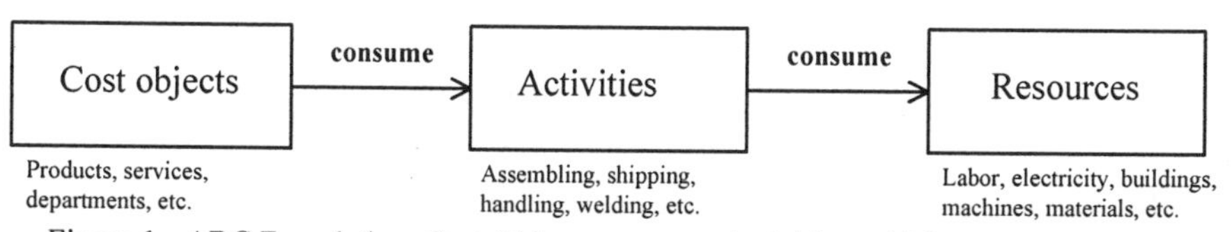

Figure 1 - ABC Foundation: Cost Objects consume Activities, which consume Resources.

As shown in Figure 1, the fundamental principle of ABC is that the performance of activities causes resources to be consumed. For example, labor (a resource) is consumed in assembling (an activity) products (cost objects). Electricity, gas, and welding rods (resources) are consumed when welding (an activity) a product (a cost object). This is quite different compared to a volume-based costing system, where allocations are made directly from the General Ledger to the cost objects using direct labor and the Bill Of Materials (BOM) as the allocation basis, see Figure 2. In volume-based costing systems the process/activity view is ignored.

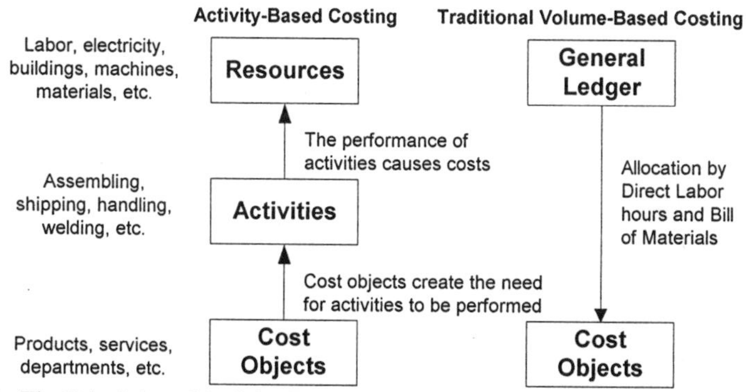

Figure 2 - The Principles of Activity-Based Costing versus Volume-Based Costing Systems. Based on (O'Guin and Rebiscke 1997).

As indicated in Figure 2, ABC can be seen as having an upward cost assignment as it estimates how the *actual* processes consume resources, while the volume-based costing systems have a downward cost assignment because they allocate cost from resources directly to objects. Another way of looking at it is that ABC is process-oriented (formulation of activities), while volume-based costing systems are structure-oriented (cost classification according to organizational structure).

Because the cost assignment is now done in two stages (first from resources to activities, and secondly from activities to objects, see Figure 2), ABC is said to be a *two-stage* costing system (Cooper 1990a). However, developments of ABC have led to so-called multiple-stage ABC systems that allow activities to be consumed by other activities, not just objects. The effect of this is that cost can now be traced from activity to activity as well, resulting in a so-called multi-level cost flow, and increasing the accuracy of the modeling and cost assessment. An example of this activity to activity tracing is the consumption of a factory's overall maintenance activity by individual machine maintenance activities. This means that the modern multiple-stage ABC approach results in a hierarchy of activities. As we will show in Chapter 4 and subsequent case studies, we use the concept of activity hierarchies and interactivity consumption flows as well, although we recommend any (over)zealous modeler to first remember the KISS-principle; Keep It Simple, Stupid. In other words, if a simple model suffices for the purpose at hand, why build a complex one?

The issue of complexity also is key in the choice of what are called the resource and activity drivers, which are the actual measures of how activities and objects consume resources and activities, respectively, which is the topic of the next section.

1.2.2 Resource and Activity Drivers Trace Consumption and Cost

A volume-based costing system uses what are called allocation bases to assign costs to products. Allocation bases are arbitrary, unit-level characteristics of the product, e.g. labor hours per product unit. Many managers and accountants have realized that a major distortion is often introduced in the cost assessments when overhead costs are allocated using these unit-level characteristics. See also our examples in Section 1.1 and Section 1.3. In (O'Guin 1990) it is reported that the difference in product costs can shift several hundred percent. According to (Cooper 1990a):

> *Conventional cost accounting systems systematically undercost small, low-volume products and overcost large high-volume products.*

ABC is a multiple-stage costing system where costs are first traced/assigned from resources to activities and then from activities to other activities and finally to cost objects. To do this tracing, three different options exist, as shown in Figure 3, namely

1. Direct attribution,
2. Causally assigned using drivers, or
3. Allocated.

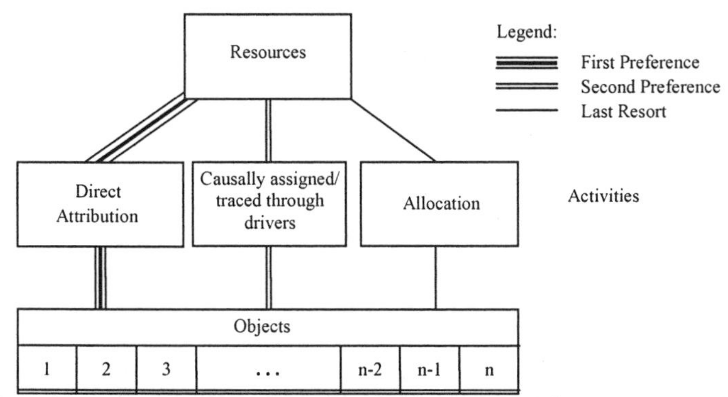

Figure 3 - Preferences for Driver Identification. Adapted from (Ostrenga 1990).

Direct attribution, also referred to as 'direct charge' or 'direct tracing', is the ideal situation that occurs when the resources match directly activities and objects or activities match directly objects. For example, imagine you are driving a car. Driving is the activity, and the car is the cost object. Fuel is a resource that is being consumed. You can measure exactly how much fuel your car is consuming by monitoring how much gas you put into the car at the gas pump, and trace it directly to your car. Direct attribution ensures a 100% correct tracing due to the one-to-one matches. This situation always occurs with material costs that can be traced to the exact product unit.

The other extreme, the last resort, is *allocation* in the traditional sense which is simply a way to distribute costs in a fairly arbitrarily manner. For example, consider allocating production planning costs using, e.g., number of units. Clearly, this is allocation since the planning costs are not related to the number of units produced. Thus, allocation must never occur for resource intensive activities because that will lead to significant distortions.

In between these two extremes, we have assignment by what are called *drivers* (the second preference in Figure 3). Drivers link resources to activities and objects in a causally correct manner. For example, imagine you are driving a car again and you need to know how much fuel you spent. Although 'miles driven' is not an exact measure of fuel consumption (ideally, you need to know speed, mass and drag of car, etc., as well), the number of miles driven is a direct cause for fuel consumption and can be used as a driver to trace fuel consumption.

In ABC, costs are first traced/assigned from resources to activities and then from activities to other activities and finally to cost objects. Hence, the drivers used in ABC are classified into *resource* and *activity* drivers (see Figure 4), which are defined as follows (see also Glossary):

- A resource driver is a measure of the quantity of resources consumed by an activity. Examples of resource drivers are the amount of electricity, area of floorspace, amount of labor, etc.
- An activity driver is a measure of the consumption of an activity by a) another activity or b) a cost object. Activity drivers that measure the consumption by an object are also referred to as 'final' activity drivers, whereas activity drivers that measure consumption of activities by other activities are also called 'intermediate' activity drivers. Examples of activity drivers are the amount of labor, the weight of a product, the number of products, etc.

In Activity-Based Management, the term 'cost driver' is also used, but as will be explained in Section 2.3, cost drivers are a broader, super-set of activity drivers. Activity drivers must be measurable, whereas cost drivers may not be measurable, but point to a true root cause for why work is done in an activity.

In the above definitions, note that, e.g., labor is both an example of a resource driver and an activity driver. However, even if the words are the same, the physical meaning is different because 'amount of labor' used as a resource driver may have a total different quantity than the 'amount of labor' that is used as an activity driver. It is very important to keep in mind that *all activity and resource drivers have a numerical quantity associated with them* that is the 'measure' or amount of consumption (see next section).

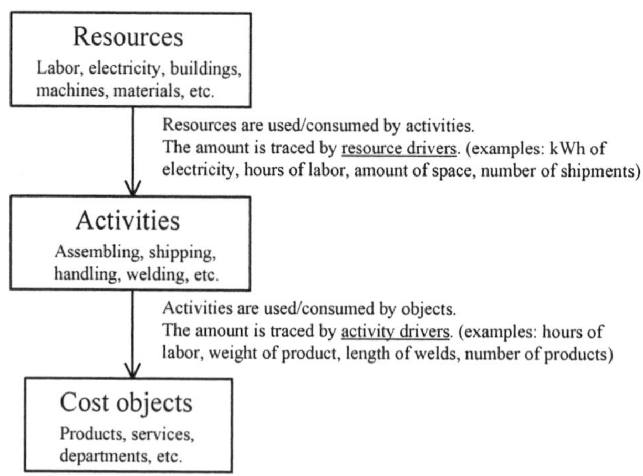

Figure 4 - Tracing of Resource and Activity Consumption using Resource and Activity Drivers.

It is important to be aware of the fact that sometimes drivers cannot be identified using a single measurable variable such as 'labor hours'. Then, one must simply be creative and come up with a combination of various measurable variables and combine them to form a driver in a way that gives as high correlation between driver and the actual consumption as possible. This can be difficult particularly for support activities, i.e. activities not directly related to production. In the Westnofa Industrier case study (Chapter 8) this was done for a purchasing activity. We knew that the ideal driver would be number of purchase orders per product, but this information was not readily available, so we had to come up with a second best solution. This was achieved by identifying the total number of purchase orders and how many articles were checked for every order. This information was combined as 'Purchase Order Articles'. Then, we identified how much of the various articles a product consumed and thereby we had created a reliable driver that in fact had no physical meaning, yet, served the purpose.

In ABC, the term 'multi-level drivers' is also used. This is simply a classification of drivers into meaningful manufacturing process related categories. The use of multi-level drivers instead of traditional (unit-level) allocation bases was promoted early on by many, see e.g. (Cooper 1990a; Cooper 1990c), to reduce distortion. Typically, the following levels of drivers are considered in ABC:

- *Unit-level drivers* - these are triggered every time a unit of a product is produced, e.g. drilling a hole and painting a surface.
- *Batch-level drivers* - these are triggered every time a batch of products is produced, e.g. machine set-up time and transportation of production lot. The effect on unit costs are significant, see Figure 5.
- *Product-level drivers* - these are triggered by the fact that products are produced, e.g. design changes and maintaining the Bill Of Materials.
- *Factory-level drivers* - these are triggered by the fact that production occurs, e.g. the amount of electricity for lighting and cleaning hours for the factory.

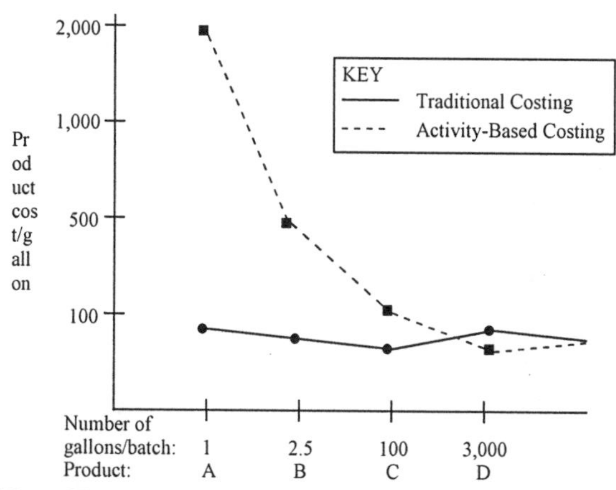

Figure 5 - The Effect of Batch Size Diversity for a Traditional Costing System vs. ABC (Maisel and Morrissey 1997).

This is a simplification of reality because product line levels, brand levels, department levels and so on are not included. In our activity-based cost and environmental models, more than four levels are employed *if needed.*

1.2.3 Driver Levels and Consumption Intensities

Activity and resource drivers are used to trace costs from resources to activities to objects. This means that drivers need to be quantified. In general, the following terms are used:

1. The driver value, amount or 'level', which is the numerical quantity of the driver. Examples are 12 hours of labor, 10 tons of fuel, 5 inspections, 6 shipments, etc.
2. The driver (consumption) 'intensity', which is the unit-price of a driver (Cooper 1990a), for example 20 dollars per hour of labor, 1000 dollar per ton fuel, 10 dollar per inspection, 50 dollars per shipment, etc. We have also seen some people refer to this as the driver 'rate'.
3. The driver cost, which is the resulting cost value obtained from multiplying the driver amount (or level) times the corresponding consumption intensity.

We have noted that some ABC articles and examples only focus on the driver cost and its value. We, however, recommend that one always quantify the driver level and consumption intensity as separate quantities first, and then calculate the driver cost. There are two fundamental reasons for this:

- The tracing and understanding of where the costs come from and where the 'dollars' go to is enhanced. For example, suppose we have an activity driver called 'labor' that cost $1000. Knowing whether it is a 100 hours at $10/hour or 10 hours at $100/hour adds understanding that the first is probably unskilled and the latter highly skilled labor.
- The information for determining the level versus intensity typically comes from separate people/groups. For example, it usually takes manufacturing and engineering people to determine how many inspections are needed on particular products. However, it takes the finance/accounting group to identify what the costs of (purchased) equipment and other resources needed are, hence, they are in the best position to quantify the driver intensity, or at least provide the basic cost data for this purpose.

In all our case studies, we have always used this approach. Moreover, it allows the easy set-up of spreadsheets. For example, Table 1 shows how the consumption of three resources by one activity (A11) can be represented in a spreadsheet table. Similar tables can be used to calculate the resource consumption for other activities. However, the number of columns can be reduced significantly by realizing that a driver's consumption intensity (or 'price') is almost always the same for all activities in which the same driver is used. In that case, we only have to write down the consumption intensity once and a table as given in Table 2 can be defined. Table 2 is representative of how we develop our spreadsheet models for our case studies. Notice how the units match in both tables and provide resulting costs in dollars per year. Matching units are extremely important. Non-matching units are an automatic signal for trouble, i.e., a bad ABC model. For completeness and increased control, we typically tend to include units in our spreadsheets.

Table 1 - Driver Cost Calculation Table Example.

		Activity A11 - Delivery to Atlanta		
Resource	**Resource Driver**	**Resource driver value**	**Consumption Intensity**	**Cost [$/year]**
Maintenance	Occurrence	5 [occ./year]	150 [$/occ.]	750
Fuel	Fuel Consumption	113 [ton/year]	1,750 [$/ton]	197,750
Drivers	Labor Hours	2,705 [hrs/year]	20 [$/hrs]	54,100
				252,600

Table 2 - Driver Cost Calculation Table with Constant Consumption Intensity.

			Activity A11 - Delivery to Atlanta		Activity A12 - Delivery to Miami	
Resource	**Resource Driver**	**Consumption Intensity**	**Resource driver value**	**Cost [$/year]**	**Resource driver value**	**Cost [$/year]**
Maintenance	Occurrence	150 [$/occ.]	5 [occ./year]	750	6 [occ /year]	900
Fuel	Fuel Consumption	1,750 [$/ton]	113 [ton/year]	197,750	2,292 [ton/year]	4,011,000
Drivers	Labor Hours	20 [$/hrs]	2,705 [hrs/year]	54,100	2,678 [hrs/year]	53,560
				252,600		**4,065,460**

1.2.4 Selecting Resource and Activity Drivers

Choosing drivers is one of the most important aspects when implementing any activity-based system because it affects the accuracy as well as the cost of the implementation. Not surprisingly, the (correct) selection of resource and activity drivers is a widely discussed subject and many guidelines exist, especially for selecting activity drivers. Because the modern ABC systems are multiple stage systems (which model interactivity consumption), identifying activity drivers (both intermediate and final) typically requires more effort than the resource drivers. When choosing drivers there are particularly three issues to always consider (Cooper 1989):

1. The ease of obtaining data required by a particular driver.
2. The correlation between the consumption of an activity implied by that driver and the real consumption.
3. The behavioral effects a driver can induce.

An expanded set of guidelines for selecting activity drivers is found in (Cokins 1996)[2]:

---

[2] According to (Cokins, 1996), the original source is Angela Norkiewicz, Manager, Cost, Pennsylvania Blue Cross.

- Avoid activity drivers for immaterial/insignificant activity costs.
- Pick activity drivers that match the type of activity.
- Pick drivers that have a high correlation to the actual consumption rate of its activity.
- Minimize the number of unique activity drivers because there will be diminishing returns in accuracy
- Find activity drives that encourage performance improvements
- Pick activity drivers that are economical to measure, and avoid activity drivers that require new methods of measurement.

Basically, to identify *activity* drivers, one should always ask:

- *What causes differences in level of effort in the activity?*
- *Why do we perform this activity, that is, what is the root cause?*

Similarly, to identify *resource* drivers, one should ask,

- *What causes differences in level of use/consumption of the resource?*
- *Why do we need this resource?*

As stated in (Cokins 1996), inexperienced ABC designer believe their model will require hundreds of activity drivers. However, simply increasing the number of activities (and hence increase the accuracy of the process model) will also segment cost diversity and will alleviate the need for a wide variety of activity drivers. In fact, in all our case studies (see Chapters 5 through 9), we also used a (surprisingly) small number of activity and resource drivers, yet we were able to capture the actual consumption.

A sound approach is to phase in the level of sophistication; start simple and start increasing the level of sophistication and accuracy as needed. But remember *there is a cost of information!* The more information needs to be gathered, the higher the cost of the modeling effort.

Even when only a few drivers are used, quantifying the drivers must be both possible and practical. When considering alternative drivers, weigh the relative cost of collecting the data against the relative precision each would offer. Before deciding on drivers, it is recommended that input be obtained from people directly involved in the activities.

The opposite extreme, however, can also occur, that is to name something an activity or resource driver just because related data is available. This temptation should be resisted as well because it will result in a 'cheap', but inaccurate (and hence useless) driver that can give significant distortion.

In a nutshell, it is advisable to restrict the number of drivers to a manageable few while always testing the trade-offs between the effort to collect data, the accuracy, and the amount of precision that the end-user of the model needs for decision-making. Searching for the best drivers is sometimes a bit like a treasure hunt. Good data often turns up in department-level or private files. Only by soliciting information about available data from people actually engaged in the day-to-day work can these sources of information be discovered.

1.3 ABC versus Traditional Costing - An Example

The numerical example in this section is a based on an example found in (Cooper 1990a), but changed in terms of wording and context. The purpose, however, is the same; to show the superiority of ABC over volume-based costing when it comes to handling and tracing overhead costs.

1.3.1 Problem Statement

Let's consider a hypothetical company (named JEBB) which wants to manufacture four products: P1, P2, P3 and P4. Although the products differ in physical size and in production volume, they are all produced on the same equipment and use similar processes. In Table 3 the breakdown of direct costs is given for each product in terms of material cost, labor hours and machine hours for producing a batch of each product. The labor rate is $20/hour. Note that the batch volumes are different for each product. We are told that the aggregated overhead cost for the manufacture of these four batches of products is $9,924. JEBB's management wants to know what price to charge for their products and wants to know what the overhead charge is for each product.

Table 3 - Costs Related to Direct Costs by Product.

			Costs related to direct production	
Product	**Size**	**Volume**	**Material costs**	**Direct labor hours**
P1	Small	10	$ 60	5
P2	Small	100	$ 600	50
P3	Large	10	$ 180	15
P4	Large	100	$ 1800	150
	Total:	220	$2640	220 hours

1.3.2 Traditional Costing Approach

In a traditional approach, we would first determine the overhead rates. Using direct labor as the allocation base (as usual), we would get the following overhead rate for JEBB:

Overhead rate = Total overhead costs / Total direct labor hours
= $ 9,924 / 220 hours = $ 45.11 per hour

Next, we would allocate the total overhead to each product by multiplying the overhead rate times the number of direct labor hours needed to produce each product. The unit overhead cost is obtained by dividing by the production volume. This is given in Table 4. From Table 4, we learn that the volume-based costing system assesses the unit overhead costs of P1 and P2 to be equal, and roughly three times lower than the unit costs for P3 and P4.

Products P3 and P4 are also estimated to cost the same. If we add up the direct and overhead (indirect) costs, we obtain the unit prices for the products as listed in Table 5 (the direct labor rate is $20/hour).

Table 4 - Overhead Costs Reported by Traditional Costing System.

Product	Direct labor hours	Overhead rates [$]	Overhead costs allocated [$]	Volume	Reported unit overhead cost [$]
P1	5	$45.11	$225.55	10	**22.56**
P2	50	$45.11	$2255.50	100	**22.56**
P3	15	$45.11	$676.65	10	**67.67**
P4	150	$45.11	$6766.50	100	**67.67**
Total	220		$9924.20	220	

Table 5 - Total Costs by Product.

Product	Volume	Costs related to direct production		Overhead cost	Total cost	Unit cost
		Material costs	Direct labor costs			
P1	10	$60	$100	$225.55	$385.55	$38.56
P2	100	$600	$1,000	$2255.50	$3,855.50	$38.56
P3	10	$180	$300	$676.65	$1,156.65	$115.67
P4	100	$1,800	$3,000	$6766.50	$11,566.50	$115.67
Total:	220	$2,640	$4,400	$9924.20	$16,964.20	

This looks great, but the limitations become clear very quickly if management starts to look into how to reduce costs (for example). The preceding tables only have data on material, labor, machines, and aggregated overhead - not much to start from. Pictorially, the above cost allocation is shown in Figure 6. Note that resources are allocated directly to products, similar as in Figure 2.

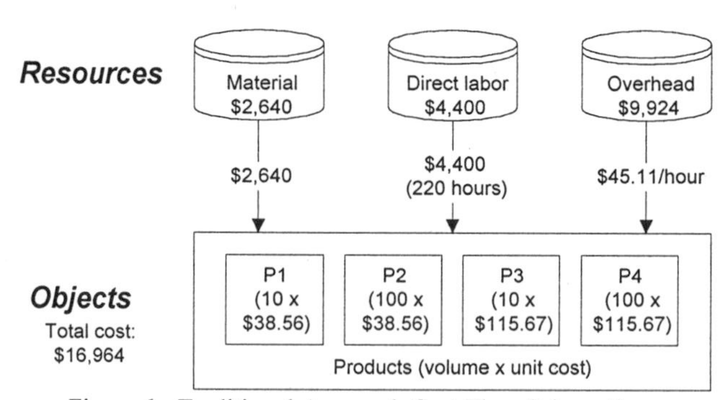

Figure 6 - Traditional Approach Cost Flow Schematic.

1.3.3 Activity-Based Costing Approach

Now let us work the same example using ABC. We know the cost objects already: Products P1, P2, P3, and P4. Let's assume that we keep the same categorization of resources as before: material, direct labor, and overhead.

Now we need to define the activities. Management agrees that there are basically two activities: manufacturing and shipping. However, the manufacturing activity can be split up further into two lower level activities:

direct production and set-up. Thus, we have three basic activities for which we want to calculate the cost: direct production, set-up, and shipping. The cost of the activity 'manufacturing' is simply the sum of the cost of direct production and set-up. Note, however, that this is already an example of a multiple stage ABC model, albeit a very simple one.

Next we need to figure out how much resources the activities consume. The activity 'direct production' consumes all materials and, hence, we can directly trace (or charge) those to the direct production activity. Remember that direct tracing is the preferred approach (see Section 1.2). Similarly, all direct labor hours are consumed by 'direct production' and again we can charge all those directly to the 'direct production' activity. The overhead, however, is consumed by all three activities. But let's assume that JEBB's management and accounting are able to trace the aggregated $9,924 overhead <u>directly</u> to the three activities, namely $5,764 for 'direct production', $2,160 for 'set-up', and $2,000 for the 'shipping' activity. The total cost of each activity can now be calculated as shown in Table 6.

Table 6 - Cost of Activities.

	Activities			
Resources	**Direct production**	**Set - up**	**Shipping**	**Total**
Material	$2,640	-	-	$2,640
Direct Labor	$4,400	-	-	$4,400
Overhead	$5,764	$2,160	$2,000	$9,924
	$12,804	$2,160	$2,000	$16,964

Now we need to trace the cost from the activities to the (cost) objects, i.e., the products. In other words, we need to find out how much of each activity is consumed by each product. Activity drivers are used to do this. Thus, we need to identify and quantify activity drivers for the 'direct production', 'set-up', and 'shipping'. A good activity driver is one that points to the root cause for a change in the cost. With that in mind, we have identified the activity drivers listed in Table 7 for the following reasons:

- We know that both material amount (or cost) and direct labor hours (or cost) affect the cost of the products. So these are almost obvious choices for activity drivers.
- Furthermore, closer investigation of management has indicated that the products need machine time in different amounts (see Table 7) and that the machines are paid for out of the overhead. So, a good driver to trace the direct production overhead to the products would be the number of machine hours that each product needs. The values are listed in Table 7.
- For the set-up activity, we assume that the cost is directly proportional to the number of set-ups. As shown in Table 7, the products require different number of set-ups.

- For shipping, management indicates that the number of shipments is a good measure for the cost, so we take that as a driver for 'shipping'. One shipment is needed for each product. This driver is in other words completely volume (annual production volume) unrelated.

Table 7 - Activity Drivers and Values per Product Batch.

		Direct production drivers			Set - up driver	Shipping driver
Product	Volume	Material	Direct labor hours	Machine hours	No. of set-ups	No. of shipments
P1	10	$60	5	5	1	1
P2	100	$600	50	50	3	1
P3	10	$180	15	15	1	1
P4	100	$1,800	150	150	3	1
Total	220	$2,640	220 hours	220 hours	8	4

Now we need to determine the consumption intensities (i.e., unit price) for each of the activity drivers. Notice that material is already expressed in dollars, so no consumption intensity is needed. We know that the direct labor consumption intensity is fixed at $20 per hour. However, we do not know the consumption intensity of the remaining activity drivers. However, we can calculate their respective consumption intensities by dividing the amount of the activity cost that they are tracing by the driver values, see Table 8. Machine hours, for example, traces $5,764 of the direct production cost, whereas no. of set-ups and no. of shipments trace the total costs of the 'set-up' and 'shipping' activities, respectively. Dividing the activity cost traced by the driver values (or levels) gives the consumption intensities listed in Table 8 for these three drivers. Note that these consumption intensities actually may vary if the activity cost remains the same, but driver values change.

Table 8 - Consumption Intensities.

	Machine hours	No. of set-ups	No. of shipments
Activity cost traced	$5,764	$2,160	$2,000
Total activity driver value	220 machine hours	8 set-ups	4 shipments
Consumption intensity	$26.20/ hour machine	$270/set-up	$500/shipment

Now we can trace the cost of each activity directly to each product by multiplying each activity driver value times the driver's consumption intensity. This is shown in Table 9 for all three activities.

In a schematic form, the basic layout of the ABC model and the cost flow can be shown as in Figure 7; there are three resources that are consumed by three basic activities. Note that the 'manufacturing' activity embodies 'direct production' and 'set-up'. The cost of the manufacturing activity is simply the sum of the cost of those two activities. Also shown in Figure 7 are the activity drivers that trace the activity costs to the objects (products P1 through P4). Note the differences between Figure 7 and Figure 6.

Table 9 - Activity Costs Traced to Products.

			Objects								
			P1		P2		P3		P4		
Activity	**Activity Driver**	**Consump. Intensity**	**Act. driver value**	**Cost**	**Act. driver value**	**Cost**	**Act. driver value**	**Cost**	**Act. driver value**	**Cost**	**Total**
Direct Production	Material	1	60	$60	600	$600	180	$180	1,800	$1800	$2,640
	Direct labor	$20/hour	5	$100	50	$1,000	15	$300	150	$3,000	$4,400
	Machine hours	$26.20/hour	5	$131	50	$1,310	15	$393	150	$3,930	$5,764
Set-up	No. of set-ups	$270/set-up	1	$270	3	$810	1	$270	3	$810	$2,160
Shipping	No. of shipments	$500/ shipment	1	$500	1	$500	1	$500	1	$500	$2,000
		Total cost		$1,061		$4,220		$1,643		$10,040	$16,964
		Unit cost		$106.10		$42.20		$164.30		$100.40	

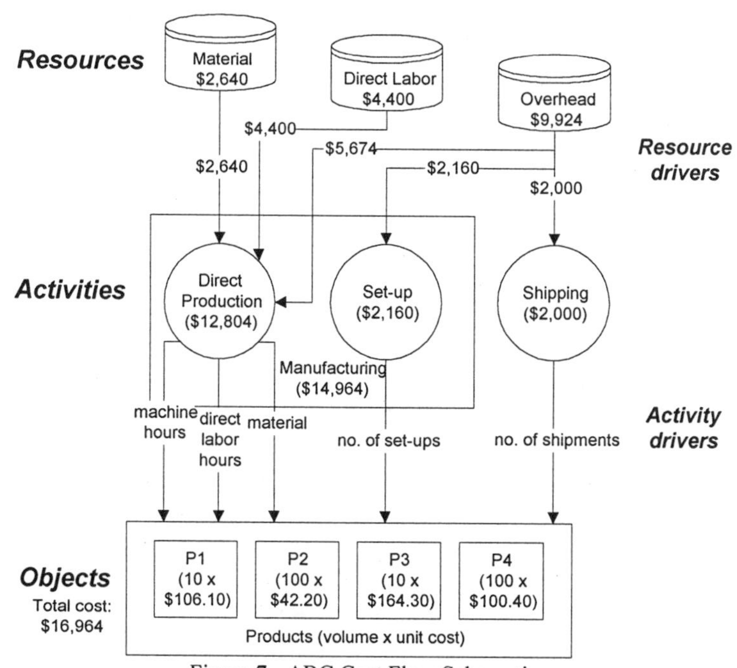

Figure 7 - ABC Cost Flow Schematic.

The difference between the two approaches is even more striking when we compare the reported unit costs between the traditional costing approach and ABC is presented, which is shown in Table 10. Clearly, it is substantial, especially for product P1. The major difference is the overhead allocation. In Table 11, the amount of overhead traced to each product using the ABC approach is shown and in Table 12 the difference in overhead allocation between the traditional costing approach and ABC is given. The difference is huge; up to 300%. Given this example, it is easy to understand why volume-based costing approaches are being criticized and why ABC is gaining ground. Making decisions based on a volume-based costing system are simply not wise simply because they grossly mistreat overhead costs.

Table 10 - Reported Unit Cost Difference.

Product	Traditional Costing Unit Cost [$]	ABC Unit Cost [$]	Difference
P1	$38.56	$106.10	275 %
P2	$38.56	$42.20	9 %
P3	$115.67	$164.30	42 %
P4	$115.67	$100.40	-15 %

Table 11 - ABC Unit Overhead Costs.

Product	Machine hours [$]	Set-up [$]	Shipping [$]	Total overhead traced [$]	Reported unit overhead [$]
P1	131	270	500	901	90.10
P2	1,310	810	500	2,620	26.20
P3	393	270	500	1,163	116.30
P4	3,930	810	500	5,240	52.40

Table 12 - Reported Unit Overhead Cost Difference.

Product	Traditional Costing Unit Overhead Cost [$]	ABC Unit Overhead Cost [$]	Difference
P1	22.56	90.10	299.4%
P2	22.56	26.20	16.1%
P3	67.67	116.30	71.9%
P4	67.67	52.40	-22.6%

1.4 Another Example - The Hidden Costs of Pollution

ABC has also been 'discovered' by those who want to promote pollution prevention activities as a useful tool. Because ABC is superior in tracing costs, it can be used to identify costs associated with pollution and waste management much clearer than the volume-based costing systems. For example, the Pollution Prevention and Assistance Division of the State of Georgia is providing short-courses in ABC to educate companies on the true costs of their wastes. Many companies think that the disposal/removal cost is the only cost for waste. But once it is made clear that, for example, an administrative assistant has to schedule a certified waste hauler's pick-up and prepare paper-work, and a forklift driver has to transport it, and other personnel has to be involved as well, then they see that the costs are actually higher.

1.4.1 Problem Statement

To show how ABC can be used to identify the true environmental costs, we use an example based on an example given in (Brooks, Davidson *et al.* 1993).

Consider a hypothetical manufacturer that produces two types of furniture: Unfinished (UFIN) and finished (FIN) chairs. The production is 1 million chairs a year - divided equally between the two product lines. The manufacturing of the unfinished chairs results in very little waste that has environmental consequences, only sawdust and residual glue. The manufacturing of the finished chairs, however, involves paints, stains, solvents and other toxic adhesives - in addition to sawdust and residual glue.

Raw material costs are $15 for each FIN Chair and $10 for each UFIN Chair. Direct labor costs for the FIN Chair is $10.50 per chair and for the UFIN Chair $10 per chair. Furthermore, the company's total overhead cost is $30 million. A large part of the corporate/plant administration costs are attributable to environmental costs, namely $5.4 million.

Management views the finished chairs as the more profitable of the two lines. Recently, management has given serious thought to phasing out the unfinished chair line. But is phasing out the unfinished chair line a good idea? To answer the question we will analyze the costs for the products using both a volume-based approach and ABC.

1.4.2 Traditional Costing Approach

If we use a traditional costing system, then the overhead costs are allocated to each chair proportional to the direct labor spent. For this example, we distinguish two separate overhead accounts: the environmental overhead ($5.4 million) and the remaining other overhead ($24.6 million).

To get the other overhead cost per chair, we divide the annual overhead cost by the annual direct labor and multiply the result times the direct labor needed to produce one chair. This gives the following other overhead costs:

UFIN other overhead	= ($24,600,000 / $10,250,000) x $10.00	= $24.00 per chair
FIN other overhead	= ($24,600,000 / $10,250,000) x $10.50	= $25.20 per chair

Similarly, the environmental overhead per chair is:

UFIN other overhead	= ($5,400,000 / $10,250,000) x $10.00	= $ 5.27 per chair
FIN other overhead	= ($5,400,000 / $10,250,000) x $10.50	= $ 5.53 per chair

Using this information, the unit cost for the UFIN chair is as follows:

Raw Materials	$ 10.00
Direct Labor	$ 10.00
Other Overhead	$ 24.00
Environmental Overhead	$ 5.27
Cost	$ 49.27

If the sales price is $50.00, this gives a profit of $0.73 for the UFIN Chair. The cost for a FIN chair is as follows:

Raw Materials	$ 15.00
Direct Labor	$ 10.50
Other Overhead	$ 25.20
Environmental Overhead	$ 5.53
Cost	$ 56.23

If the sales price is $60.00, this gives a profit of $3.77 for the FIN chair, which clearly supports the management decision. What is wrong?

If we look carefully, we see that the two processes are treated in the same way. The volume-based costing system ignores the difference in activities in the two processes - that UFIN triggers few environmental costs, while FIN triggers most of the environmental costs. Notice that the company develops the costs *without* considering the related environmental expenditures.

1.4.3 Activity-Based Costing Approach

If we follow an ABC approach, then we would split up the manufacturing process into different activities, and trace the cost to the products according to the activities that they consume. For the FIN and UFIN chairs, by identifying and separating the finishing activities and assigning the related costs accordingly to each type of chair, the cost assignment changes dramatically. For the UFIN chair, only sawdust and residual glue have to be disposed. It turns out that is disposal cost is $30,000 per year. That means that the environmental costs for the UFIN chair are only a fraction of the $5.4 million total environmental costs for the company. The actual annual UFIN environmental cost is $30,000/500,000 = $0.06/unit. Using this information, the cost of the UFIN Chair now becomes $44.06, see below:

Raw Materials	$ 10.00
Direct Labor	$ 10.00
Other Overhead	$ 24.00
Environmental Overhead	$ 0.06
Cost	$ 44.06

With the same selling price of $50, this as earlier, this gives a $5.94 profit.

Clearly, the FIN Chair must be responsible for the remainder of the environmental overhead which is ($5,400,000 - $30,000)/500,000 = $10.74/unit. If we recalculate the cost of the FIN Chair, we now get a cost of $61.44, see below.

Raw Materials	$ 15.00
Direct Labor	$ 10.50
Other Overhead	$ 25.20
Environmental Overhead	$ 10.74
Cost	$ 61.44

With a $60.00 selling price we get a $1.44 loss! The situation has changed dramatically, and the management decision is - according to ABC - wrong!

Again we see that the traditional approach is seriously flawed and could have put the company out of business if management decided to boost the production of the finished chair over the unfinished chair. No wonder that in (Fox 1986) the question is raised whether cost accounting is an asset or a liability.

1.5 Summarizing the ABC Approach

In the preceding sections, we discussed ABC and outlined its advantages over volume-based costing systems. In summary, the basic differences between traditional costing systems and ABC lie in the basic assumptions:

- Traditional volume-based costing;
  ⇒ Products consume resources.
  ⇒ Costs are allocated using unit-level allocation bases (e.g., direct labor).
- Activity-Based Costing;
  ⇒ Products (objects) consume activities; products do not directly use up resources, but activities consume resources.
  ⇒ Costs are allocated/traced using (multi-level) activity and resource drivers.

Although ABC may sound like a completely new way of costing, it is really not that radical. In fact, for engineers who think in terms of (physical) processes, ABC sounds very obvious and perhaps even trivially simple. The key thing to remember is that *the main purpose of ABC is to increase the accuracy and tracing of cost accounting*.

2. FROM ABC TO ABM

There is nothing so useless as doing efficiently that which should not be done at all.

Peter F. Drucker

So far, we have compared ABC to volume-based costing systems and hopefully demonstrated the superiority of ABC, but two issues remain:

1. *What drives/triggers costs in ABC?*
2. *How can we reduce costs, or how can resources be better utilized?*

The latter is especially important for managers and has led to the development of what is called Activity-Based Management (ABM). Its relation with ABC warrants discussion, because as will be clear, our work falls under the ABM heading.

2.1 What is ABM?

There are a large number of 'Activity-Based' acronyms being used in management nowadays (and we added one more with this book). ABC, ACBM, ABM, ABCEM - what do all those acronyms mean and how do they relate? To explain the differences, we use the same definitions as found in (Cokins 1996):

- In a narrow sense, *Activity Based Costing* (ABC) can be considered the mathematics used to reassign cost accurately to cost objects, that is, outputs products, services, customers. Its primary purpose is for profitability analysis.
- *Activity-Based Cost Management* (ABCM) uses the ABC cost information to not only rationalize what products or services to sell, but more importantly to identify opportunities to change the activities and processes to improve productivity
- *Activity-Based Management* (ABM) integrates ABC and ABCM with non-cost metrics such as cycle time, quality, agility, flexibility, and customer service. ABM goes beyond cost information.

It is not surprising after reading the above definitions that ABC and many other activity-based approaches, including ABCM, are often collectively grouped under the term Activity-Based Management (ABM). It is also easy to realize that many variations to ABM are possible. As an example, Figure 8 shows how ABC can be connected with cost planning and control.

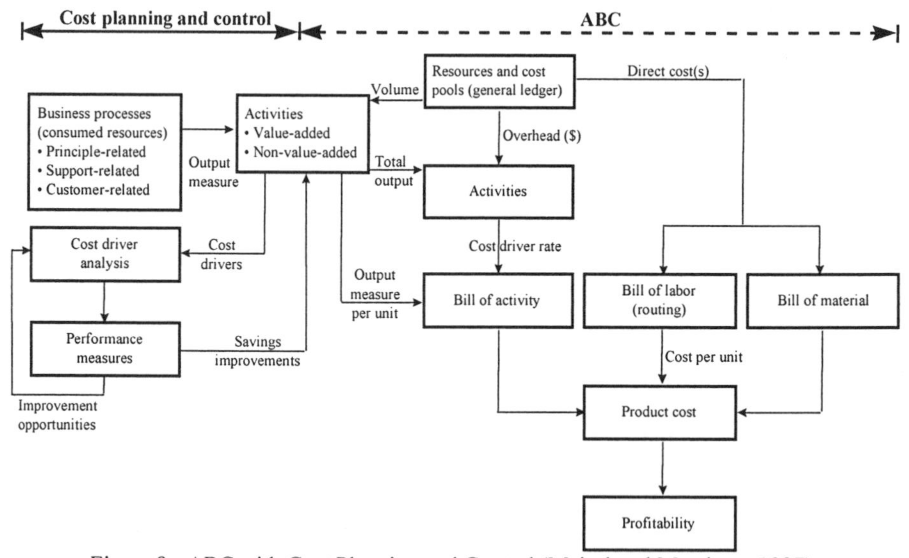

Figure 8 - ABC with Cost Planning and Control (Maisel and Morrissey 1997).

From a management perspective, the difference between ABM and traditional cost management is significant. Traditionally, cost management is concerned with managing costs. In ABM, however, it is recognized that costs cannot be controlled. Rather, one controls activities that in turn cause costs (Youde 1992). The fundamental principle that all ABM approaches have in common is that they focus on managing processes (that consist of activities) rather than costs *per se*. The foundation to this new thinking is based on what is called the 'two-dimensional' ABC concept or 'second generation ABC architecture'.

2.2 Managing Processes versus Managing Costs

In Figure 9, the so called 'two-dimensional' ABC/ABM concept is shown, where the second dimension is the process view used to do non-economic performance measurement. This is also referred to as a 'second generation ABC architecture' (Turney 1991a) because it is an improvement over the older version shown in Figure 2 that was articulated by e.g. (Cooper 1990c).

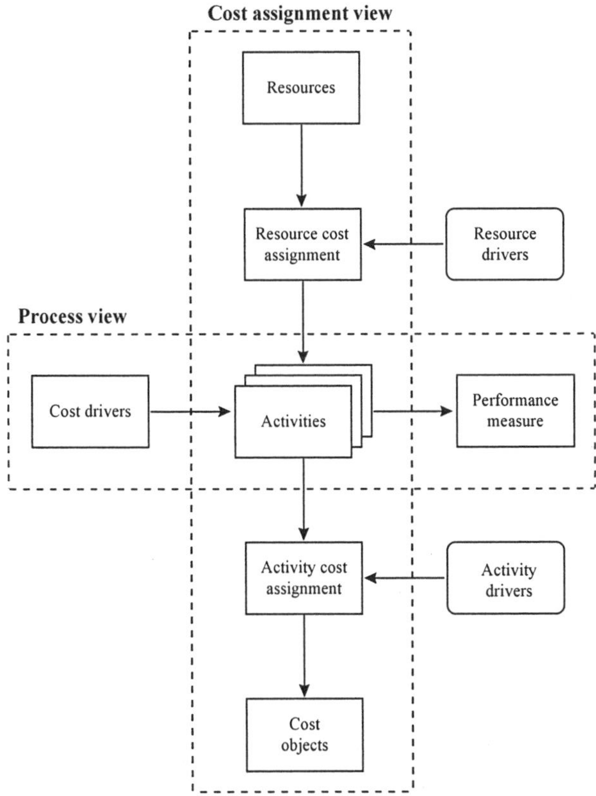

Figure 9 - Two-Dimensional ABC System (Turney 1991a).

The core of the second generation ABC architecture is still that resources are consumed by activities, which are consumed by objects. This is the same as before and is referred to as the cost assignment view in Figure 9. What is new is that the second generation includes an enhanced and distinct process view centered on the activities. This is the horizontal part called the process view in Figure 9. Together, these two views form a cross, often referred to as the CAM-I cross because Peter Turney and Norm Raffish created the diagram for the not-for-profit Consortium for Advanced Manufacturers-International

(CAM-I). The two views serve different purposes (Turney 1992; Cokins 1996):

1. The cost assignment view focuses on *what things cost* and deals with issues such as pricing, product mix, outsourcing, product design, and setting priorities for improvement efforts.
2. The process view focuses on *why things cost* and deals with issues like
   - *Why* work takes place (the root cause) and *how much* effort must be expended to carry out the work. This is measured by cost drivers. Cost drivers are any factors that change the cost of an activity. For example, the quality of goods received at an activity determines the amount of work required by that activity. Cost drivers include factors relating to the performance of prior activities in the value chain, as well as factors internal to the activity.
   - *How well* the work was performed. This is measured with the performance measures, which are indicators of the work performed and results achieved in an activity. Performance measures can include quality (e.g., number of defective parts per million), time (e.g., time to perform an activity), and efficiency (e.g., number of products per ton material).

This CAM-I cross is widely used to convince managers that they should be managing processes rather than costs. Again, this is something that may seem obvious for the engineers in the trenches of manufacturing processes and even for us it seemed so logical that we wondered what MBAs are taught in school. Nevertheless, ABM simply forces managers (and others) to take a closer look at what is actually going on in their business. Taking a high-level helicopter view using the traditional accounting books is not enough anymore; knowledge and understanding of what is actually going on is needed.

2.3 Cost Drivers versus Activity Drivers

We have noticed that the use of the word ‘cost drivers’ can be confusing in explaining ABC and ABM to novices in the area because people immediately start wondering what the differences are between activity drivers and cost drivers. Well, an activity driver is a measure of change in an activity’s costs caused by an object (product) or another activity. Cost drivers are broader measures that go beyond products and other activities as the root cause of a change in an activity’s cost. Activity drivers may not be the true drivers of cost in the sense of triggering or being the root cause. The true drivers are called ‘operational cost drivers’, or simply cost drivers (Cokins, Stratton *et al.* 1992). Activity drivers are consequences of what has happened, whereas cost drivers reveal what is making it happen. For example, workers spend ten minutes on assembling product A and fifteen minutes on assembling product B. The activity driver simply is the time spent assembling: 10 minutes for A and 15 minutes for B. But think for a moment *why* product A takes 10 minutes and *why* product B takes 15 minutes. One reason may be that product B is more complex than product A. In that case, 'product complexity' is the cost driver.

Because activity drivers are integral to product costing, they must be (easily) quantifiable and measurable, e.g., minutes spent assembling, number of set-ups. Cost drivers, on the other hand, may be less measurable but more insightful. In Figure 10, an overview of factors that drive costs is given. All can be viewed as cost driver examples. Clearly, some are more easily measured than others.

Volume complexity	**Size complexity**	**Product design complexity**	**Distribution complexity**
Number of units Process time Direct labor ($) Direct material ($)	Space Head count Equipment	Number of part numbers in system Number of material structure levels Number of part numbers per product Number of engineering change orders	Number of customers Number of customer orders Number of internal orders
Procurement complexity	**Process complexity**	**Product range complexity**	**Non-value-added complexity**
Number of vendors Number of purchase orders Number of part inspected Number of "A" parts	Subcontracting Material transactions Cycle time Process flow Schedule changes	Number of products Number of options Number of accessories Number of new products	Inventory levels Rework Scrap

Figure 10 - Factors that Drive Costs (Eiler and Ball 1997).

2.4 Cost Reduction in ABM

So, if a manager should not manage costs, how can he/she reduce cost in ABM? The answer lies again in the transactions.

2.4.1 Transactions Drive Costs

In general, *transactions* drive costs in ABC/ABM. A transaction is basically a transmission of information regarding something. So, if you buy something, there is a transaction (the purchase order, for example.). If you send an e-mail, there is a mail transaction (the mail itself, for example). An activity, on the other hand, is the actual thing that is being done. Can you have an activity without at least one transaction? No, an activity will always be associated with at least one transaction. Most activities will be associated with several types of transactions. Transactions are what cause the consumption of resources in an activity and are basically what we are trying to model using the drivers. Hence, a driver is a measure of a particular transaction that we believe is representative for the overall activity, or a significant part of it. This is why ABC is sometimes referred to as transaction-based costing.

That transactions drive costs is old news. In an insightful paper (according to (Johnson and Kaplan 1987)), (Drucker 1963) observed:

> *While 90% of the results are being produced by the first 10% of events, 90% of the costs are being increased by the remaining and result-less 90% events. Economic events are, by and large, directly proportionate to revenue, while costs are directly proportionate by number of transactions.*
> *Furthermore, ...efforts will allocate themselves to the 90% of events that produce practically no results.... In fact, the most expensive and potentially productive resources (i.e., highly trained people) will misallocate themselves the worst.*

Manufacturing transactions (see also Figure 10) can be grouped into four types of transactions (Miller and Vollmann 1985):

1. Logistical transactions to order, to execute and to confirm materials movements. Personnel busy with such transactions include indirect shop floor workers as well as people engaged in receiving, shipping, data entry, Electronic Data Processing (EDP) and accounting.
2. Balancing transactions to match supply of materials, labor and machines with demand. These transactions are typically performed by people doing purchasing, materials planning, production control, forecasting and scheduling.
3. Quality transactions to validate that production is conforming to specifications. People in quality control, indirect engineering and procurement perform quality transactions.
4. Change transactions to update manufacturing information. Manufacturing, industrial and quality engineers involved with Engineering Change Orders (ECO), schedules, routings, standards, specifications and bills of materials perform change transactions.

A well designed ABC system can per definition handle these transactions well because ABC is transaction/process-based. Note that one group of transactions are related to quality. This explains the large potential for quality driven organizations to implement ABC and Total Quality Management (TQM). Therefore, ABC is a quality enforcing system if it is used as such, and not just a 'cost-cutting tool', see (Johnson 1992).

2.4.2 Reducing Costs by Reducing Transactions

A costing system should not only reveal the state of affairs by tracing transactions, but also be useful in reducing cost by identifying better ways to use the resources by reducing the number of total transactions. In an ABC system this is done in four ways (Turney 1991b):

1. Activity reduction is one of the key elements in continuous improvement. This implies that the elapsed time and effort required to perform activities must be reduced.
2. Activity elimination is based on the fact that changes in the production process or products can eliminate the need to perform certain activities. Many activities in an organization do not contribute to customer value, responsiveness or quality (non - value - added activities). However, one should be careful to assume that those activities can be eliminated in general. An example of this is all the cashiers, they perform non-value - added activities, but their jobs are not eliminated. Activity elimination is the only way to affect the fixed activity costs, and it is therefore the most effective way to reduce cost/increase resource utilization. This is what Business Process Reengineering (BPR) relies on, see (Cokins 1993).
3. Activity selection is applicable when a product or a production process can be designed in several ways, with each alternative carrying its own set of activities and costs.
4. Activity sharing provides economies of scale as the designer of a product or process can choose design alternatives that permit products to share activities.

As we can see in Table 13, it is necessary to focus on both product design and process design to minimize the costs or improve resource utilization. It is interesting to note that (Turney 1991b) does not realize that product design has significant impact on activity reduction and elimination.

Table 13 - Cost Reduction Opportunities in ABC (Turney 1991b).

How to reduce costs	Process design	Product design
Activity reduction	Reducing setup time	
Activity elimination	Eliminating material handling activities	
Activity selection		Choosing an insertion process
Activity sharing		Using common components

This is something, however, more and more people are becoming aware of and it is an important part of our work. In Table 14 a fairly complete list of possible cost reduction methods is given. ABC supports most of them, while a traditional costing system gives little aid. In

Table 15, it is shown what (in general) the traditional costing system's ways to reduce costs are, compared to an ABC system. The differences stem from the different basic assumptions (see Section 1.2).

Table 14 - Cost Reduction Methods. Based on (Shields and Young 1991).

Design and Manufacturing Methods
- Design to manufacture
- Group technology:
  - Standardizing and reduction of the number of parts
  - Standardizing manufacturing process
  - Manufacturing cells
  - CAPP
- Design to cost
- Design for assembly
  - Taguchi methods
  - Boothroyd & Dewhurst's DFA
  - Hitachi's assemblability evaluation method
- Concurrent and simultaneous engineering
- Reliability engineering
- Value analysis and engineering
- TQC of design and development

Design and Manufacturing Organization Structure
- Manufacturing sign-off
- Integrator
- Cross-functional team
- Concurrent engineering team
- Simultaneous engineering
- Product-process design department
- Early manufacturing involvement

Material Sourcing
- Vendor selection and certification
- Electronic data interchange
- Purchasing of materials
- TQC of incoming materials before arrival

Inventory Management
- Manufacturing Resource Planning (MRP II)
- Just-in-Time (JIT)

Advanced Manufacturing Technology
- Computer-Aided Design (CAD)
- Robotics
- Flexible Manufacturing Systems (FMS)
- Computer-Integrated Manufacturing (CIM)
- MRP

Capacity Utilization
- Optimized Production Technology (OPT)
- CIM
- Total preventive maintenance
- MRP II

Manufacturing Costs
- Economies of scale
  - Dedicated technology
  - Standardization
  - High volume/experience curve
- Economies of scope
  - Flexible technology
  - Focused factories
  - Elimination of changeovers

Activity and Cost Driver Analysis
- Eliminating non-value-adding activities
- Reduction of value-adding cost drivers

Total Quality Control
- Statistical process control
- Cost of quality
- Quality circles

Customer Consumption Costs
- Design for maintainability
- Design for reliability
- Design for serviceability

Performance Measures of Continuous Improvement
- Constant flow of inventory
- Standing inventory
- Simplicity
- Quality
- Productivity
- Flexibility
- Time

Motivation
- Target costing
- Motivational standards
- Ratchet productivity standards
- Design target accountability
- Design productivity standards
- Management by objectives
- Employee ownership
- Employee training
- Suggestion box systems
- Performance contingent compensation
- Skill contingent compensation

Accounting control
- Budget planning and control
- Cost planning and estimation
- Actual cost accounting
- Standard cost accounting

Table 15 - Examples of Messages Communicated by a Traditional and ABC System (Turney 1991b).

Examples of Reducing Costs	Volume-Based Cost System	ABC System
Reducing setup time	Ignore or reduce direct labor	Reduce setup time to achieve low cost diversity
Eliminating material handling activities	Ignore or reduce direct labor	Eliminate activities to reduce the cost of handling materials
Choosing an insertion process	Pick alternative with lowest unit level activities	Pick lowest cost alternative
Using common components	Using common components yields no cost savings, using non common components creates no cost penalty	Use common components wherever possible

2.5 ABM and Total Quality Management

The main difference between ABC and ABM is that in ABM non-cost performance measures are also used. Quality measures are an important sub-set of those non-cost performance measures. It is therefore not surprising that ABM and TQM are often used together: ABC/ABM is process-oriented, and the process continuous improvement is a cornerstone of TQM. Furthermore, TQM is focusing on quality, which is one of the four transaction types (see Section 2.3).

People have realized that ABC and TQM support each other in a complimentary fashion. For TQM, the ABC/ABM cost perspective is important to give guidance and focus on *what* matters from a financial bottom-line perspective, see (Kaplan 1992). Many Baldridge Award winners have encountered severe financial difficulties. Hence, focusing on continuous improvement, etc., is not enough by itself.

For ABC/ABM some of the basic TQM notions are vital to avoid the 'business as usual' syndrome attacked by (Johnson 1992). The times that ABC implementations have failed are often attributable to a lack of management's understanding that fundamental changes are needed and that ABC is not just another cost cutting initiative. TQM requires people to think of fundamental changes. Hence, ABC aids TQM and TQM puts ABC in the right context.

The importance of focusing on quality in a long-term perspective is illustrated by a study of 187 European manufacturers, see (Ferdows and DeMeyer 1991; Shields and Young 1991), in which it was shown that long-term cost improvements result from having first achieved improvements in quality, then dependability, and finally in speed (i.e. time). Cost efficiency can only contribute up to 20% of possible advantages, according to (Skinner 1986). Others, such as (Dutton and Marx 1997), estimate that traditional cost efficiency improvement strategies can only affect roughly 10% of manufacturing costs. In (Brinker 1994; Edwards 1998) many of these issues are discussed in-depth by many well-renown authors, and we refer to these books.

3. IMPLEMENTING ABC - WHAT CAN GO WRONG

There are risks and costs to a program of action, but they are far less than the long-range risks and costs of comfortable inaction.

John F. Kennedy

The development of a good cost management system does not happen over night; it goes in stages. Figure 11 gives a good representation of such stages, and many readers can probably relate to these from their own experiences. Clearly, the objective of the management system drives how far (i.e., to what stage) the development has to go.

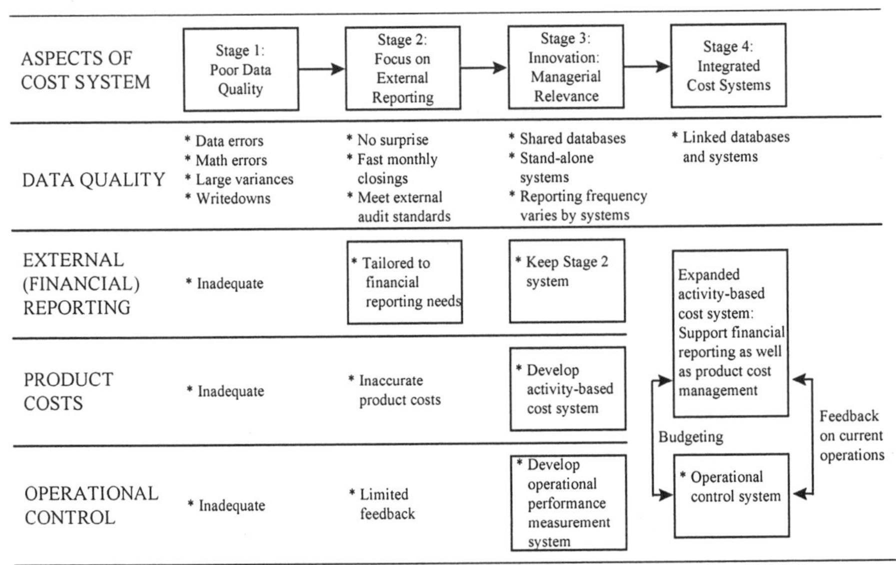

Figure 11 - The Development of Cost Management Systems (Kaplan 1990b).

But developing a management system is not easy...

In (Player 1993) a top ten list of things that typically can go wrong, and how to avoid them, is presented, see also (Turney 1990). This list is applicable to most activity-based framework implementation, and the main points are presented here.

Did not ask 'The Five Whys' is problem number one, which refers to the problem of not defining the scope of implementation properly. Also, the general lack of asking why is a problem in itself. The Five Whys refer to a technique attributable to the developer of Just-In-Time (JIT) concept Taichii Ohno (Tatikonda and Tatikonda 1991), see (Schneiderman 1996) for example, which focuses on finding the root cause and not the symptoms. The Fishbone diagram and Pareto Analysis are also used for this purpose.

The second most important problem is forgot the 'Three Cost Views' that were presented in (Johnson and Kaplan 1987); i.e. strategic, operational and financial. The problem arises when practitioners try to do all of them *simultaneously*. The three cost views are presented in Table 16.

Problem number three is looked at only manufacturing costs; not total costs, and thereby omitting large areas of potential improvements.

The fourth problem is related to the team set-up; used team solely from finance; not cross-functional, which is rooted in three issues:

1. A group consisting solely of finance people will get the stigma of being 'just another accounting' project, and therefore not get the necessary company-wide buy-in necessary for success.
2. A model implemented solely by finance people will inevitable get too much focus on the financial perspective.
3. Finance people do not have the knowledge to model the business processes accurately enough. This is crucial since ABC is process-oriented. In fact, when Taichii Ohno implemented JIT at Toyota he kept cost accountants out of the plant to 'prevent the knowledge of cost accounting from entering the minds of his people' (Lee 1987).

Table 16 - The Three Cost Views (Player 1993).

	Operational	Strategic	Financial
User of Information	• Line managers • Process improvement teams • Quality teams	• Strategic planners • Cost engineers • Capital budgeters • Product sourcing	• Financial controllers • Tax managers • Financial accounting • Treasury
Uses	• Key performance indicators • Value/non-value added indicators • Activity analysis for process improvements	• Activity-based product costing • Target Costing • Investment justification • Life-Cycle Costing • Make/buy analysis	• Inventory calculation • External reporting • Shareholders • Lenders • Tax authorities
Level of Aggregation	• Very detailed • Little aggregation	• Plan or product line aggregation • Detailed based on what is needed for a specific decision	• High aggregation • Often company-wide data
Reporting Frequency	• Immediate • Sometimes hourly or daily	• Ad Hoc, as needed • Usually a special study	• Periodic; often monthly • Probably could be quarterly or annually if other needs were met

The tenth problem in the (Player 1993) list is in our opinion among the most important; did not address the changes. There is an inherent resistance to change in organizations, and this can be very damaging since changes typically are necessary after, e.g., ABC model implementation. In fact, changes occur constantly and an ABC model is capable of assessing this. In an organization where this is not appreciated, the benefits of an ABC analysis will ultimately be severely reduced.

4. ABC ADOPTION AND IMPLEMENTATION

The only ones who welcome change are wet babies.

Proverb

The Cost Management Group of the Institute of Management Accountants (IMA) conducted a major survey in 1996, discussed in (Krumwiede 1998). Here, we are only presenting some highlights that are pertinent to this book. The first issue is how large portion of the polled companies is considering adopting ABC and for what reasons. The second issue is concerning the status of those companies that have adopted ABC. These issues are discussed subsequently.

4.1 ABC Adoption

First of all, adoption is defined as the stage when companies decide to commit necessary resources to ABC implementation. Of all the companies that responded to the survey, the following results were identified:

- 49% adopted to ABC. This is 41% up from earlier.
- 5% considered seriously ABC and then rejected it.
- 25% is seriously considering ABC.
- 21% has not yet seriously considered ABC.

If we divide the companies into manufacturing and non-manufacturing companies, the adoption is 45% and 61% (up 42%), respectively. Clearly, the ABC concept is spreading fairly rapidly. The reasons are many. In Table 17 the various adoption factors are listed for both adopters and non-adopters. The results in Table 17 indicate that companies that feel they have distortion problems are most motivated regarding an ABC implementation. Also, adopters are typically large organizations with more overhead costs, which may also explain the problems of distortion. Of the least important factors is the lack of system initiatives. Basically, a significant proportion of the polled companies want to either wait with an ABC implementation or not implement it at all. Two reasons for this were identified; 1) ABC implementations take considerable amount of time and resources, which are not available because other systems are implemented and 2) many companies want to wait with an ABC implementation because ABC requires better information systems.

Finally, the above-average usefulness of cost information in both non-adopters and adopters is high, yet the non-adopters reject ABC. This is because some companies believe they cannot use improved cost information in decision-making. These companies are typically in a commodity type market or they viewed ABC as overkill.

Table 17 - ABC Adoption Factors.

Adoption Factor	Non-adopters	Adopters
Potential for distortions	39% above-average potential.	71% above-average potential.
Percentage of overhead costs to total production costs	28.7%	32.8%
Decision usefulness of cost information	54% had above-average usefulness.	65% had above-average usefulness.
Lack of system initiatives	15% report major system or software initiatives occurring.	7% report major system or software initiative occurring.
Size of the organization	\$51M - \$100M (average)	\$101M - \$500M (average)

4.2 ABC Implementation

For companies that adopted ABC, (Krumwiede 1998) found the following:

- 54% have reached the stage at which ABC is used somewhat for decision making outside the accounting function.
- 16% have approved ABC for implementation.
- 14% are looking for organizational buy-in.
- 14% are still in the activity analysis stage.
- Only 2% implemented ABC and then abandoned it.

These numbers suggest that companies that implemented ABC are overall satisfied with ABC (only 2% abandoned ABC). However, it is clear that an ABC implementation is not a straightforward thing. Organizational acceptance is crucial since an ABC system not only requires information from everywhere in the organization, but also the results can be quite controversial and cause discussions unless people have ownership in the implementation.

5. WHAT IS NEXT?

An invasion of armies can be resisted, but not an idea whose time has come.

Victor Hugo

Due to the capabilities of reducing costs and altering the general resource consumption, ABC is being used in more and more areas (see Figure 12). In our work, this wide range will be extended to also include environmental issues. Activity-based approaches are becoming more and more all-encompassing and judging from the literature, it does not seem that this trend will stop in the near future.

Intuitively ABC seems, and is, far more logical and more useful than the traditional costing systems. However, it seems that a) direct labor as allocation base is still dominant in product costing, b) these product costs are used for a wide range of strategic and competitive decisions, and c) non-manufacturing costs are rarely (if ever) included as a part of product or product line costs. Worse, however, is that many people seem to be confident with this situation. Even faced with a remarkable amount of information, 'business as usual' is often still prevailing. This gives innovative companies an advantage to not just sustain their competitive edge, but to also sharpen it.

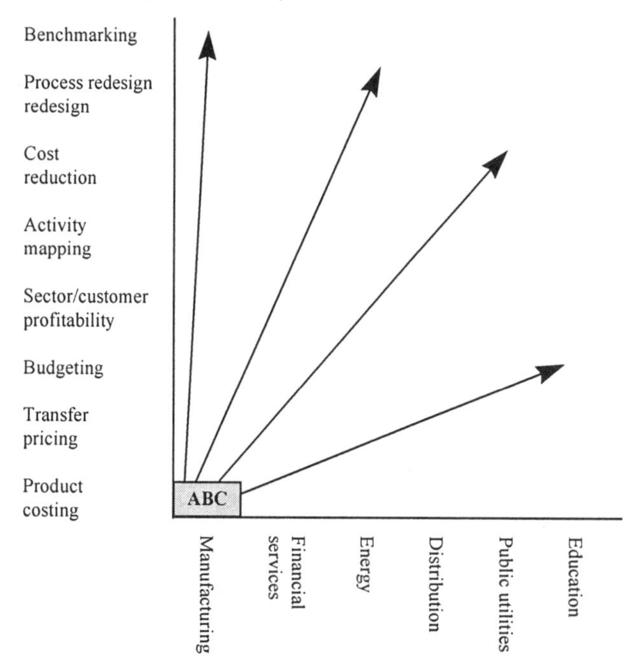

Figure 12 - The Increasing Areas of ABC Application (Morrow and Ashworth 1994).

We believe that ABC will, and probably should, be an integral part of any company's performance measurement system since 'doing the right thing right the first time' is the recipe for success.

However, ABC is no miracle cure, as discussed by (Johnson 1992) in a provoking paper where he attacks all the arising activity-based 'panaceas'. What is required is a change in mindset. Implementing ABC without promoting a culture of continuous improvement, bottom-up management, process thinking and so forth is almost a wasted effort. Focusing on continuous improvement is not enough by itself either. What is needed is a balanced approach, which can be easily achieved by realizing that ABC and ABM are process-oriented and that it requires a paradigm shift to process thinking to be really successful, see (Emblemsvåg and Bras 2000) for more on process thinking.

With that in mind, we turn ourselves back to environmental management that is (in a broad sense) focused on Nature and its processes. How do ABC and ABM fit in environmental management? This brings us to our expansion of ABC with environmental dimensions, and Activity-Based Cost and Environmental Management - the topic of the next chapter.

Chapter 4

ACTIVITY-BASED COST AND ENVIRONMENTAL MANAGEMENT

The economy is a wholly owned subsidiary of the environment.

Ray C. Anderson

In this chapter we present the principles behind our integrated Activity-Based Cost and Environmental Management (ABCEM) approach, as well as how to develop Activity-Based Cost, Energy, and Waste models. This chapter is organized as follows:

- In Section 1 we introduce the basic principles on how to expand Activity-based Costing and Management into the environmental domain.
- In Section 2 we introduce steps for developing Activity-Based Cost, Energy, and Waste assessment models along with a simple, yet illustrative, running example.
- In subsequent sections, we discuss some issues related to developing such models in more detail.

First, we start with how to expand Activity-Based Costing (ABC) and Activity-Based Management (ABM) into the environmental domain.

1. EXPANDING ABC INTO THE ENVIRONMENTAL DOMAIN

Money brings you food, but not appetite; medicine, but not health; acquaintances, but not friends.

Henrik Ibsen

Clearly, ABC is a powerful tool, also in pollution prevention, as shown in the previous chapter. ABC tracks how costs flow from resources to activities to objects. But, ABC is focused on only one dimension: monetary cost. And, cost may not be indicative at all of the severity of the actual environmental impact. In this section, we will show how environmental measures can be integrated into ABC and ABM. It is actually so simple, that we often wonder why more people have not done it.

1.1 Expanding ABC with Environmental Dimensions - Basic Concepts

1.1.1 Tracing Dollars versus Guilders versus kiloWatthours versus kilograms

As described in Chapter 3, resources are consumed by activities which are consumed by objects, and the consumption of activities by objects is traced by activity drivers, whereas the consumption of resources by activities is traced by resource drivers. This is schematically represented in Figure 1. From a numerical perspective, this consumption is measured in terms of (monetary) costs, expressed in, e.g., US Dollars. And as shown in Figure 1, the costs flow from the resources to the activities and objects.

ABC follows the conservation of total cost theorem: *no cost can be created or destroyed.* That means that, if we know the cost of the resources *and* the resource and activity driver values (levels) as well as their corresponding consumption intensities, then the cost of the activities and objects are (uniquely) known as well in an ABC model. In other words, *given* the resource cost, resource and activity driver values (levels) and their corresponding consumption intensities, the activity cost and object cost are known. From an ABC modeling perspective, we can view the resource cost, the resource and activity drivers, and the consumption intensities, as *inputs* for the ABC model, whereas the activity cost and objects cost are the *outputs*, (see Figure 1).

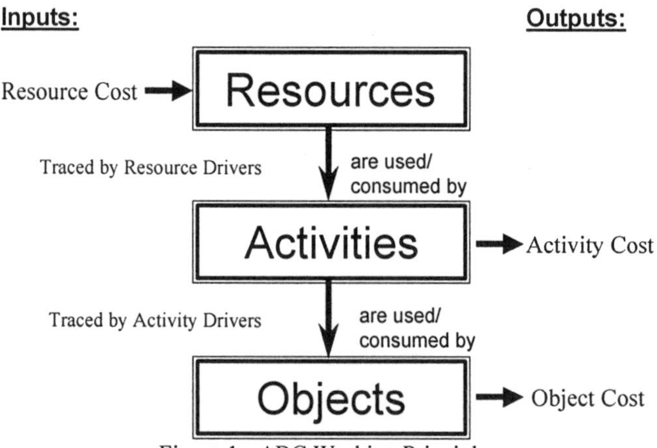

Figure 1 - ABC Working Principle.

In this book, we mostly use US Dollars ($) to measure costs and in Chapter 3 we gave two numerical examples where resources have a cost in terms of US Dollars. In Chapter 6, we have case studies in which costs are measured using Norwegian Kroner (NOK). Clearly, the principles of ABC do not change if we used British Pound Sterling or Dutch Guilders, or any other currency. Even if we invented a new currency (e.g., the JEBBCU[1]), the principles of ABC still remain the same.

Now let's take this one step further. Let's say that resources have a cost in terms of kiloWatthours (kWh) instead of dollars, and that we now want to trace the consumption of resources by activities and objects in terms of kWh, instead of dollars. Do you think the principles of ABC would change? Absolutely, and clearly, not.

What if we expressed the 'cost' of a resource in kilograms, could we trace the consumption of a resource by activities and objects in terms of kilograms? Yes, because from an ABC method's point of view, it is simply just another 'currency' and the principles remain unchanged. Mathematically, it's just a number like any other number.

1.1.2 Tracing Dollars AND kiloWatthours AND kilograms - Multi-Dimensional ABC

Clearly, changing cost currency of a resource and even changing the 'cost' to a non-monetary unit (like kWh or kg) does not change the ABC method. But let's take this another step further. *What about tracing dollars AND kiloWatthours AND kilograms?* This is something new, because what we are saying is that a resource has not a single value cost ($), but a multi-value cost ($, kWh, kg). Traditional ABC only uses a singular value for cost. What we are saying is that we want to use multiple values for 'cost'. This may seem radical, but it really is not. Imagine that we had three exactly the same ABC models, but in one we would express the cost of a resource in dollars, in the second we would express the cost in kWh and in the third we would express the cost in kilograms. Each model would give the cost of activities and objects in terms of what unit was used for the resource cost; dollar or kWh or kg. Again, the principles of ABC remain the same.

But making three models can be rather cumbersome. Could we combine the three models with different resource 'cost' measures into one model? Yes, we can, because mathematically it is no problem at all. From a mathematical point of view, we are merely saying that the cost of a resource is now mathematically expressed as a multi-dimensional $1 \times n$ vector, where n is the number of cost measures for a resource. If we use $, kWh, and kg, then $n = 3$. Those who remember basic linear algebra remember that vectors are easy to deal with; we can add and subtract them from each other as long as they have the same dimensions. Furthermore, they can be multiplied and divided by scalars (single numbers).

---

[1] Jan Emblemsvåg and Bert Bras Currency Unit.

What does this do for the ABC method? Well, what we are proposing is that the basic ABC method is expanded from a single dimensional method that traces monetary cost to a multi-dimensional method that traces monetary cost AND other non-monetary (environmental) measures. Mathematically, it is no problem. Schematically, the multi-dimensional ABC method is represented in Figure 2. Note the changes with respect to Figure 1; instead of a singular cost, we consider multiple types of 'costs' in Figure 2.

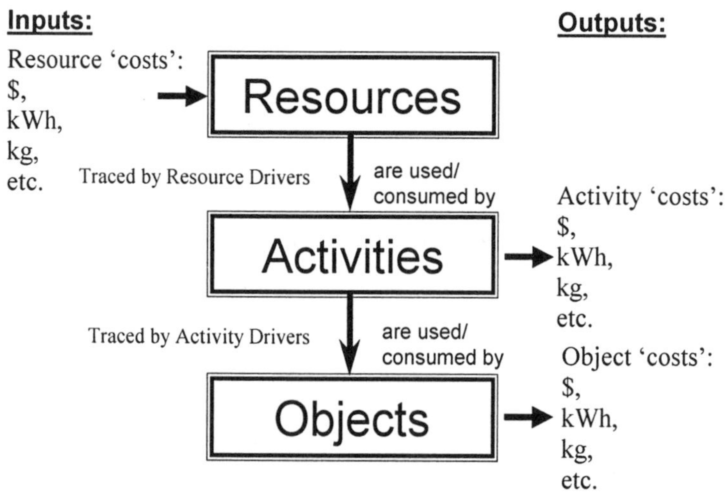

Figure 2 - Multi-Dimensional ABC Working Principle.

The fact that ABC can be expanded with additional non-monetary dimensions forms the core of our approach. Although we focus on adding environmental dimensions, many of our readers will realize that this concept is not limited to environmental issues. In fact, whatever dimension you want can be added. What to add from an environmental perspective is the topic of the next section.

1.2 Choosing the Environmental Dimensions

As we showed in the previous section, you can expand ABC without limits to include other non-monetary dimensions. So in principle, the sky is the limit and we can add whatever we want. However, information has a price and the more data we need for our models, the more it is going to cost. Thus, it is important to use as few as possible dimensions without losing relevancy.

Typical ISO 14042 LCIA studies list a whole set of environmental impact categories such as acidification, eutrophication, ozone depletion, global warming potential, etc. If needed, we could attach these dimensions to the resources in an ABC model without problem. However, as discussed Chapter 1 we like to start from a couple of simple measures that:

- Are comparable.
- Are not open to subjective influences at the practitioner level.
- Can be easily measured (just like money is today).
- Complement money.

With this in mind, we have chosen to go back to basics (see Chapter 1) and include *energy consumption* and *waste generation* as the two core environmental dimensions in our approach. *This means that in our Activity-Based Cost and Environmental Management approach, we use (as a minimum) three basic 'cost' dimensions:*

1. *Cost*, which captures the costs from the economic perspective.
2. *Energy consumption*, which captures the 'costs' of the energy perspective.
3. *Waste generation*, which captures the 'costs' of the waste (i.e. materials) perspective.

Note that we use accountable performance measurement dimensions. This is to distinguish these three dimensions from the huge variety of possible dimensions and to signify that one can in principle, and preferably, keep accounts of these dimensions. Also note that our core dimensions are not the same as the impact categories in the ISO 14042 standard, which represent actual areas of environmental problems such as ozone depletion. However, energy and waste data are also needed for ISO 14041 inventory assessments (cost is not included at all in ISO 14040). Our two core dimensions <u>can</u> be used as environmental performance indicators (EPIs) in an ISO 14031 Environmental Performance Evaluation. We always emphasize that the dimensions we propose are purely relative measures that do not represent *actual* environmental problems:

- Costs/revenues represent wealth (revenues) and general resource conversion efficiencies,
- Energy consumption represents energy conversion efficiencies, while
- Waste generation represents material conversion efficiencies.

Note that the three perspectives are complimentary. None of them are sufficient to describe, e.g., a process, but combined they describe all the accountable aspects of a process that are related to either the economic or the environmental status.

1.3 The Energy and Waste Content of Resources

In the preceding section, we explained how ABC can be expanded with additional non-monetary dimensions, specifically energy consumption and waste generation. Conventional ABC states that every resource has a monetary cost. Quantifying a resource's cost is relatively easy, because it is what you pay (or paid) for it. But how do you quantify or express the 'energy consumption' or 'waste generation' of a resource? It is not something you find on a resource's price sheet. But, as we will see, there exist something analogous to a resource's monetary cost (or price) from the energy and waste perspectives, namely energy content and waste content, respectively.

1.3.1 Energy Versus Embodied Energy/Energy Content

Energy is something we are all familiar with. We all got exposed to its basic concepts in school. The units to quantify energy are well known: Joule, kiloWatthour, calorie, Btu, etc. But there are many different energy terms in the literature, such as exergy (energy available for useful work), anergy (energy unavailable to useful work), internal energy, Gibbs free energy, kinetic energy, potential energy, electric energy, etc. Electricity is probably one of the most well known and easiest to measure types of energy; one can simply measure the electricity consumption with a meter and that is it. But electricity is not the only source of energy we use.

So how do we define a resource's energy consumption? To answer that question, we need to discuss the nature of energy first. There are two types of 'energy' associated with a resource:

1. Energy: With this we mean the *actual* energy released when using/consuming a resource. For example, fuel (a resource) can be burned and heat and/or work are obtained. This type of energy is often simply referred to as the 'energy' of a resource and it can be in different forms (e.g., chemical energy, potential energy, kinetic energy, etc.).
2. Embodied energy or energy content: With this we mean the energy used to produce, process, and transport the resource to where it is actually needed. For example, a lot of energy was expended in making gasoline before we put it in a car and burn it. The embodied energy, or energy content, of a resource is in essence the 'cost' for this resource in terms of units of energy spent.

Clearly, energy and energy content are not the same. The energy released when burning gas does not reflect the energy *expenditures* up to that point in the value chain. It is for that reason that one should not just use chemical heat values to measure true energy consumption, but should include upstream energy content as well. More on this will be given in Section 2.3.2.

1.3.2 Waste versus Embodied Waste/Waste Content

Analogous the preceding discussion on energy, we can also define two types of 'waste' associated with a resource:

1. Waste: With this we mean the *actual* waste created when using/consuming a resource. For example, when burning fuel (a resource) we create emissions (wastes) such as CO_2 and NO_x.
2. Embodied waste or waste content: With this we mean the waste created to produce, process, and transport the resource to where it is actually needed. For example, a lot of waste is created in making steel. As another example, according to a study referred to in (Anderson 1998) the production of a 10 pound laptop personal computer creates 40,000 pounds of waste throughout the entire value chain. If we consider the laptop to be a resource, then its waste content would be 40,000 pounds. The embodied waste, or waste content, of a resource is in essence the 'cost' for this resource in terms of units of waste created.

Clearly, like energy and energy content, waste and waste content are also not the same. But, wastes are more tangible than energy. Wastes can be solids, liquids, and/or gases. Waste and waste content can be measured in kilograms, pounds, etc. However, intuitively, a kilogram of plutonium waste is not the same as a kilogram of steel waste from an environmental point of view. For that reason, we prefer to use a measure that reflects to some extent the environmental impact of a waste. As discussed in Chapter 2, numerous impact measures exist and can be used, but we prefer to use the Waste Index (WI). In our opinion, it forms a good compromise between simple mass units and controversial impact measures. Nevertheless, we foresee that many of our readers will choose to use mass units (e.g., kilograms) for measuring the waste because it is by far the easiest data obtain. Plus mass data forms the foundation of all other measures, including the WI. More on waste and waste content will be given in Section 2.3.3.

1.4 Tracing Energy and Waste from Resources to Activities to Objects

Let's assume that we are able to quantitatively express a resource's energy and waste content, in addition to its monetary cost. ABC helps us find the cost of an activity cost and object, but how do we find the waste and energy of an activity and object?

The answer is simple; resource and activity drivers are used to trace the waste and energy from resources to activities and to objects in exactly the same manner as we use resource and activity drivers to trace cost. Note that the definition of a resource and activity driver is still the same as in Chapter 3, Section 1.2.2:

- A resource driver is a measure of the quantity of resources consumed by an activity.

- An <u>activity driver</u> is a measure of the consumption of an activity by a) another activity or b) an object.

The only change we made was changing the word 'cost object' to 'object' in the definition of activity driver to make it more general.

In a way, it is obvious that the same principles can be used to trace energy consumption and waste generation from resources to activities and objects. If we 'drive a car' (an activity), we use 'fuel' (a resource). The 'amount of fuel' (the resource driver) we use, e.g., 15 liters, is independent of what the cost, energy content, or waste content of the fuel is. Thus, the driver amount or level is independent of the assessment dimension, in this case. However, we do need to expand/change the definition for 'consumption intensity'.

As defined in Chapter 3, a driver's cost is the level (or amount) of a driver multiplied by its corresponding (cost) consumption intensity, resulting in a numerical cost value. In keeping with the definitions for driver consumption intensity and cost, we define the following terms:

- <u>Energy consumption intensity</u>, which is the energy consumption per unit energy driver, for example, 10 kWh per hour, 5 BTUs per square yard of carpet, 1000 calories per serving, etc.
- <u>Waste generation intensity</u>, which is waste generation per unit waste driver, for example, 10 kilograms per hour, 5 liters per square yard of carpet, 1.5 pico Waste Units per ton, etc.
- <u>Driver energy consumption</u>, which is the level (or amount) of a driver multiplied by its corresponding energy consumption intensity, resulting in a numerical energy consumption value, e.g., in Mega Joules.
- <u>Driver waste generation</u>: The level (or amount) of a driver multiplied by its corresponding waste generation intensity, resulting in a numerical waste value, e.g., in tons or Waste Units.

Returning to our 'driving a car' activity, let us assume that the cost consumption intensity is \$1 per liter and the energy consumption intensity is 47 MJ per liter and the waste generation intensity is 3 kilograms per liter. In Table 1, it is shown how the activity's cost, energy consumption and waste generation are calculated.

Table 1 - Activity Cost, Energy, and Waste Calculation.

Resource	Res. Driver	Driver Amount	Cost Consumption Intensity	Energy Consumption Intensity	Waste Generation Intensity	Activity Cost	Activity Energy Consumption	Activity Waste Generation
Fuel	Amount	10 liters	\$1/liter	47 MJ/liter	3 kg/liter	\$10	470 MJ	30 kg

1.5 The Result: Activity-Based Cost, Energy, and Waste Assessments

The result of all the preceding discussion is schematically represented in Figure 3. Basically, we have developed an integrated Activity-Based Cost, Energy, and Waste assessment.

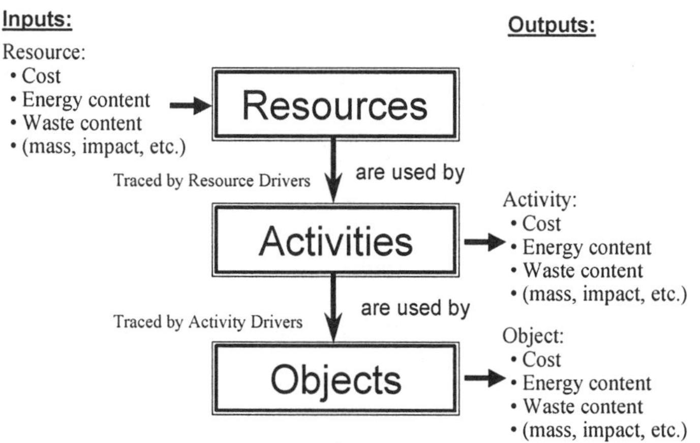

Figure 3 - Activity-Based Cost, Energy, and Waste Assessment Working Principle.

In Figure 3, it is important to note that the data gathering is concentrated around the resources. Once we know the cost, energy content and waste content (and, if desired, the mass and impact) of each resource, then the cost, energy content, waste content of each activity and object are also known as soon as the resource and activity drivers with corresponding intensities are defined and quantified. As mentioned, other assessment dimensions such as mass and environmental impact measures (such as the Eco-Indicator 95) can be added as indicated in Figure 3.

Thus, once the environmental impact of a resource is assessed, the environmental impacts of the activities and objects are also known because the impact is propagated from the resources to the activities and objects via the resource and activity drivers. In our opinion, this view simplifies ISO 14042 Impact Assessment because the environmental impact of activities and objects can only change when changes in drivers or resources occur.

Nevertheless, it all works the same as with assigning costs using standard ABC. The difference is that in Activity-Based Cost and Environmental Management, we are also assigning environmental 'costs' from resources to activities to objects. But the principle is exactly the same. This is the beauty of Activity-Based Cost and Environmental Management: it is completely transparent with ABC/ABM with a similar mind-set and framework.

1.6 Activity-Based Cost, Energy, and Waste Management

In this section we discuss how it is possible to extend ABC/ABM into the environmental domain by expanding the overall framework with energy and waste drivers, as well as cross-dimensional performance measures.

1.6.1 Expanding the CAM-I Cross

So how does discussion in the preceding sections relate to ABM? Well, recall the CAM-I cross from Figure 9 in Chapter 3, which represents the 'classical' two-dimensional product versus process view of ABM. Now look at Figure 4 below. Notice the similarity, except that Figure 4 has multiple assessment dimensions, namely, cost, energy and waste, instead of only monetary cost. In fact, more could be added, if needed, as reflected by the performance/impact measures in Figure 4.

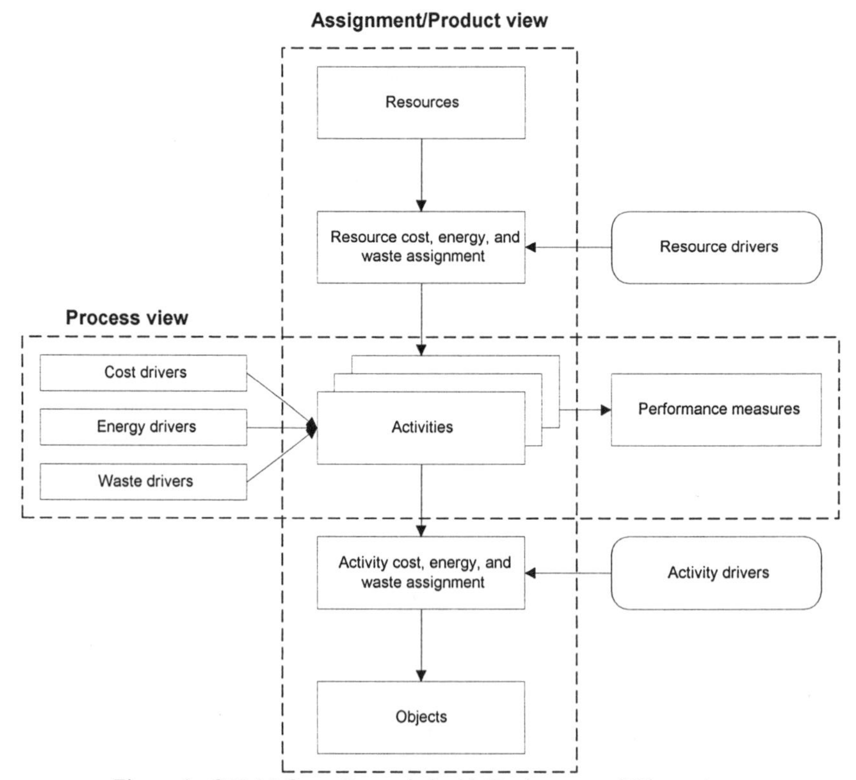

Figure 4 - CAM-I Cross Expanded with Environmental Dimensions.

1.6.2 Energy and Waste Drivers

Analogous to the use and definition of cost drivers (see Chapter 3, Section 2.3), we can now also define energy and waste drivers:

- An energy resource/activity driver is a measure for how much energy an activity consumes, while energy consumption intensity is the energy consumption per unit energy resource/activity driver.
- An energy driver is any factor that causes a change in the energy consumption of an activity.
- A waste resource/activity driver is a measure for how much waste an activity generates, while waste generation intensity is the waste generation per unit waste resource/activity driver.
- A waste driver is any factor that causes a change in the waste generation of an activity.

Other types of environmental drivers can be defined as well, if desired. Cost drivers and environmental drivers (such as the energy and waste drivers) may be the same. For example, the quality of incoming material may be the root cause for differences in an activity's cost, but it may also be the root cause for a fluctuating amount of waste that is generated in the activity (e.g., lesser quality material may result in higher scrap rates). We have found that in many cases there is such a one-to-one link between cost and environmental drivers, which indicates that focusing on managing environmental drivers can have a cost benefit as well, and vice versa.

1.6.3 Environmental Performance Measures

The traditional activity performance measures can also be expanded into the environmental domain. Basic process efficiency is one such measure that is already widely used. In Chapter 7, we show how quality loss, a classical performance measure, can be expressed in terms of energy and waste and serve as an environmental performance measure. In the case studies following this chapter, we use two other environmental performance measures that we have developed:

- <u>Economic Energy Efficiency (EEE)</u>, which is defined as the total energy consumption of an assessment object (incl. activities) divided by the total costs for the same object; inventory effects are eliminated. The EEE should be as low as possible. The EEE is therefore a relative measure of how well an organization utilizes energy in relation to the resource usage.
- <u>Economic Waste Efficiency (EWE)</u>, which is defined as the total waste generation of an assessment object (incl. activities) divided by the total costs for the same object where inventory effects are eliminated. The

EWE is a relative measure of how much waste an organization generates in relation to the resource usage and should also be as low as possible.

Analogous to these measures, we also define the Profitability Resource Efficiency to capture a purely monetary point of view:

- Profitability Resource Efficiency (PRE), which is defined as the profitability of an assessment object divided by the costs, where the costs are determined in a full absorption cost model. Cost of inventories is not included, and financial returns are not included. The PRE should be as high as possible as it measures relatively how effectively the assessment object generates profit.

Many other environmental performance measures exist and ISO 14031 - Environmental Performance Evaluation also provides some examples. But we have noticed that companies and their managers and engineers are many times reluctant to apply them, partly out of fear for the unknown, partly out of fear for the (perceived) effort involved. We hope to show with this book that environmental assessments can become an integral part of cost assessments, thus reducing workload. With that, we turn now our attention to how to actually create models for Activity-Based Cost, Energy, and Waste assessments.

2. DEVELOPING ACTIVITY-BASED COST, ENERGY, AND WASTE MODELS - SPECIFIC STEPS

The expectations of life depend upon diligence; the mechanic that would perfect his work must first sharpen his tools.

Confucius

Creating Activity-Based Cost, Energy and Waste models consists basically of three main steps:

1. Identify and quantify the resources, activities, and objects
2. Identify and quantify the resource and activity drivers
3. Perform the numerical analyses

In the following, we break these steps down in further steps and more detail, and in Figure 5 an overview of the process is given.

The specific steps have been divided into modeling and calculation steps (steps 1-6 and 7-10, respectively). Notice that the first three steps deal with the definition of objects, activities, and resources. Although we list identifying the objects as the first step, one can also reverse the order and start first with the resource perspective. Both should end up with the same result - the difference being that the first takes a bottom-up approach whereas the latter (starting from resources) takes of top-down approach.

We also want to point out that if we were only interested in the *overall* result of a company (e.g., in terms of cost, energy consumption and waste generation), we only would have to quantify the resources, and hence only Step 3 would be strictly needed. The reason for going through all the other steps is so that we can assess and trace the consumption of resources to the activity and product/service levels. This also allows forecasting and simulation.

Warning. You can spend enormous amounts of time trying to find embodied energy and waste values. Data is important, but remember that a chain is never stronger than its weakest link. Our philosophy is that **we prefer you spend your time on developing the structure of a good model, rather than chasing numbers**, especially up-front. After having gone through some initial modeling iterations, you may realize that some numbers are irrelevant anyway. Associated with this, we emphasize that **stating assumptions is very important throughout,** especially for the environmental dimensions because we do not know of any good environmental accounting systems (yet). For the costing dimension of the model this is a lesser problem because of the widespread use of monetary accounting systems.

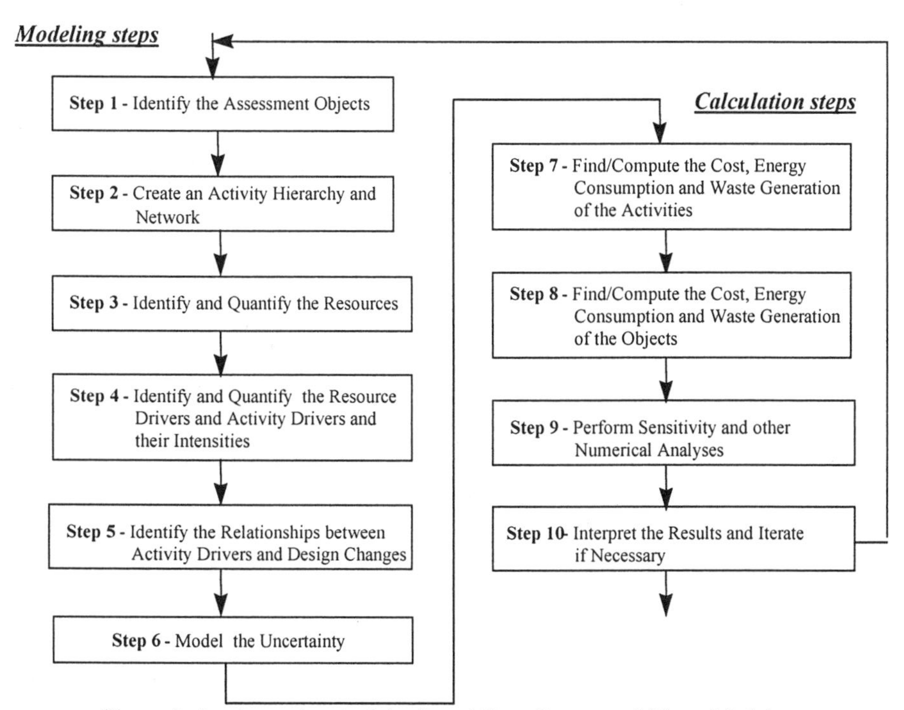

Figure 5 - Implementing Activity-Based Cost, Energy and Waste Models.

Next, we explain every step in more detail. However, first we introduce a running example that we use to illustrate each step.

Running Example

Our running example deals with a company XYZ that produces two products, i.e. assessment objects: hammers and scissors. Management at XYZ has asked us to create an Activity-Based Cost, Energy, and Waste model, and the running example illustrates how it can be done in a simple fashion. The goal of this implementation is to assess the unit cost of the hammers and the scissors. The systems boundaries are defined as everything Company XYZ must account for, as identified by their accounting system.

It should be noted that we only discuss the cost dimension completely and the manufacturing stage here for brevity, but that the inclusion of the two other dimensions and other life-cycle stages are similar in principle. However, the quantification of the waste and energy content of the resources, see Section 2.3, is discussed separately.

2.1 Step 1 - Identify the Assessment Objects

The first step to do is to identify what should be assessed, i.e. the assessment objects. These could be products, facilities, services, processes, distribution channels, suppliers, etc. When many product lines are involved they should be defined, and possible assumptions stated.

Typically, we recommend the generic part of any assessments, namely, goal and system boundary definition, to be part of this step because obviously *what* needs to be assessed is directly related to the goal of the assessment as well as the system boundary of the assessment.

When it comes to system boundary definition, it is important that the boundaries are defined in a meaningful way. This is directly in line with the goal and scope definition of conventional ISO 14040 LCA approaches. The system boundaries should at least be equal to the system boundaries chosen for a cost assessment because cost is the most important performance measure of a company. However, the boundaries may be much broader if life-cycle stages that are currently externalities for a company are included.

Running Example

With respect to the running example of Company XYZ, the assessment objects are the two products: hammers and scissors.

2.2 Step 2 - Create an Activity Hierarchy and Network

Having identified the assessment objects, i.e., what we want to assess, and the corresponding system boundary definition, we now need to identify what activities these objects are consuming.

To identify the activities, we take every process within the system boundary and break it down into more and more detailed processes, i.e. activities, and thereby create an activity hierarchy. The activities should be defined as detailed as possible, yet not beyond what is possible to get reliable information about. *It is important to relate the level of detail to the goal of the modeling*. For example, if a model for strategic analysis is the goal, then we do not have to define the activities as detailed as if the model is for operational usage. Likewise, a model not intended for any design purposes can be less detailed that a model intended for design purposes.

Naming the activities in a meaningful way is important. The names should be expressed in terms of verbs. For example, an activity should be called 'Package Product' and not 'Product Packaging'.

Furthermore, (in addition to naming) we recommend labeling each activity in the special manner as shown in Figure 6. This way of labeling the activities makes it easy to see where the activities belong in the activity hierarchy. Plus it saves a lot of space in diagrams.

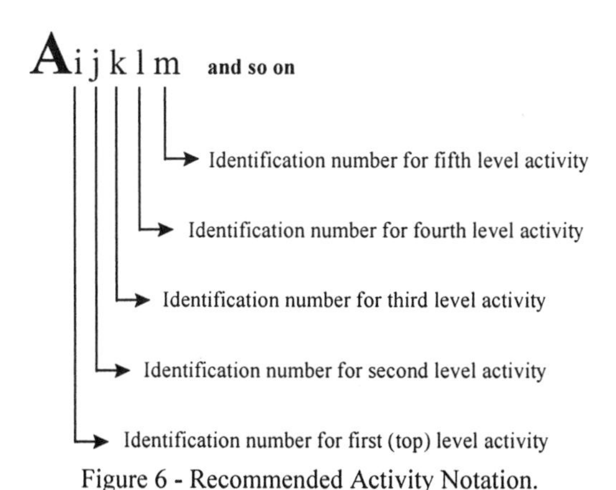

Figure 6 - Recommended Activity Notation.

How an activity hierarchy and an activity network looks like in general is shown in Figure 7 and Figure 8, respectively. The activity hierarchy shown in Figure 7 is not easy to use for large models simply because the hierarchy will often not fit on a single sheet of paper. Then, we typically present the activity hierarchy as a table as shown in the case studies in this book. Such tables, however, should be interpreted exactly as a hierarchy.

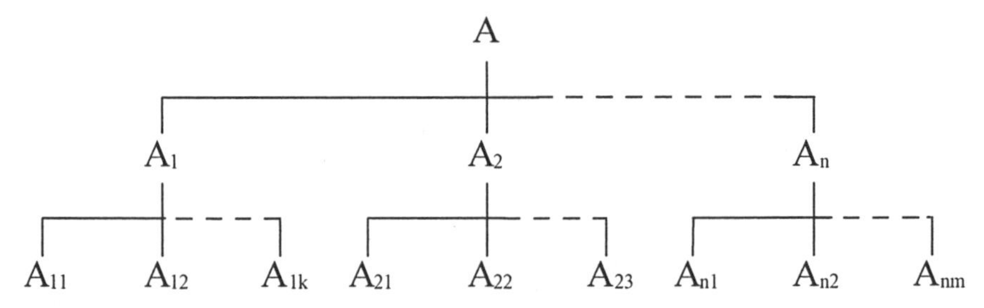

Figure 7 - General Activity Hierarchy.

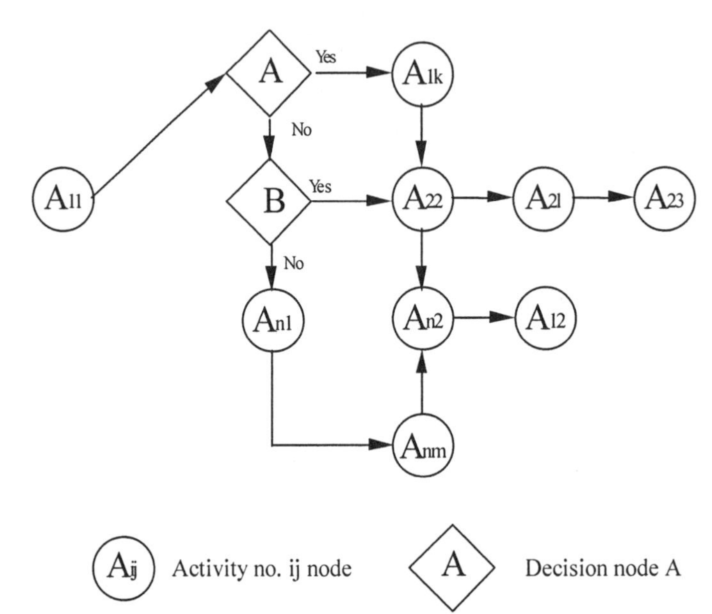

Figure 8 - General Activity Network.

An activity network shows the connections between activities and can be identical to a process network, but it does not have to be. *Every process network can be seen as an activity network, but not every network is a process network.* The purpose of an activity network is mainly to show what products consume which activities and the order of consumption. This is important to know when designing a model because otherwise the wrong products may be associated with the wrong activities, which would cause large modeling errors. We recommend that when the activity hierarchy has been created, at least the lowest level of activities, A_{11} through A_{nm}, be represented in an activity network as shown in Figure 8. We use the same icons as in (Greenwood and Reeve 1992).

The diamond-shaped nodes in Figure 8 are decision nodes used to graphically display how various design options will affect the activity network. They are also used to simply distinguish between the activity consumption of various products. For example, if Product A is associated with a 'Yes' in Decision Node A, we see immediately that Product A will incur Activity A_{lk} and then activities A_{22}, A_{21}, A_{22}, A_{n2} and A_{l2}.

Running Example

With respect to Company XYZ, the easiest way of representing an activity hierarchy is as a table, and not as shown in Figure 7. However, we illustrate here both ways. In Figure 9 the activity hierarchy for Company XYZ is presented graphically. Clearly, when the number activities become large this approach soon becomes very space consuming. Therefore, we prefer to represent the activity hierarchy as shown in Table 2, because it is far less space consuming and just as easy to understand. The shading is to signify that these activities are the lowest level activities, which are also shown in the activity network in Figure 10.

Normally, we would not write activity names in the activity network as done in Figure 10, again due to space limitations. Here, we have done it to make the example as clear as possible. Note: the 'Routing' decision node in Figure 10 simply directs scissors and hammers to the appropriate activities.

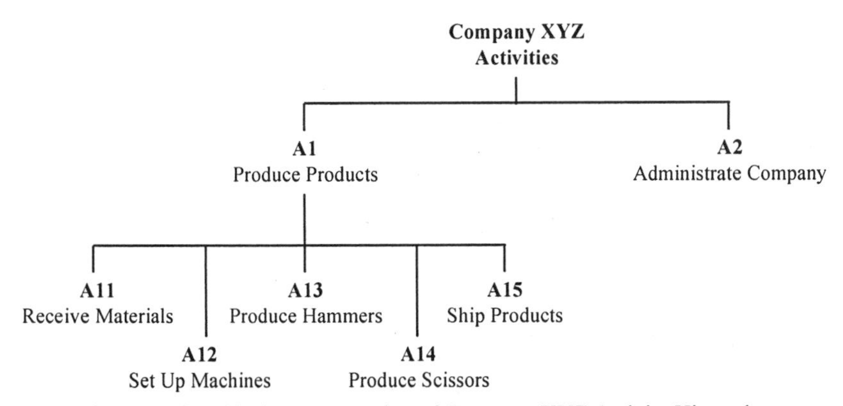

Figure 9 - Graphical Representation of Company XYZ Activity Hierarchy.

Table 2 - Tabular Representation of Company XYZ Activity Hierarchy.

Level 1 Activities		Level 2 Activities	
Produce Products	A1	Receive Materials	A11
		Set Up Machines	A12
		Produce Hammers	A13
		Produce Scissors	A14
		Ship Products	A15
Administrate Company	A2		

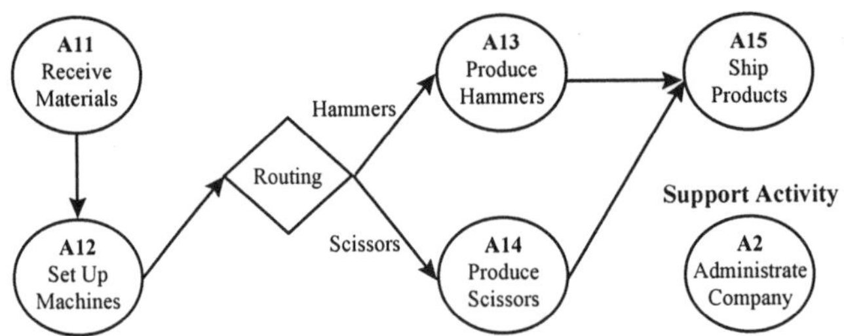

Figure 10 - Company XYZ Activity Network.

2.3 Step 3 - Identify and Quantify the Resources

Identifying the resources means basically listing everything within the system boundaries that is associated with costs (and revenues), energy consumption and waste generation. The identification only requires the name of a resource and, preferably, a type or category, e.g. depreciation, office-rent, insurance and so forth. Like activities, resources can be refined into smaller and smaller resource elements. E.g., a 'building' is a resource that can be split up into resource elements such as 'rent', 'cleaning', 'heating and air-conditioning', 'depreciation' and 'insurance', which can be further refined into even smaller resource elements depending on what is useful regarding the goals of the implementation.

The resources should also be quantified at this stage, *if possible*. The reason for emphasizing 'if possible' is that for some resources, especially those directly related to production volume, the exact amounts may not be completely known until some of the other steps have been completed. Nevertheless, every attempt should be made to quantify monetary cost, energy consumption and waste generation values of a resource.

2.3.1 Quantifying Resource Cost

Quantifying the monetary cost is simply the same as for regular ABC. Elements of the general ledger can be used for this.

2.3.2 Quantifying the Resource's Energy Consumption

Quantifying the resource's energy consumption can be a bit tricky (mostly due to lack of data), but it gets easier if we think about it analogous to monetary costs.

In our Activity-Based Cost, Energy and Waste models, there are several distinctly different ways how we can define a resource's energy consumption. Consider for example a car (a resource). A car has a certain energy content from fabricating it, etc. But when we use the car for transportation (an activity), it also consumes energy in the form of fuel (another resource). There are (at least) two approaches to modeling this situation.

- The first approach is to treat the car and fuel as separate resources that are being consumed by the activity 'transportation'. Thus, the activity 'transportation' would consume two resources: car and fuel. Clearly, a car has an energy content, but a car does not have actual energy. Fuel, on the other hand, has both energy content and an actual energy component that is released when burning it, so if an activity consumes fuel, it consumes in essence both types of energy, i.e., the sum of the two.
- The second approach is to aggregate the resources. In the car example, it is clear that we are not using fuel without using the car, so we can aggregate the fuel's energy content and actual energy with the car's energy content. In that case, the activity 'transportation' consumes only one resource; 'car'. However, this resource now has a higher energy content because it contains fuel as well. It is the same principle as aggregating the car and fuel from a monetary perspective; by aggregating the annual fuel into the car, the annual cost of the car has gone up.

Both approaches have their advantages and we have used both depending on the purpose of the modeling effort. The advantage of the first approach is that it allows for more accurate tracing of resource consumption. The second advantage is that it is creates a simpler model (less resources to trace).

In both cases, it is important to include the (upstream) energy content and its actual energy, but quantifying this can be tricky. One either needs to talk to one's suppliers and get the data from them, or see whether the data can be found in a handbook or database. With the advent of LCA, many LCA studies, tools, and databases have been developed that contain such data. We have used the IDEMAT database from the Technical University Delft, The Netherlands, quite extensively in our case studies. But other (and better/more comprehensive) tools and databases exist.

What if it is impossible to find energy content data? If no data related to the embodied energy can be found, or it is too expensive and time-consuming, we recommend that you start your model without it and simply use the actual energy consumption of a resource. For example, the actual energy consumption (e.g., electricity) of machinery (a resource) can often be found in their manuals. Similarly, chemical and engineering handbooks have data (heating values) on the (chemical) energy released if fuels are burned. These numbers underestimate the true energy consumption of a resource (you are basically assuming that the resource came for free), but it is better than nothing. Again, our philosophy is that up-front we would rather see you spending your time on developing the structure of a sound, useful, model rather than chasing numbers, which may turn out to be insignificant anyway.

2.3.3 Quantifying Resource Waste generation

Like for the resource's energy consumption, quantifying the resource's generated waste (or total waste content) can be a bit tricky (mostly due to lack of data). In our Activity-Based Cost, Energy and Waste models, there are different ways to model the waste. Consider the example of driving a car from the preceding section again. Both a car and fuel have embodied (upstream) waste. The burning of fuel to power the car generates air emissions (wastes) such as CO, CO_2, and NO_x. The following modeling approaches are possible:

- Aggregate the resources and calculate the aggregated generated waste. Basically, we would treat the car and fuel as an aggregated resource, and sum the amount of embodied waste of the car and the embodied waste fuel with the amount of waste generated when burning the fuel.
- Consider the car and fuel as separate resources with separate waste contents. The advantage of this approach is that now we can better separate and trace actual wastes from using the car versus using the fuel. For example, when using a car, the changing of oil, wiper blades, etc. causes a waste. Separating fuel and car as separate resources provides a more detailed view on where actual waste generation occurs.
- Consider each waste element to be a resource. For example, CO_2 is generated when burning fuel. However, there may be other resources that generate CO_2 when they are being consumed by activities. Given the debates over global warming, a company may want to trace where CO_2 is generated in its process. This can be achieved by considering CO_2 to be a resource that is generated by the activities. This is sort of counter-intuitive to the classical ABC approach where resources are (always) consumed. However, generation is simply the opposite of consumption, and mathematically it simply means a multiplication by -1.
- Consider each waste element to be an object. Waste is a tangible item (more so than energy) and sometimes considered to be a by-product of a process. In that case, it could be useful to consider the waste element (e.g., CO_2) as an object. This approach is especially useful if disposal costs are associated with the waste and we want to trace the cost of the activities involved in the disposal, as well as their energy consumption and waste generation, to the waste.

Again, what approach to use depends very much on the purpose of the modeling effort and there is no clear-cut recommendation.

We emphasize that it is important to include the (upstream) waste content. Quantifying this can (again) be tricky, but it is often easier than for energy because waste is more tangible and many companies maintain mass balances for their manufacturing processes. Often a resource's waste content is somewhat proportional to a resource's mass. Again, LCA studies, tools, and databases contain such data, but we recommend obtaining the information directly from the supply chain because many of the LCA tools and databases use very generic data that may not be applicable if a high degree of precision in the assessment is required. Furthermore, LCA database invariably only contain direct product related waste, no waste related to overhead activities.

What if it is impossible to find waste content data? If all fails, start with zero waste content, but assume that the resource is going to be a waste down the road itself. So, for a car, take its mass and assume that this mass is going to be the resource's waste, in addition to the fuel and other wastes it generates. Basically, in this case you make the assumption that the mass of a resource is directly proportional to the (upstream) waste content. However, many resources do not have a mass, like electricity. In such cases, as a last resort, you could multiply the amount of the resource times some kind of "inefficiency" or "waste" factor. For example, for every kWh of electricity, you could assume that you generate 2.5 units of waste.

Again, focus on model structure first before blindly chasing numbers.

Running Example

If we represent the resources that Company XYZ possesses we get Table 3. If we add all the resources together we get $530,000.

Table 3 - Company XYZ Resource Hierarchy.

Level 1 Resource Elements		Level 2 Resource Elements	
Transportation Resources		Forklift Costs	$10,000
		Truck Costs	$20,000
Materials	$300,000		
Production Equipment		Assembly Cost	$50,000
		Final Inspection Costs	$25,000
Information Technology	$10,000		
Labor	$60,000		
Administration Resources		Advertising Cost	$20,000
		Marketing Costs	$35,000

Drawing resource hierarchies is extremely difficult for large models simply due to the shear amount of information. However, when setting up an activity-based model, it is important that to keep this type of structure in mind in order to ensure that the various resource elements can be easily and meaningfully aggregated so that, e.g., all the costs associated with transportation can be easily found and verified.

As noted earlier, the energy and waste contents must be calculated. Here we will illustrate the calculation of the forklift resources. The calculation consists of two steps:

1. The upstream energy consumption and waste generation, and

2. The waste generation/energy consumption in the current stage, i.e., the stage in the life-cycle/value chain for which the assessment is done.

To calculate the upstream energy and waste, we need to identify the energy and material streams prior to Company XYZ's stage in the life-cycle. This can be done by using existing software, such as IDEMAT. However, we will (and must) make some simplifications concerning the forklift itself. We assume that the mass of a forklift is about 2,000kg, it consists of 100% steel (for simplicity) and that the forklift has an electric motor.

We first start with the upstream energy and waste content:

- *Energy content*: In IDEMAT we can find the unit energy content of automobile grade steel (31.57 MJ/kg) and by multiplying it by the mass of the forklift (2,000 kg) we get an energy content of 63,140 MJ/forklift. Using Generally Accepted Accounting Principles (GAAP) we depreciate the forklift over, say, 5 years, yielding an annual energy consumption of 12,628 MJ/forklift.
- *Waste content*: IDEMAT also provides information about most emissions cause by the production of 1 kilogram automobile grade steel (e.g. 1.229 kg NO_2 to the air). These emission data are used in conjunction with the data in Appendix A and the waste content is calculated according to the WI. Like for the energy content, the annual waste generation is found by depreciating the total waste generation over a certain time period.

To calculate the energy consumption of the current stage we use data gathered by Company XYZ. For an electric forklift we assume that the electricity is generated from various sources and that the data in IDEMAT is representative for Company XYZ. For the total energy consumption and waste generated by the forklift, we now get the following:

- *Energy consumed*: We find the electricity consumption of the forklift is 10,000 kWh/year, and we convert this information to 36,000 MJ/year. Data from IDEMAT suggests that for every 1 MJ of electricity, 2.79 MJ are required (implying a 35% power generation efficiency), thus the upstream energy content is 100,440 MJ. Combined with the (depreciated) upstream energy consumption of 12,628 MJ/forklift/year, the total annual forklift energy consumption becomes 149,068 MJ/year.
- *Waste generated*: The waste generation is computed the same way as before except that we now use the emissions for the electricity generation as found in IDEMAT as input. The waste generation for this stage in the life-cycle is added to the upstream waste generation, yielding the total annual waste generation for the forklift.

How these calculations are done is shown in detail in the following chapters where we present some of our actual case studies.

2.4 Step 4 - Identify and Quantify the Resource Drivers and Activity Drivers and their Intensities

This step is one of the most crucial steps because it establishes the relationships between resources, activities, and objects. The purpose of resource drivers is to trace resource consumption by the activities, while activity drivers trace activity consumption by the assessment objects. Basically, the resource and activity drivers define the causal relationships between resources, activities and assessment objects.

When identifying these drivers it is crucial that they are chosen as closely as possibly to represent the *actual* consumption (see also Chapter 3, Section 1.2.4). But, as explained before, we inherently run in a trade-off between accuracy and the cost of (perfect) information.

For example, consider the activity 'driving a car'. A resource that is being consumed is 'fuel'. Let's assume that the resource is quantified in terms of gallons (or liters). The best driver to measure how much fuel is being consumed would be to define a driver called 'fuel consumption' in terms of gallons (or liters). However, as we all know, mostly we have no idea about the actual fuel consumption, but we do know the amount of miles (or kilometers) we drive. Hence, a less accurate, but easier (i.e., cheaper) way to measure the driver would be 'miles driven'. By multiplying this driver by the average miles per gallon and a (average) cost per gallon of fuel, we can now calculate the amount of resource 'fuel' (in terms of cost) consumed by activity 'driving a car'. Clearly, we lose accuracy since more average information is used by using the average miles per gallon number, but the information for quantifying the driver is much easier, or less costly, to obtain.

In cases where only one assessment object is studied, only resource drivers are needed if the object wholly consumes all activities. In such a case, the cost, energy consumption and waste generation of the object is the sum of all the cost, energy consumption, and waste generation of all activities.

When the drivers are identified, the consumption and generation intensities, i.e. driver intensities, must be identified. There are two types of intensities, namely, fixed and variable:

- A fixed intensity is fixed no matter the magnitude of the drivers, e.g. direct labor and material prices, energy content per liter fuel, and waste content per kilogram steel.
- A variable intensity varies as the magnitude of the drivers varies, e.g. the machine hour price, set-up price, and shipment price in the example given in Chapter 3, Section 1.3.3. These intensities are typically found by dividing the total resources associated with that driver by the total amount of the driver.

The last issue in this step is to distinguish between design dependent drivers and design independent drivers. This is only applicable when designing relative models for comparison, which would require only design dependent activity drivers. For example, if two designs are to be cost evaluated, then it is only necessary to investigate the relative cost for both designs and choose the design with lowest relative cost.

Running Example

The activity and resource drivers are listed in Table 4. In this simple example, there are only two fixed consumption intensities; the unit material cost for the hammers and the scissors found in the Bill-Of-Materials (BOM). We assume that the BOM unit material cost for a hammer is $5 per unit and for scissors $4 per unit.

Table 4 - Company XYZ Drivers.

Activity Drivers	Resource Drivers
Product Involved	Activities Involved
Product Volume	Number Of Units
Bill Of Materials Costs	Labor Hours
Number Of Batches	Information Technology (IT) Cost
Market Share	

2.5 Step 5 - Identify the Relationships between Activity Drivers and Design Changes

In this step we need to distinguish between two very different approaches. The simplest is when we just want to assess something and to understand how activities and objects consume resources without *changing* a design.

The other approach is when we consider making a design change in the process (activities) or products (objects). In this case, often we want to know what the effect of the design change is on the consumption of resources in terms of cost, energy and waste. For this purpose we need to identify the causal relationships between these design changes and the activity drivers. For example, assembly time might be an activity driver that directly relates an activity called 'assembly' to an object called 'car door'. Clearly, if a designer is going to redesign the configuration of the car door, then most likely the assembly time will be affected, and thus the consumption of the 'assembly' activity by the 'car door' will change.

We recommend identifying these kinds of relationships explicitly so that a clear understanding is gained of how products affect activity consumption. Note that not every design change will cause necessarily a change in activity consumption. If, for example, the designer had changed the material of the door from say one type of steel to another type of steel, then the assembly time would not have changed.

Relationships between activity drivers and design changes can be anything from explicit mathematical functions to what we call 'action charts'. Mathematical functions are very accurate but equally difficult to establish. Action charts are a listing of detailed actions associated with an assessment object, aggregated into a single performance measure. In 'action charts', there are no explicit relations between design parameters and activity drivers, which gives much higher flexibility. Action charts are therefore good for directing *attention* towards design changes. Action charts are also good tools for performance measurements such as quality and time because action charts are primarily time-based and quality measures such as scrapping percentage and the like are easily included. Interestingly, we derived our action charts from disassembly charts found in, e.g., (Beitz, Suhr et al. 1992) and our own work on de- and remanufacture assessments. We modified them to better fit our purpose, and they are used in the Westnofa Industrier case study in this book. We therefore do not show the layout of the action charts here.

Action charts can be incorporated into a model or used as sub-models. The latter is probably best in very large models because it reduces the size of the model. Reducing the model size reduces simulation time, but by using action charts as sub-models, some tracing is lost. A tradeoff must therefore be made between simulation speed and tracing which ultimately is a question of implementation budget.

Running Example

Since no design related tasks are identified in this model we omit Step 4.

2.6 Step 6 - Model the Uncertainty

In our models, we always make a point of modeling the inherent uncertainty in an explicit manner. There are two main motivators for including uncertainty:

- When there is a known uncertainty associated with a specific variable in the model. For example, we know that a particular activity driver value may vary between 1.5 and 2.3 hours with a normal distribution.
- When we want to know what the effect of a change in a variable value is on the overall assessment results, in other words, we perform a sensitivity analysis. In this case, we induce a deliberate uncertainty to see how this will change the end-result.

In either case, the known or induced uncertainty is simply modeled by assigning appropriate uncertainty distributions to the corresponding variable(s) in the model. Uncertainty can be modeled in a variety of ways depending on what kind of uncertainty is to be modeled. Because of their flexibility, we typically use fuzzy intervals in our models (see also Section 4). This means that we have to define the type of distribution to use, as well as the mean and a left and right deviation. Typically, one can use past experience for this. For example, in Figure 11, an uncertainty distribution for dismantling a car is given. This figure illustrates the fact that typically it takes 2.5 hours to dismantle a car and it could be as low as 2 hours, but it is more likely that it will take longer although never more than 4 hours. Clearly, Figure 10 is based on *some* experience in car dismantling. If, however, such experience is unavailable, i.e., the uncertainty itself is uncertain, then we often defer to using normal distributions, where the standard deviation is 10% of the mean. In other words, we use a wide distribution.

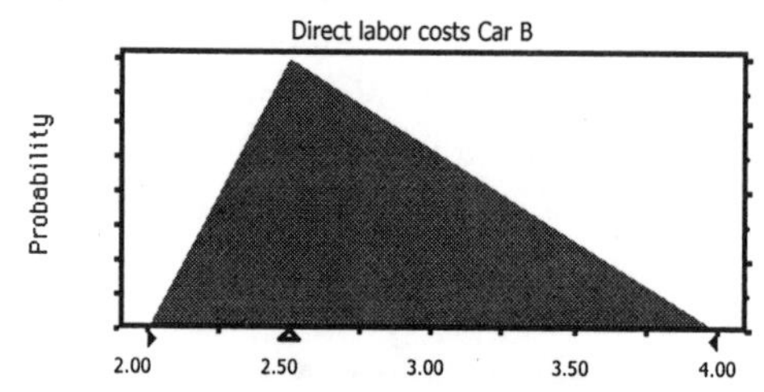

Figure 11 - Example of a Fuzzy Interval in a Model (Emblemsvåg and Bras 1997b).

The other mentioned purpose for including uncertainty is to perform sensitivity analyses and to trace the critical success factors. In this case, we typically use bounded and symmetric uncertainty distributions, such as triangular distributions, where the upper and lower bounds are ±10% of the mean. The uncertainty distributions allow us to use Monte Carlo simulation techniques (see also Section 4) for numerically identifying the effects of these uncertainties on the costs, energy consumption and waste generation, which leads us to the next step in our method.

2.7 Step 7 - Find/Compute the Cost, Energy Consumption and Waste Generation of the Activities

The cost, energy consumption and waste generation of a activity are found by multiplying the resource drivers by the consumption and generation intensities, followed by summing up all resulting values for all resources that the particular activity consumes. For activities with multiple drivers, the driver's cost, energy consumption and waste generation must be determined before the drivers are summed up to yield the overall activity cost, energy consumption and waste generation.

In general, we recommend implementing the model in a spreadsheet or so. All models listed in this book are implemented in Microsoft Excel. In fact, many tables are directly taken from the Excel spreadsheet files. If the model has been implemented in a spreadsheet, then finding/computing the cost, energy consumption and waste generation becomes almost a trivial step.

Running Example

In Table 5 we show how to calculate the consumption of resources by the activities.

Table 5 - Company XYZ Resource Consumption by Activities.

Resources	Resource Cost [$]	Resource Driver Name	Consumption Intensity	A11 Res. Driver Val.	Activity Cost [$]	A12 Res. Driver Val.	Activity Cost [$]	A13 Res. Driver Val.	Activity Cost [$]	A14 Res. Driver Val.	Activity Cost [$]	A15 Res. Driver Val.	Activity Cost [$]	A2 Res. Driver Val.	Activity Cost [$]
Fork Lift	10 000	Activities Involved			10 000										
Materials	300 000	Activities Involved			300 000										
Assembly	50 000	Number Of Units	0.77 [$/units]					40 000 [units]	30 769	25 000 [units]	19 231				
Labor	60 000	Labor Hours	5.66 [$/h]	900 [h]	5 094	500 [h]	2 830	3 500 [h]	19 811	3 000 [h]	16 981	900 [h]	5 094	1 800 [h]	10 189
Information Technology	10 000	IT Cost	1.00 [$/$]					4 000 [$]	4 000	3 500 [$]	3 500	2 500 [$]	2 500		
Final Inspection	25 000	Number Of Units	0.38 [$/units]					40 000 [units]	15 385	25 000 [units]	9 615				
Advertising	20 000	Activities Involved													20 000
Marketing	35 000	Activities Involved													35 000
Truck	20 000	Activities Involved											20 000		
Total Costs	**530 000**				**315 094**		**2 830**		**69 965**		**49 327**		**27 594**		**65 189**

From Table 5 we see that activity A11 annually consumes $315,094 worth of resources. Similarly, we can identify the cost for all the activities. This information is useful, e.g., when it comes to improving the process efficiency because we want to improve first and foremost the activities that are most costly. Calculations involving energy consumption and waste generation would work exactly the same way as for costs except the units and the resource definitions would be different.

2.8 Step 8 - Find/Compute the Cost, Energy Consumption and Waste Generation of the Objects

The cost, energy consumption and waste generation of a particular assessment object is found by multiplying the activity drivers by the consumption and generation intensities, followed by a summation of the resulting values for all activities that the particular assessment object consumes. For activities with multiple drivers, the driver's cost, energy consumption and waste generation must be determined before the drivers are summed up to yield the activity cost, energy consumption and waste generation.

Running Example

To actually calculate the unit costs of the two products, which was the goal of the model, we proceed as shown in Table 6. Recall that in Table 5 we show how to calculate the consumption of resources by the activities. In Table 6 we show how the products consume the activities. The products are the reason why the activities consume the resources. Note that in both tables the total cost adds up to $530,000. This is important to check, because it ensures correct cost assignment in the model

Table 6 - Company XYZ Activity Consumption by Products.

Activity	Activity Cost [$]	Activity Driver Name	Consumption Intensity		Hammer 40 000 Act. Driver Val.		Product Cost [$]	Scissors 25 000 Act. Driver Val.		Product Cost [$]
A11	315 094	BOM Cost	1.05	[$/$]	200 000	[$]	210 063	100 000	[$]	105 031
A12	2 830	Number Of Batches	7.55	[$/Batches]	200	[Batches]	1 509	175	[Batches]	1 321
A13	69 965	Product Involved					69 965			
A14	49 327	Product Involved								49 327
A15	27 594	Product Volume	1 103.77	[$/m3]	20	[m3]	22 075	5	[m3]	5 519
A2	65 189	Market Share	372 506.74	[$/%]	10.0 %		37 251	7.5 %		27 938
	530 000						**340 864**			**189 136**
Product Unit Cost [$/unit]							**8.52**			**7.57**

From Table 6 we see that the unit costs are $8.52 per unit and $7.57 per unit for the hammer and the scissors, respectively. Next to the assessment object names 'Hammer' and 'Scissors' there appears a number. These are the annual production numbers and we see that 40,000 hammers and 25,000 scissors were produced.

The energy consumption and waste generation calculations would look and work exactly the same as in Table 6. The units would of course be different, as well as the numerical values for the resources and intensities. We would also use 'BOM Energy Content' instead of 'BOM Cost' for activity A11 in the energy consumption calculations and similarly a 'BOM Waste Content' for the waste generation calculations. But the overall framework and the principles behind it would remain exactly the same for the energy consumption and waste generation calculations. For the Hammer we could, e.g., find that the unit energy consumption is 11.5 MJ/unit and the unit waste generation is 2.5 pWU/unit. In other words, the resource consumption attributable to the realization of 40,000 hammers annually would be quantified in three dimensions that would tell us the overall resource consumption costs, the energy consumed by converting raw materials into 40,000 useful hammers, and the waste generated in the processes.

2.9 Step 9 - Perform Sensitivity and Other Numerical Analyses

The power of having a computer model is that numerical analyses can be performed efficiently. More specifically, we always recommend performing sensitivity and trend analyses, and if uncertainty is modeled as per Step 6, the effect of these uncertainties on the resulting cost, energy and waste must be determined. For these kinds of analyses, we have found it useful to use the Monte Carlo simulation technique, which is a very simple, yet powerful numerical approximation method where we perform a controlled and virtual experiment within a model. Although numerous different simulation software programs exist, we have found the Crystal Ball® software to be very good for this purpose. In fact, all the case studies in this book have been implemented in MS Excel® spreadsheets and use the Crystal Ball® software to do the Monte Carlo simulations.

Crystal Ball®, which is an MS Visual Basic® application, adds into the MS Excel spreadsheet software (hence we talk about 'cells'). It allows the definition of 'assumption' and 'forecast' cells in a spreadsheet computer model. An assumption cell can be viewed as a source variable (any design, process, or driver value can be an assumption cell), whose variability (modeled as uncertainty distributions) inflict changes in the forecast cell(s). A forecast cell can therefore be viewed as a response variable (any cost, revenue and/or profitability in the model you want to investigate) whose response is measured statistically.

Consider the example in Figure 12 where product cost is modeled as a simple linear function of material and direct labor cost, i.e. 'Product Cost = Direct Labor + Material'. Based upon our *assumptions* with respect to material and direct labor, we want to *forecast* the product cost. In each assumption cell, an uncertainty distribution is defined as is appropriate for the particular value in that cell. In our example, the 'Direct Labor' assumption cell is distributed as a triangular distribution with a mean of $12 - and a lower and upper bound of $4 and $20 respectively. The 'Material' assumption cell is distributed elliptically with a mean of $10 - and a lower and upper bound of $5 and $15 respectively. The key issue now is how these two assumption cells affect the forecast cell. To find out, we run a Monte Carlo simulation.

We see the three first trials: $4+$6=$10, $12+$8=$20 and $20+$15=$35. The numbers in the assumption cells are picked randomly within the modeled distribution. That is, if we picked an infinite numbers from 'Direct Labor', and afterwards drew a histogram - the histogram would look *exactly* like the triangular distribution.

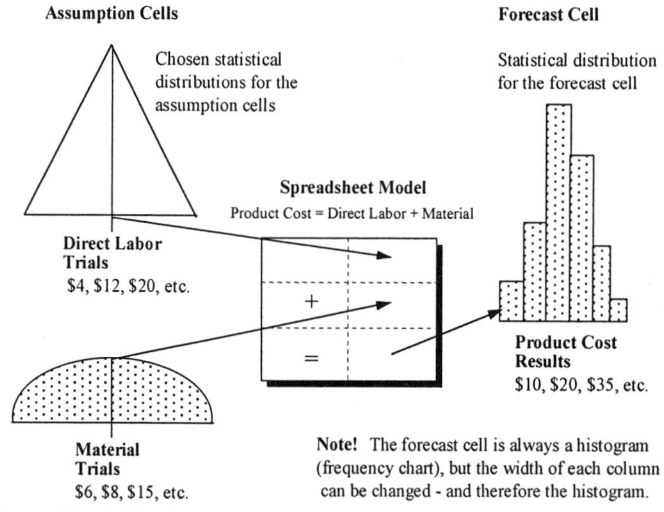

Figure 12 - A Simple Random Sampling Monte Carlo Simulation Example.

After all the trials have been completed, a histogram of the forecast cell is created - this is a graphical view of the forecast cell (response variable) and how *it is estimated numerically to vary - given the variability of the assumption cells*. Each estimate for the 'Product Cost' is stored in the computer and statistical analyses are performed on these stored values as if they were obtained by a real-life experiment. In other words, we are running an experiment in a virtual world that is defined in the model as assumption cells, forecast cells, and model relationships. In Figure 13, an example of such a histogram that is generated using the Crystal Ball® software is shown.

Due to the randomness, the numbers that have propagated through the model can be used in ordinary statistical analysis as if we were running a real experiment, e.g., to construct confidence intervals, perform T-tests, etc. The problem with measuring the response statistically is that the number of trials in the simulation affects the reliability of the model. So care must be exercised, especially in large models - say, models with more than 100 assumption cells

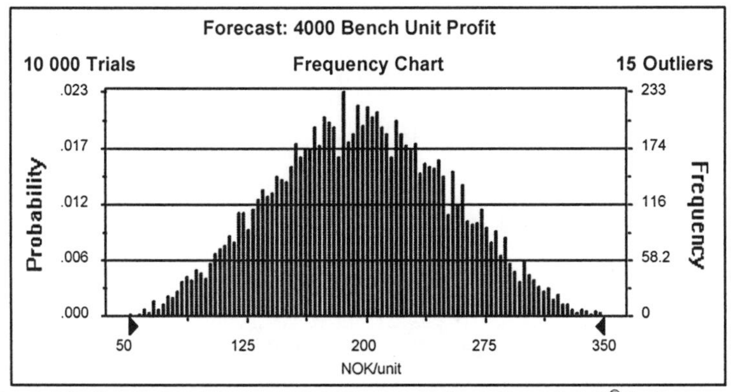

Figure 13 - Example of Forecast Cell Histogram from Crystal Ball® Software.

When running these virtual experiments, it is crucial to know *what* the purpose of the Monte Carlo simulations is. We use Monte Carlo simulations in two different ways:

1. In the case of uncertainty analyses, there exists real uncertainty in the assumption cells and the choices of uncertainty distributions must therefore reflect the actual uncertainty in the assumption cells. That is, if we have reason to believe that the uncertainty is distributed as a Weibull distribution, as in the cases of failure of components, the assumption cell should be modeled as such.
2. In the cases of tracing, sensitivity analyses, and information management one simply chooses a distribution that is bounded, symmetric and easy to model, such as the triangular uncertainty distribution because we are not interested in uncertainty *per se*, but more in its potential effects.

In Figure 14, we give an example of a sensitivity chart that is automatically generated from the Crystal Ball® software. Bars to the right mean that an increase in the associated factor will increase the 'Target Forecast'. In Figure 14, therefore, am increase in 'Price 4000 Bench' will cause an increase in '4000 Bench Unit Profit', whereas an increase in 'Sales rebate Wonderland' will cause a decrease in '4000 Bench Unit Profit'. In the following chapters, where the case studies are presented, many of these sensitivity charts will be shown and used to identify the critical success factors in Activity-Based Cost and Environmental Management. Figure 13 and Figure 14, in fact, are from our Westnofa case study.

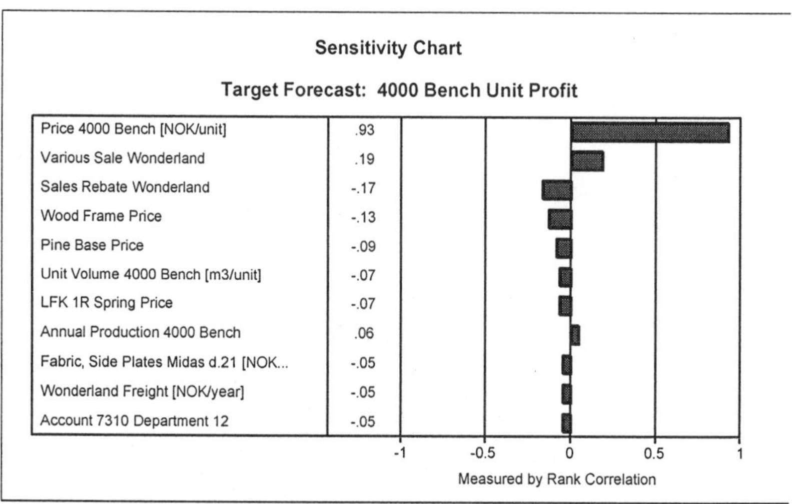

Figure 14 - Example of Sensitivity Chart from Crystal Ball® Software.

Running Example

How the simulation works in this example is not explained here. Basically, the software allows us to take any combination of source variables in the model and perform a Monte Carlo simulation with sensitivity analyses with respect to the forecast/response variables.

2.10 Step 10 - Interpret the Results and Iterate if necessary

Whether the model is satisfactory or not is a matter of how it responds to changes and the degree of meaningfulness of the results, and that it actually meets the goals specified in the beginning of the implementation. There are three issues to consider here:

1. Check if the model meets the goals specified in the beginning. This is obvious, but nevertheless extremely important.
2. Go trough the model and check if there are any *computational* errors. A simple check is always to use control sums on the object level, activity level, and resource level. If all control sums are equal, there are no computational errors.
3. Check for *logical* errors. These are much more difficult to spot, and here we utilize the sensitivity charts to see if there are any illogical effects that we notice. More importantly, however, is the way the implementation is

done. It is very important, when implementing activity-based systems in general, that the implementation procedure is systematic and that the people implementing the model *know* what they are doing. Experience is important.

If the model is found *unsatisfactory*, then (depending on what failed in the model) the appropriate steps must be iterated until model is satisfactory. If the model is satisfactory, then we should also check, based on the interpretation, whether the results are satisfactory or further iterations are needed from a product- and/or process design perspective.

Running Example

We assume that the model is satisfactory for company XYZ and in Figure 15, the layout of the complete (hypothetical) Activity-Based Cost, Energy, and Waste model for this running example is given.

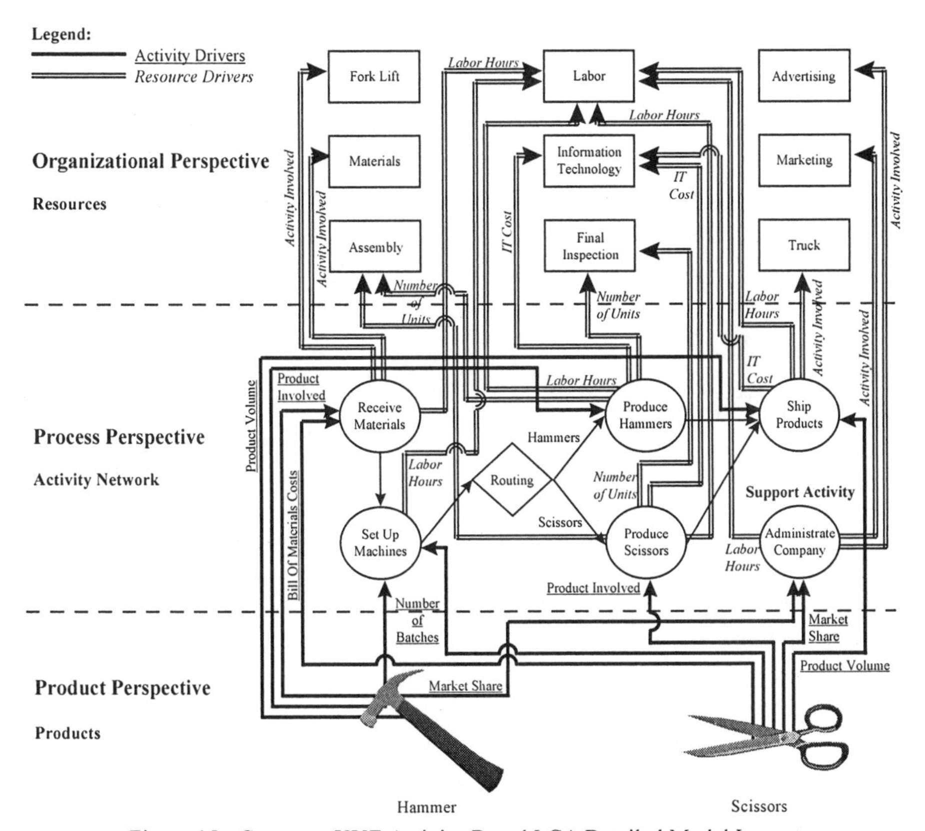

Figure 15 - Company XYZ Activity-Based LCA Detailed Model Layout.

The thick arrows show the direction of the consumption of the activities and the resources. We see, e.g., that Hammers consume the 'Set Up Machines' activity, among others, which is traced by the 'Number Of Batches' activity driver. The 'Set Up Machines' activity in turn consumes, among others, the 'Labor' resource, which is traced by the 'Labor Hours' resource driver. We note that when the two products consume a certain activity, the same activity driver always measures the consumption, although the magnitude of the driver is different. This is only a general rule as long as there is only one driver per activity and/or resource like here. If there were multiple drivers of the 'Set Up Machine' activity, then two products *may* consume in different ways the two drivers. This is also the case when several activities consume a resource, as can be seen from the calculations in Table 5 and Table 6.

3. ISSUES RELATED TO RESOURCE DEFINITION AND QUANTIFICATION

To be successful do not be rigid and immobile in your thinking but always keep abreast of the time and change with it.

Fu Hsi
I Ching (The Book of Changes)

The definition and quantification of resources is a crucial step in the Activity-Based Cost, Energy, and Waste modeling effort. In this section we provide some additional information on issues related to resource definition and quantification.

3.1 The Role of the General Ledger

The secret of success is to know something nobody else knows.

Aristotle Onassis

The general ledger can be very helpful in identifying the resources that are consumed in a company. Ideally, the general ledger in a company should be object-oriented and not function oriented as today. For example, today, one account may be 'depreciation'. Depreciation should, however, have been related to the various processes (e.g. 'Assemble engines'). This is, however, not possible due to the Generally Accepted Accounting Principles (GAAP).

Another preferred solution is an object-oriented general ledger where depreciation is accounted for by the various resources to be depreciated, e.g. a machine or computer. Hence, when implementing Activity-Based Cost and Environmental models, the general ledger should be reorganized to fit an object-oriented resource hierarchy. This will minimize distortion and the process/activity assessments become more correct. Furthermore, it has the advantage that we do not have to use trace keys to link the various accounts in the general ledger to the physical resources. The usage of such trace keys is usually crude because the general ledger is too aggregated and the implementers lack the necessary understanding of both the local accounting practices and the physical resources and processes. The result is significant and unnecessary distortion, which in turn impairs the overall reliability of the implementation.

Nonetheless, the approach chosen by several commercial ABC software vendors *is* to take general ledger directly as is, see e.g. CMS-PC 4.0 for Windows by ICMS, Inc. (Albright, Sharpe et al. 1998) and ACTIVA by PriceWaterhouseCoopers (Borden 1994). This is clearly not an ideal approach, but it has the advantages of being fast, cost effective and easy to implement as it requires almost no process understanding by the implementers. In our opinion, that, however, negates a fundamental principle of ABM - process thinking.

3.2 Renewable Energy also has Embodied Energy and Waste

When I see an adult on a bicycle, I have hope for the human race.

H.G. Wells

One issue we want to point is the inherent confusion about the cleanliness of renewable versus nonrenewable energy because it relates to a resource's energy and waste content. Often, nonrenewable energy is equaled to being 'dirty' and renewable is called 'clean'. However, these judgements may be seriously flawed.

First of all, what is renewable and what is nonrenewable is a matter of perspective. Typically, renewable energy is energy arising from phenomena in Nature with short cyclic time up to a year or so, while nonrenewable energy has long, e.g. geological, cyclic time. So clearly, there is a time difference between the two categories of energy.

The second issue, concerning what is "clean"' and what is "dirty", is also a matter of perspective. No known form of energy is 100% clean because there are <u>always</u> resources used in/needed for converting the inherent energy into exergy (the amount of energy available to do work). The degree of dirtiness/cleanliness is simply a matter of the amount of waste generated in the process of converting that energy into exergy. In energy analyses, overhead energy consumption and waste generation throughout the entire value chain are often ignored, leading to erring conclusions that solar panels are (per definition) environmentally friendly because there are no releases associated with converting solar energy to exergy. However, this totally ignores the fact that the production of solar panels uses energy as well, and even worse, acids and other nasty chemicals could have been used in the production of the electronic circuits. As long as we do not know the amount of energy spent earlier on in the value chain to manufacture the solar panels, extract the raw materials, etc., we cannot really tell whether solar panels or say natural gas power is better.

Consequently, to determine what is the least dirty energy we need to investigate the entire value chain. It is striking that this discussion is missing even in highly acclaimed publications such as (Brown, Flavin et al. 1999), where modern technology is praised in the context of making energy management more efficient and more sustainable. But what about all the energy expenditures needed to make the technology in other links of the value chain? That is not mentioned. We are not saying that technology is necessarily bad, but because the entire value chain has not been accounted for, we cannot really know.

3.3 Indirect Energy Consumption, Waste Generation and Environmental Return on Investments

The only green car
Is skeletally rusted and overgrown

Heathcote Williams
In "Autogeddon"

In accounting, it would be a major error to exclude overhead or indirect costs. For example, the heating of a production facility is indirect cost. Yet, in environmental assessments, indirect environmental impacts are often completely ignored. Clearly, the heating of a production facility consumes energy and generates emissions. However, if one reads the ISO 14040 Life-Cycle Assessment standards, when quickly realizes that ISO 14040 primarily focuses on direct environmental impact for a specific product. Needless to say, everybody knows from cost accounting that excluding indirect cost can cause severe distortions. Similarly, in environmental 'accounting', the exclusion of indirect environmental impact may result in strategically bad decision-making.

As an example, one of the authors was told by a major electronics manufacturer how they actually designed and developed a 'green' television that consumed 5-10% less energy than a conventional unit, plus was better recyclable, and used less materials in the first place. However, the company decided not to produce it. Why not, you may ask, clearly the TV looks to be much more environmentally benign product from a life-cycle perspective. Well, the company realized that in order to produce the TV, they would have to build a new production line. The environmental cost of that production line would not have been regained by the environmental savings of the TVs produced. Hence, there was no 'environmental return on investment'.

Return-on-Investment (ROI) is another term well known in cost accounting. In the environmental assessment domain, however, it is astonishingly infrequently used. Yet, it makes perfect sense to think we may create a large environmental impact that will result in a lower environmental impact in the long run. Creating public transportation networks are examples of major developments that need to pay themselves back through reduced car use and corresponding less fuel related emissions.

There are more terms that are widely used in cost accounting, but still have not been 'discovered' by the environmental assessment specialist. Depreciation is another such term.

3.4 Depreciating Energy Content, Waste Content, and other Environmental Impacts

That which costs little is less valued.

Miguel De Cervantes

You may have noticed that we did something in our running example that seemed quite normal from a monetary perspective, but yet immensely strange from an environmental perspective: we depreciated the energy content and waste content of a forklift (see Section 2.3). Depreciation is a very well known concept in accounting and many, many depreciation methods exist. But basically, depreciation ensures that the 'fixed' costs of an asset are spread out over its life. Without depreciation, if a company would buy, say, a truck for $50,000, the company would suddenly see a $50,000 cost and profitability would go down. Assuming the car lasts 10 years and depreciate linearly, the company would see only an annual cost of $5,000 for 10 years.

Clearly, depreciation works in cost accounting. Why would it be needed in environmental accounting?

Well, consider you are a manufacturer happily producing products. You even became ISO 14000 certified and did your ISO 14031 environmental performance evaluation. Now suppose you want to expand and started building a new factory. Clearly, you are causing a major environmental impact by building this new facility. Think of all the concrete and steel you need, with its energy- and waste content. If you would do an annual ISO 14031 environmental performance evaluation, and if you would include the impact caused by the new facility, then you could suddenly look like the biggest and worst polluter on the block. But if the facility is supposed to be there for 30 years, then the energy- and waste content can be spread out over 30 years. Basically, you are depreciating the environmental impact over time. It works in accounting, it can easily also work in environmental accounting, as long as we all agree on the rules, i.e. similar to Generally Accepted Accounting Principles (GAAP).

3.5 Waste of Capacity

Muda. It's the one word of Japanese you really must know. It sounds awful as it rolls of your tongue and it should, because *muda* means 'waste', specifically any human activity which absorbs resources but creates no value.

James P. Womack and Daniel T. Jones
In "Lean Thinking"

There is another, more subtle, aspect to waste that most LCA methods totally ignore or cannot capture due to their lack of connection to the economic system, namely 'waste of capacity'. Some types of capacity waste are, as discussed by (McNair 1997) and others:

1. *Definitional waste*, which is the abandonment of theoretical capacity in favor of a more 'practical' capacity. This type of waste may account for up to 20 - 30 percent of total theoretical capacity. For example, plants that do not operate on weekends automatically have a definitional waste of 2/7 (i.e., 28+%).
2. *Structural waste*. The structure of an organization is made up of many different resources whose capacity varies. To generate the overall required output a set of resources are needed for which the capacity often will not match the actual capacity required to generate the output. The difference represents structural waste. There are also other sources of structural waste such as overhead.
3. *Technical waste*. As technology replaces labor, the causal relationships between input and output are increasingly difficult to trace, this leads to excessive imbalance in the system and waste builds up.
4. *Management-based waste*. The assumptions management makes in defining capacity can actually build waste into a company's cost structure.
5. *Accounting-based waste*. Assessing the impact of (new) practices on profitability is difficult and we have accounting systems to aid us. Some accounting systems rely on 'standards' and here we have two inherent sources of waste;
   - The waste factored into the standards. This arises easily whenever the standards are set to take what is *perceived* as inevitable into account.
   - The waste hidden in inventory evaluation. Full cost absorption and standard costing have the unfortunate effect that it can easily promote production of work-in-progress inventory to balance the books and adjust for the discrepancy of budgeted production volume and actual production volume.

How do we handle capacity waste? Again, we tie into modern management practices such as Total Quality Management (TQM) but above all ABC. In TQM and ABC the focus is on creating value for the customers and continuously improving this value. Everything that does not contribute towards this goal is sought eliminated and/or reduced. This is by itself a strong move towards reducing the various types of wastes. On top of that, ABC also focuses on the idle capacity of activities (Johnson and Kaplan 1987) and then seeks to eliminate this source of waste as well. In total, the TQM-ABC continuum is a very powerful combination that attacks all sources of capacity waste. Unfortunately, ABC is often combined with standard costing principles to give benchmarks for cost drivers, i.e. one compares for example last year's activity level (as measured by the cost drivers) with this year's activity level, and thereby violates a TQM cornerstone - the search for perfection. This means that one runs into the so-called accounting-based waste. The way we implement ABC and Activity-Based Cost and Environmental Management, in contrast, effectively eliminates the need for any standards because the use of Monte Carlo simulations and statistical sensitivity analyses allows us to constantly work on the overall most critical success factors to constantly lower resource consumption. Over time, this process will approach perfection just as a TQM process constantly pushes towards perfection.

4. MORE ON UNCERTAINTY MODELING AND SIMULATION

Do not put your faith in what statistics say until you have carefully considered what they do not say.

William W. Watt

In the next sections more background material on our use of uncertainty distributions and Monte Carlo simulations is provided.

4.1 Uncertainty Distributions

Uncertainty can be modeled in a variety of ways depending on what kind of uncertainty is to be modeled. Typically, there are three main approaches for handling uncertainty:

1. Simplify the problem from handling uncertainty to handling *risk*. The difference is that in a situation with risk we have additional information and can for example do a risk tree analysis, see e.g. (Park and Sharp-Bette 1990). Hence, we can use probability estimates to simplify the problem. However, in the complex models we are developing this is rarely feasible

because we do not have the required additional information. Furthermore, because of the computational strengths of modern days computers it does not make much sense to simplify a problem that is already fairly easy to handle, by making several additional assumptions that only will reduce the usefulness of the output from the models.

2. Bayesian statistics or what most people simply call statistics. The problem with Bayesian statistics is that we need historical data, which is impossible to find when performing original design and hard to find in many other situations. Consequently, we often cannot use Bayesian statistics.
3. Fuzzy numbers and fuzzy intervals. This is a very flexible way of handling uncertainty, and it has two great advantages:
   a) Fuzzy numbers and fuzzy intervals can be used with or without hard data. Bayesian statistics, on the other hand, need quite a large sample of hard data. However, it should be noted that it is always preferred to have as much relevant hard data as possible.
   b) The use of fuzzy numbers and fuzzy intervals has very few restrictions. A fuzzy number has to be convex and bounded, while a fuzzy interval only has to be bounded, see (Dubois and Prade 1978; Bandemer and Näther 1992). According to (Kaufmann 1983), this is the perfect tool of numerical theories and applications.

As mentioned, we typically use fuzzy intervals in our models, but the most flexible way to handle uncertainty is simply called 'uncertainty distributions'. Uncertainty distributions are a combination of risk, Bayesian statistics and fuzzy numbers and intervals. Uncertainty distributions have therefore no restrictions. You can, for example, cut off a normal distribution anywhere you like as in Figure 16. This is an approach provided by (Decisioneering 1996), although the theoretical foundation is not articulated, probably due to its all-encompassing simplicity.

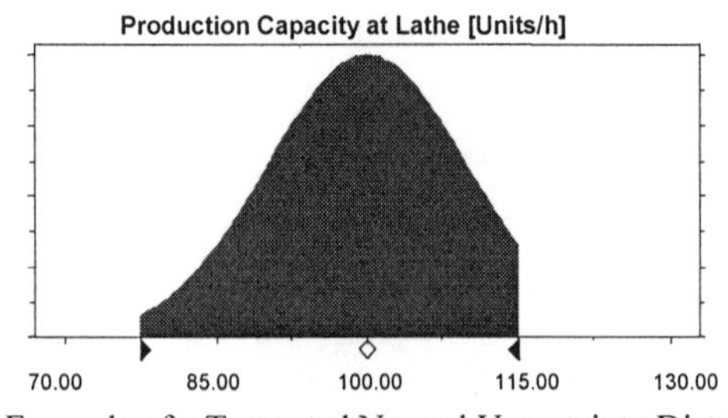

Figure 16 - Example of a Truncated Normal Uncertainty Distribution.

The choice of how to handle uncertainty has significant bearings on what numerical methods are chosen. It turns out that Monte Carlo methods allow the most versatile handling of uncertainty.

4.2 Why Monte Carlo Methods?

Solving the models we create exactly, as opposed to numerically, is not recommended because of the large number of variables and the need to handle uncertainty in a flexible fashion. We have therefore concentrated on numerical methods, more specifically on numerical and statistical methods that rely on *n*-point evaluations in *m*-dimensional space to produce an approximate solution. In plain English, *m* is the number of variables while *n* is the number of points used to assess. According to (Fishman 1996):

> "Among all numerical methods that rely on *n*-point evaluations in *m*-dimensional space to produce an approximate solution, the so called Monte Carlo methods have an absolute error of estimate that decreases as $n^{-1/2}$ whereas, in the absence of exploitable special structure, all others have errors that decrease as $n^{-1/m}$ at best".

Furthermore, interestingly enough,

> "Monte Carlo methods were considered mathematical blasphemy until as recently as 1970, today however, Monte Carlo methods are the *only* approaches capable of providing useful insights in several problems in physics".

Rubinstein (1981) asserts that the Monte Carlo method is now *the most* powerful and commonly used technique for analyzing complex problems. Others are less enthusiastic about Monte Carlo methods. In (Liebtrau and Scott 1991), for example, they claim that "the computational burden precludes the use of standard, i.e., Simple Random Sampling (SRS) Monte Carlo analysis". This statement must be interpreted in relation of climate models that comprise of hierarchical computer codes, because then the Monte Carlo analyses must be performed at several levels and provide input to each other, which by the way is not the case in our work (at least so far). A SRS Monte Carlo method will be time consuming in such a situation as argued in (Liebtrau and Scott 1991).

However, by using Latin Hypercube Sampling (LHS), the number of trials can be reduced drastically. The only problem with LHS is the difficulty in computing the mean output response. It is therefore suggested to break up the models into a hierarchy of models and then run simulation first to identify the most important variables and then run Monte Carlo simulations including only the most important variables. However, this can be a dangerous approach because it neglects the 'insignificant' variables. According to (Allvine 1996), the reason behind Delta Airlines success over twenty years was that they did all the "little things", i.e. insignificant variables, right. It should be mentioned that there are many different sampling techniques that can be used with Monte Carlo methods, each having its own certain advantages in certain situations.

In any case, when we take into account that we have problems with many variables (large *m*) that are *not* nested, we have not been able to find any numerical method that can even remotely compete with Monte Carlo methods for our purpose. Another issue is that Monte Carlo methods never go wrong, see (Liebtrau and Scott 1991). The only issue is what is faster:

1. To run a Monte Carlo simulation once and for all, or
2. To do some manipulations to reduce the size of the problem and then solve the problem wondering did you miss something?

We believe that when the entire time is included and the fact that modern software allows distributed computing, see (Joshi, Lemay et al. 1996), over a web of computers, Monte Carlo methods will take more and more over in uncertainty analysis, sensitivity analysis and optimization.

5. SOME MORE ON MODEL QUALITY

What good is speed if the brain has oozed out on the way?

Karl Kraus

A good model should be capable of handling all the different cost driving complexities of a company. But a costing model will never be 100% correct (Cooper 1990b).

It is important to choose the cost drivers so that the cost drivers in as high degree as possible reflect the way activities actually are consumed. When this has been achieved to a satisfactory degree, we can use the model to estimate costs, trace the costs and thereby identify cost reduction opportunities. However, the cost of maintaining and designing must always be taken into consideration, because it is no point spending thousands of dollars to keep track of a minor cost contributor. Simplicity versus accuracy is therefore a very important issue when designing an ABC model.

From a conceptual point of view the problem of distortion can be illustrated as in Figure 17 where it is one out three major sources of error in performance measurement. The two others are reliability problems concerning implementation and deficiencies concerning the method itself and potential lack of data. The goal is to make the three circles overlap as well as possible since the 'Best Assessment' is the intersection of the three circles.

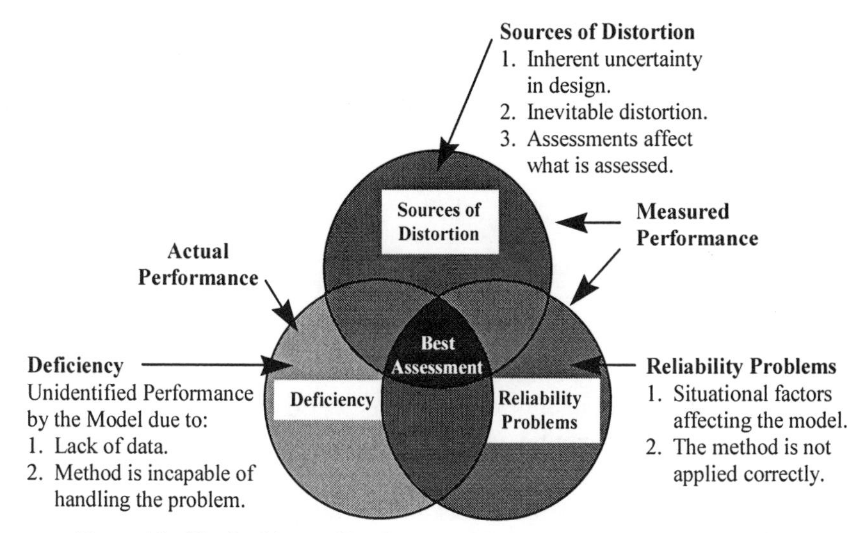

Figure 17 - The Problems of Performance Measurement (Emblemsvåg 1999).

From our own experience, we emphasize that it is extremely important that the drivers, activities and resources are defined so that the distortion in the model is as low as economically (due to budget constraints) feasible. Hence, it is vital to have the "best and brightest", see (Miller 1990), in a team implementing activity-based systems. From this we readily understand that as ABC systems, as well as Activity-Based Cost and Environmental Management systems, perform fairly differently depending on the level of skills of the implementers. The results will always be comparable, but the quality can vary.

To ensure high quality of the implementation of activity-based systems some rules of thumb can be useful, given a certain implementation budget:

1. Ensure that there is in the implementation team at least one person who have a holistic picture of the assessment objects such as a plant manager or similar. Also, persons with thorough knowledge of the various functions in the company and the processes must participate. Specifically, there should always be people from management, marketing, production and design.
2. Define activities, resources and drivers so that as many one to one matches are achieved. Then, concentrate on finding good drivers for all significant resource elements. This is crucial in order to minimize distortion of the system. For resources of little importance, say less than 0.1%, allocation is always acceptable if it is difficult to identify drivers.
3. Address the issues of maintaining the model early. That is, never implement a model that is impossible to maintain within an acceptable budget.

And always keep in mind that the premise for designing a new ABC system should be (Cooper 1990d):

"It is better to be approximately right than exactly wrong".

6. WHAT IS NEXT?

Be both a speaker of words and a doer of deeds.

Homer

We have given you quite an extensive and lengthy discussion on how to extend ABC into the environmental domain, and how Activity-Based Cost and Environmental Assessments and Management can be achieved. Although a wide variety of environmental assessment dimensions can be integrated, we recommend starting simple and starting with developing integrated Activity-Based Cost, Energy, and Waste models. We showed you what we use as basic steps in developing such models, but nothing can illustrate it better as some actual implementations. In the next four chapters, therefore, we present case studies that we hope will illustrate the development, implementation, and actual use of Activity-Based Cost, Energy, and Waste models for Activity-Based Cost and Environmental Management.

Chapter 5

THE WAGONHO! CASE STUDY

There are those who act without knowing, I will have none of this. To learn a lot, choose the good, and follow it, to see a lot and learn to recognize it; this is next to knowledge.

Confucius
Analects 7:27

WagonHo!, Inc. is a toy manufacturer. It operates in a somewhat price sensitive niche market with fairly good demand. Management faces three problems:

1. They have a $1.3 million loss for the year.
2. They expect higher energy costs due to energy shortage.
3. They face possible tougher pollution legislation.

In this chapter, we will show how to make an Activity-Based Cost, Energy and Waste model, and how to use it to help management overcome these three problems. In (Emblemsvåg 1999), this case study was also used for contrasting conventional approaches (Contribution Margin Costing and ISO 14000 LCA) to our approach. In this chapter, however, the primary purpose of this case study is to serve as a learning exercise for how to develop an Activity-Based Cost, Energy and Waste model, and how to use it for engineering and managerial decision-support.

1. DESCRIPTION OF WAGONHO!

Nature uses as little as possible of anything.

Johannes Kepler

We have to start by saying that the toy manufacturer WagonHo!, Inc. is not a real company; it is a small company that exists in the computers at Center for Manufacturing Information Technology (CMIT) in Atlanta, Georgia. CMIT uses it as a simulation company where companies in Georgia can come and test out the newest information technologies in manufacturing. However, CMIT *does* have a model factory built in the laboratory where they actually produce products.

The 'company' experienced a $1.3 million loss. This is a highly unsatisfactory result, and the management is of course in dire need of decision-support to turn this dreadful result around.

1.1 The Organization

WagonHo! has 56 employees, and the organization is simple. The Chief Executive Officer (CEO) is Samuel P. Stone and the Plant Manager is Mary Ann Chesnutt. There are 6 production teams where each of the supervisors has six employees under them. Except from these 42 employees that are directly involved with production, the remaining 14 are overhead resources.

1.2 The Products

The strategy of WagonHo! is to target the high price/quality market for upper-class children. The products are mainly made of plastics, steel screws and some wood. The specific products are as follows, see also Figure 1:

- CW1000 Wagon, referred to as CW1000. This is a wagon with four wheels and front steering. The sales price of this product was $120, corrected for 12% sales rebates and 2% provisions we get $103.20. Current production is 5,000 units/year.
- CW4000 Wheel Barrow, which is denoted CW4000. This is a single wheeled barrow, without steering. It sold for $100, but the net sales price was $86.00. Current production is 3,000 units/year.
- CW7000 Garden Cart, which we simply call CW7000. This is a two-wheeled cart also without steering that was sold for $105 giving a net sales price of $90.30. Current production is 2,000 units/year.

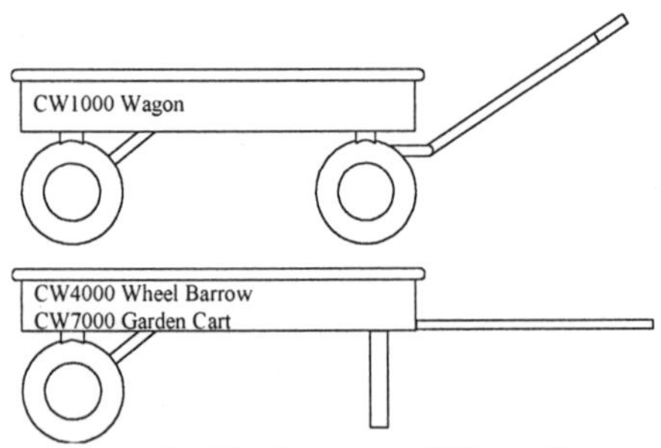

Figure 1 - The Products of WagonHo!.

1.3 The Production Facility

The shop floor is configured as in Figure 2 with; a) six lathes (L), b) six milling machines (M), c) six sub-assembly (SA) station, d) a kitting area, e) six final assembly and inspection (FA) stations and f) one central conveyer.

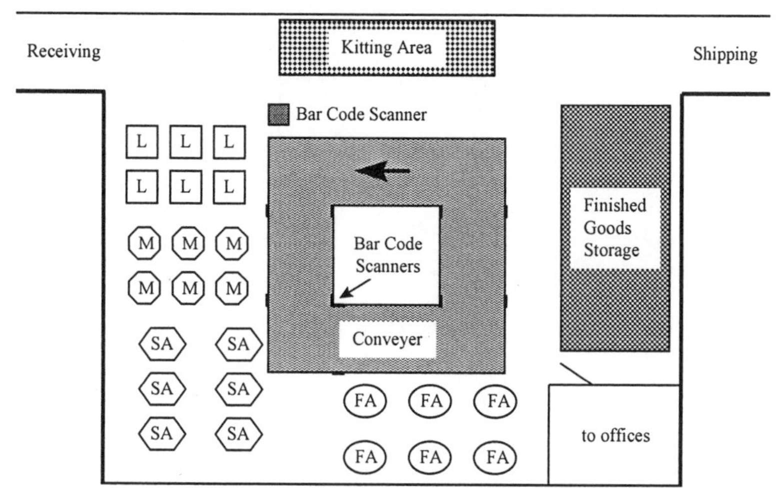

Figure 2 - The Initial Shop Floor Layout at WagonHo!.

The estimated aggregated production cycle times are 462, 247 and 259 minutes for the CW1000, CW4000 and CW7000, respectively. The cycle times are long and clearly need improvement.

WagonHo! is currently using Contribution Margin Costing, a volume-based costing system. To aid their cost management, WagonHo! has traditional standard times [h] for production in the costing system. In Table 1 the hourly labor costs of the workers are given, and by multiplying the hourly labor cost with the standard times we get the Bill of Materials (BOM).

As an example, the BOM for the CW4000 is shown in Table 2. The BOM should be read from bottom to top. Thus, for the CW4000, see Table 2, the total direct cost is $46.86 per unit while the unit production time is 4.07 hours per unit. We can also see the various product/subassembly numbers. For example, the product number for the 'bed' is 4100 and its part number is CW1373. Once the Swivel Plate is made it becomes a part for the CW4000.

In Table 3 a summary of unit BOM for the products is presented along with the aggregated production times, which serve as fixed costs allocation base. The aggregated production times is found by multiplying the unit time for each product by the annual production of each product.

Table 1 - Hourly Labor Costs at WagonHo!.

Labor Classifications	Hourly Labor Cost [$/h]
Mill Operator	9.00
Lathe Operator	8.50
Assembler	7.80
Kit Maker	7.30
Inspector	11.30

Table 2 - CW4000 Bill of Material (BOM).

Product Num.	Description	Parent Num.	Part Num.	Number Reqd	Standard Time [h]	Unit Time [h]	Unit Cost	Ext Cost	Comments
4000	**Wheel Barrow**		**CW4000**	**1**		**4.07**	**0.00**	**46.86**	**Complete**
	Labor - Final Inspection	4000		1	0.30	0.30	3.39	3.39	Labor
4100	Bed	4000	CW1373	1			6.54	6.54	Purchase Assembled
4200	Screws	4100	SP4881	10			0.02	0.20	Purchase
4210	Labor - Final Assembly	4000		1	0.45	0.45	3.51	3.51	Assembly Labor
4220	Labor - Kit purch. parts	4000		1	0.30	0.30	2.19	2.19	Kitting Labor
4300	Wheel Assembly	4100		1			0.00	0.00	Subassembly
4310	Labor - Assembly	4300		1	0.30	0.30	2.34	2.34	Assembly Labor
4320	Axle Bracket	4300	CW2019	2			0.87	1.74	Make
4330	Short Axle	4300	CW3626	1			0.12	0.12	Purchase
4340	Wheel	4300	CW2314	1			0.73	0.73	Make
4350	Cotter Pins	4300	SP6122	2			0.01	0.02	Purchase
4360	Labor - Kit make part	4300		2	0.20	0.40	1.46	2.92	Kitting Labor
4370	Labor - Make part	4300		1	0.84	0.84	7.56	7.56	Machine part
4380	Raw Material - Wheel	4300	RM5784	0.5			0.78	0.39	Purchase Raw
4390	Raw Material - Brackets	4300	RM5784	1			0.78	0.78	Purchase Raw
4400	Leg/Handle Assembly	4100		1			0.00	0.00	Subassembly
4410	Labor - Assembly	4400		1	0.15	0.15	1.17	1.17	Assembly Labor
4420	Handle	4400	CW3908	1			0.47	0.47	Purchase
4430	Leg Stand	4400	CW4240	1			0.95	0.95	Make
4440	Screws	4400	SP4881	6			0.02	0.12	Purchase
4450	Labor - Kit make part	4400		2	0.20	0.40	1.46	2.92	Kitting Labor
4460	Labor - Make part	4400		1	0.93	0.93	8.02	8.02	Machine part
4470	Raw Mat. - Leg Stand	4400	RM5784	1			0.78	0.78	Purchase Raw

Table 3 - Summary of BOM and Aggregated Production Times for the Products.

Product	Unit BOM [$/unit]	Aggregated Production Times [h/year]
CW1000 Wagon	90.04	38,500
CW4000 Wheel Barrow	46.86	12,210
CW7000 Garden Cart	57.40	8,640

2. DEVELOPING AN ACTIVITY-BASED COST, ENERGY AND WASTE MODEL FOR WAGONHO!

Without education, you're not going anywhere in this world.

Malcolm X

The steps for implementing Activity-Based Cost, Energy and Waste models are given in Figure 3, see also Chapter 4, Section 2. The Activity-Based Cost, Energy and Waste model we implement at WagonHo! is a *full absorption* model. That is, we trace all the cost to the products, except costs like gifts, newspapers, etc. Full absorption assessment is being recognized as very important for environmental efforts in general, because it affects pricing, distribution, sales promotion, cost management and ultimately everything in a company, see e.g. (Miles and Russel 1997). It is also referred to as total cost assessment (GEMI 1994a; GEMI 1994b), total cost accounting, or full cost accounting as (EPA 1992) defines it.

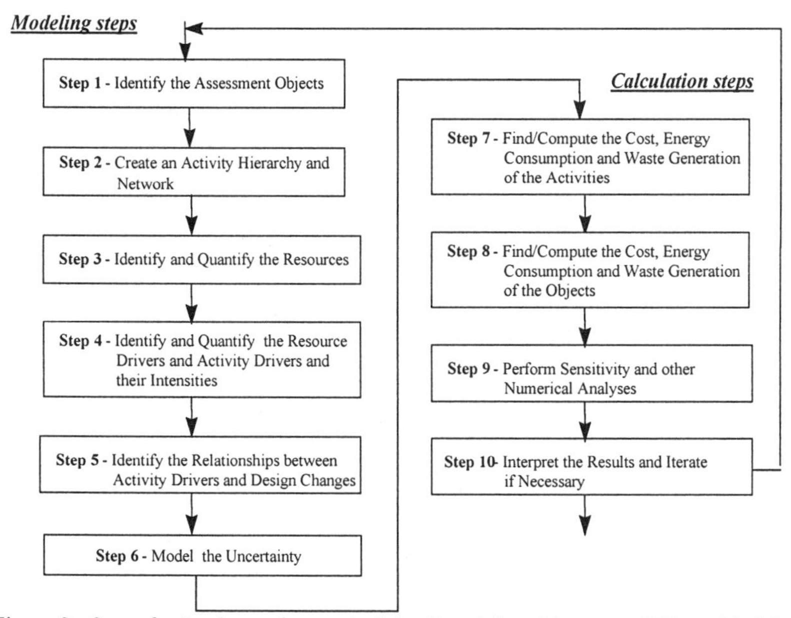

Figure 3 - Steps for Implementing an Activity-Based Cost, Energy and Waste Model.

2.1 Step 1 - Identify the Assessment Objects

In this case, there are four assessment objects: three products and the company itself.

2.2 Step 2 - Create an Activity Hierarchy and Network

To create the activity hierarchy, we simply break down all the operations performed at WagonHo! and group them into activities. How we gather them depends on what we think is most useful in terms of data gathering, accuracy, process understanding, etc. Consequently, there are different ways to create an activity hierarchy. The activity hierarchy we developed for WagonHo! is shown in Table 4.

The activities in the gray shaded cells are then linked together in an activity network, see Figure 4, but not necessarily in order of the process. The main purpose of the activity network is to give a visual, but simplified, process view. It is very important to relate the right products to the right activities in order to avoid large errors. We see for example that CW1000 is the only product that uses the lathe. If possible, however, the actual process flow should be mapped in the activity hierarchy and network.

Table 4 - WagonHo! Activity Hierarchy.

Level 1		Level 2		Level 3	
Production	A1	Logistics	A11	Receive Parts	A111
				Run Inventory	A112
				Ship Products	A113
		Produce Products	A12	Kit Parts	A121
				Run Mill	A122
				Run Lathe	A123
				Assemble Products	A124
				Inspect Products	A125
Product Support Activities	A2	Design Products	A21		
		Sell Products	A22	Sell Products	A221
				Service Customers	A222
Facility Support Activities	A3	Maintenance	A31		
Administration	A4	Lead Company	A41		
		Run Production	A42		
		Process Orders	A43		
		Accounting	A44		

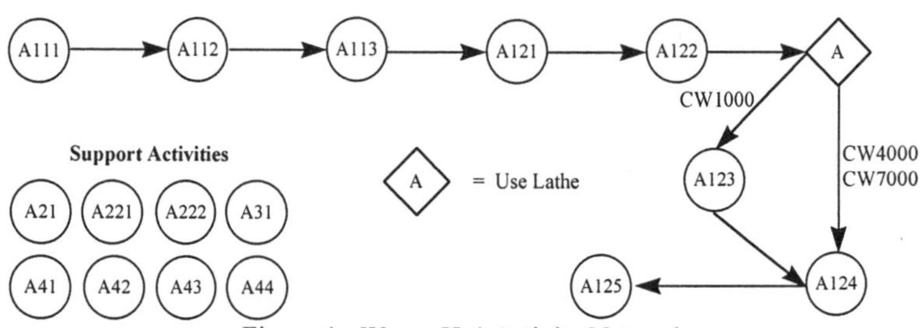

Figure 4 - WagonHo! Activity Network.

2.3 Step 3 - Identify and Quantify the Resources

Resources are defined as 'economic, energy or waste related elements that are consumed by the performance of activities'. In Table 5 a summary of the resources is presented. When it comes to waste generation, we express it by using the Waste Index (WI), see Chapter 2. We believe that both the energy and waste estimates in Table 5 are probably too low, given the lack of true environmental accounting in the value chain. Nevertheless, Table 5 shows the principle that analogous to monetary costs, we can also define an energy and waste 'cost' for the resources.

Table 5 - Resources in WagonHo!.

Resource Elements	Costs [$/year]	Energy Consumption [MJ/year]	Waste Generation [pWU/year]
Building:			
Atlanta Technology Center	261,000	9,683,144	25,183
Logistics Equipment:			
Fork Lift	2,900		
Company Cars:			
Ford Taurus	8,500	61,540	99.94
Ford F150	8,300	51,280	83.03
Production Equipment:			
Conveyer System	5,100	15,810	
Kitting Equipment	270	410	
Lathe Machines	22,500	8,990	
Milling Machines	17,000	9,500	
Assembly Equipment	1,050	1,000	
Inspection Equipment	650	500	
Design Equipment	3,500		
Material Design	2,500		
Material Maintenance	3,000		
Material	215,350	320,590	
Activity Labor Costs	1,560,800		
Office Equipment:			
Computer Systems	39,500		
Furniture etc.	4,000		
Reception	2,150		
Total:	**2,158,070**	**10,152,764**	**25,366**

2.3.1 Quantifying Resource Cost

The costs were quantified using information from the Bill of Material and the existing cost accounting system. All equipment costs are depreciated costs to which the annual maintenance costs have been added. The building is an aggregated resource that includes the annual gas and electricity costs.

2.3.3 Quantifying Resource Energy Consumption

As mentioned, the 'building' is an aggregated resource that includes annual gas and electricity. Its annual energy consumption is calculated as shown in Table 6. As shown Table 6, the building's energy content (which was estimated using the amount of concrete embodied in the building) is being depreciated resulting in an energy content depreciation post. The energy content of both gas and electricity are also taken into account, as is the actual energy from the electricity. The actual energy from the gas is not taken into account, which means that Table 6 is a low estimate of the buildings real energy consumption. However, note that the electricity energy content and amount dwarf the gas energy content. Therefore, we already know that the data for gas will not be very significant and no effort is made at this stage to find the data. We would rather recommend model developers to investigate the electricity energy content in more detail given that its amount is significant.

Similarly, the company cars' energy consumption is calculated from their fuel consumption and its associated energy content. In this case, we left out the depreciation of the cars' energy content.

The production equipment's energy consumption is only based on the equipment's energy content, which was based on the equipment's type and amount of main material, because the electricity that the equipment consumes has already been accounted for in the building energy consumption. Hence, only the annual depreciation of the production equipment's energy content is listed in Table 5.

The energy content for the materials was determined using each product's Bill of Materials combined with energy content data found in IDEMAT for the materials. Analogous to a 'regular' Bill of Materials, we can develop an 'Energy Bill of Materials' that contains energy data instead of cost for product components. In Table 7, an example of such an Energy BOM is given for the CW4000 product. In Table 8, the total direct energy consumption (EC) by product is given, based solely on direct material use. As we will see, the actual product energy consumption is higher because of the indirect energy consumption (e.g., building related) that will also be traced to the products.

Table 6 - Building Annual Energy Consumption Calculation.

Building	16,550 sq. ft.	
Energy content depreciation		82,000 [MJ/year]
Gas (amount used)	1000 kg/year	
Gas Energy Content	5.7904 MJ/kg	57,904 [MJ/year]
Electricity (amount used)	2,520,000 [MJ/year]	2,520,000 [MJ/year]
Electricity Energy Content	2.787 MJ/MJ	7,023,240 [MJ/year]
Total Energy Consumption		9,683,144 [MJ/year]

Table 7 - Energy Bill of Materials for CW4000 Product.

Product Num.		Description	Parent Num.	Part Number	Num. Required	Unit EC	Ext EC
4000		**Wheel Barrow**		**CW4000**			**27.68**
4100		Bed	4000	CW1373	1	20.60	20.60
4200		Screws	4100	SP4881	10	0.09	0.90
4300		Wheel Assembly	4100				3.17
	4320	Axle Bracket	4300	CW2019	2	0.41	0.82
	4330	Short Axle	4300	CW3626	1	0.15	0.15
	4340	Wheel	4300	CW2314	1	0.61	0.61
	4350	Cotter Pins	4300	SP6122	2	0.18	0.36
	4380	Raw Material - Wheel	4300	RM5784	0.5	0.82	0.41
	4390	Raw Material - Brackets	4300	RM5784	1	0.82	0.82
4400		Leg/Handle Assembly	4100				3.01
	4420	Handle	4400	CW3232	1	0.66	0.66
	4430	Leg Stand	4400	CW4240	1	0.99	0.99
	4440	Screws	4400	SP4881	6	0.09	0.54
	4470	Raw Material - Leg Stand	4400	RM5784	1	0.82	0.82

Table 8 - Direct Material-based Energy Consumption by Product.

Total Direct Energy Consumption (EC)		**CW1000 (5000 units)**	**CW4000 (3000 units)**	**CW7000 (2000 units)**
Material	320,590 [MJ/year]	179,550	83,040	58,000
	Unit EC [MJ/unit]	35.91	27.68	29.00

2.3.3 Quantifying the Waste

For the waste, we focused only on what waste is caused by the annual use of the cars and building. The waste content of fuel, gas, and electricity is calculated base on the Waste Index (WI) presented in Chapter 2, and expressed in picoWasteUnits (pWU). Multiplying the waste resource amount times the sum of the waste content and the waste from use gives the total annual waste generation in Table 9. Thus, Annual Waste Generation = Annual Waste x (Waste Content + Waste from Use). Note that there is no waste from *using* electricity, but electricity does have a significant waste content.

To compute the waste content in terms of pWU/kg we need the information in Table 10 which shows what compounds are being released into Nature as waste when the resources are used/consumed. Please note that the bold numbers in Table 10 are summation totals for associated waste elements.

Table 9 - Computing the Waste Index for the Waste Elements at WagonHo!.

Waste Source	Annual Waste		Waste Content	Waste Release from Use	Annual Waste Generation [pWU]
Company Cars					
Ford Taurus	Car Fuel	1,300.0 kg	0.0670 pWU/kg	0.0099 pWU/kg	99.94
Ford F150	Car Fuel	1,080.0 kg	0.0670 pWU/kg	0.0099 pWU/kg	83.03
Building					
Atlanta Technology	Natural Gas	1,000.0 kg	0.0001 pWU/kg	0.0001 pWU/kg	0.17
Center	Electricity	2,520,000 MJ	0.0100 pWU/MJ	0 pWU/MJ	25,182.00

Table 10 - Computing the Waste from Use per Unit Resource.

Resource	Waste Release	Degrades into	Unit Release	$V.T_N/2A_N$ Ratio (Appendix A)	Unit Waste Release from Use [pWu/kg]
Car Fuel					**9.90E-03**
	CO_2	Appears naturally	2.09E-01	2.15E-02	4.50E-03
	CO	Appears naturally	2.30E-03	2.21E-01	5.09E-04
	SO_2	Appears naturally	2.00E-04	6.28E-01	1.26E-04
	CH_4	Appears naturally	4.20E-05	1.72E+00	7.24E-05
	NO_x	Appears naturally	2.60E-04	1.80E+01	4.69E-03
Natural Gas					**1.15E-04**
	CO_2	Appears naturally	2.09E-03	2.15E-02	4.50E-05
	CO	Appears naturally	5.04E-05	2.21E-01	1.12E-05
	CH_4	Appears naturally	1.32E-05	1.72E+00	2.28E-05
	NO_x	Appears naturally	2.01E-06	1.80E+01	3.62E-05

By distinguishing between waste content and waste releases, a company can investigate internal and external factors much more systematically and elegantly similarly to cost management where one distinguishes between material costs and other costs. The only way to affect the overall material costs is by using less material or a cheaper material. Similarly here, the only way WagonHo! can reduce their electricity waste generation is by using less electricity or finding electricity with a lower waste content.

2.3.4 Information Assumptions

Clearly, due to the current absence of energy and waste accounting by companies, Activity-Based Cost, Energy, and Waste models must also rely on data gatherings along the lines of conventional LCA type approaches. Unlike conventional LCA data, however, we (only) need a specific set of basic data and do not need any impact categorizations, functional units definitions, etc. We obtained our information from the following sources:

- The energy content numbers for 'Natural Gas' releases and 'Electricity' are from IDEMAT'96, which was also used to calculate the waste content (see also Appendix A).
- The 'Car Fuel' release data are from: David Hart, Centre for Environmental Technology, Imperial College, London; IDEMAT; and from Günter Hörmandinger, Energy Policy Research, London.

Please note that since this is a learning case study only we have not been as diligent when it comes to information gathering as elsewhere in this book.

2.4 Step 4 - Identify and Quantify the Resource Drivers and Activity Drivers and their Intensities

In Table 11 we show the resource drivers employed in the Activity-Based Cost, Energy and Waste model at WagonHo!.

Table 11 - Resource Drivers Employed at WagonHo!.

Resources	Resource Drivers	Cost Consumption Intensity	Energy Consumption Intensity	Waste Generation Intensity
Building				
Atlanta Technology Center	Area [sq. ft.]	15.77 [$/sq. ft.]	585.08 [MJ/sq. ft.]	1.52 [pWU/sq. ft.]
Logistics Equipment				
Fork Lift	Direct Labor [h]	1.53 [$/h]		
Company Cars				
Ford Taurus	Direct labor [%]	85.00 [$/%]	615.4 [MJ/%]	0.9994 [pWU/%]
Ford F150	Direct labor [%]	83.00 [$/%]	512.8 [MJ/%]	0.8303 [pWU/%]
Production Equipment				
Conveyer System	Direct			
Kitting Equipment	Direct			
Lathe Machines	Direct			
Milling Machines	Direct			
Assembly Equipment	Direct			
Inspection Equipment	Direct			
Design Equipment	Direct			
Material Design	Direct			
Material Maintenance	Direct			
Material Production	Direct			
Activity Labor Costs	Direct Labor [h]	*varies*		
Office Equipment				
Computer Systems	Acquisition Costs	0.23 [$/$]		
Furniture etc.	Acquisition Costs	0.13 [$/$]		
Reception	No. of Communications	0.03 [$/comm.]		

The activity drivers are not shown here, but in Section 2.8, Table 18. 'Direct' means that the resource element matches the activity completely. This is preferable since the distortion introduced is zero, see Chapter 3. Since all the work is performed unit by unit we chose 'Direct labor' as a resource driver. The cars are traced by a percentage use basis rather than on an hourly basis. 'Area' (floorspace) traces the 'consumption' of the building.

2.5 Step 5 - Identify the Relationships between Activity Drivers and Design Changes

In this model there are no explicit relationships between activity drivers and design variables. Potential product design changes, however, can be found by simply identifying the weaknesses about each product. The design group can then improve the design by eliminating/reducing these weaknesses. When it comes to process design and potential organizational changes, these are directly related to the activity drivers and therefore always incorporated into an activity-based model.

2.6 Step 6 - Model the Uncertainty

In this model, we have included triangular uncertainty distributions, where the upper and lower bounds are ±10% of the mean, for all key input data. The purpose of this is to perform sensitivity analyses and to trace the effect of changes in values, see also Section 3.

2.7 Step 7 - Find/Compute the Cost, Energy Consumption and Waste Generation of the Activities

The cost, energy consumption, and waste generation of an activity are found by multiplying each resource driver value times its associated consumption- or generation intensity, followed by a summation of all resulting values for the particular activity. To facilitate this, the WagonHo! model was implemented in Microsoft Excel® as three separate spreadsheets: one for cost, one for energy and one for waste. These spreadsheets are too large to show on a page in this book. In Table 12, Table 13, and Table 14, however, the same portion of the spreadsheet tables is given for each of the three assessment dimensions. In Table 13, for example, the energy consumption calculation for activities A121 and A21 is given. The horizontal break-line denotes that some rows have been left out. Observe that the driver values are the same in all three tables, but the consumption and generating intensities are different.

Table 12 - Resource Consumption Calculation Example - Cost Consumption.

			Activity A121		Activity A21	
Resource	**Resource Driver**	**Consumption Intensity**	**Resource driver value**	**'Cost' [$]**	**Resource driver value**	**'Cost' [$]**
Material	Direct	1 [$]	215,350 [-]	215,350		
Building	Area	15.77 [$/sq. ft]	1,000 [sq. ft]	15,770	100 [sq. ft]	157.70
Conveyor	Direct	5,100 [$]	15,810 [-]			
Kitting eq.	Direct	270 [$]	410 [-]	410		
Ford Taurus	Direct labor %	85 [$/%]			10 %	8.5
Ford F150	Direct labor %	83 [$/%]			5 %	4.15
		Total:		**231,530**		**170.35**

Table 13 - Resource Consumption Calculation Example - Energy Consumption.

			Activity A121		Activity A21	
Resource	**Resource Driver**	**Consumption Intensity**	**Resource driver value**	**'Cost' [MJ]**	**Resource driver value**	**'Cost' [MJ]**
Material	Direct	1 [MJ]	320,590 [-]	320,590		
Building	Area	585.08 [MJ/sq. ft]	1,000 [sq. ft]	585,084	100 [sq. ft]	58,508
Conveyor	Direct	1 [MJ]	15,810 [-]			
Kitting eq.	Direct	1 [MJ]	410 [-]	410		
Ford Taurus	Direct labor %	615.4 [MJ/%]			10 %	6,154
Ford F150	Direct labor %	512.8 [MJ/%]			5 %	2,564
		Total:		**906,084**		**67,226**

Table 14 - Resource Consumption Calculation Example - Waste Generation.

			Activity A121		Activity A21	
Resource	**Resource Driver**	**Generation Intensity**	**Resource driver value**	**'Cost' [pWU]**	**Resource driver value**	**'Cost' [pWU]**
Material	Direct					
Building	Area	1.52 [pWU/sq. ft]	1,000 [sq. ft]	1,520	100 [sq. ft]	152
Conveyor	Direct					
Kitting eq.	Direct					
Ford Taurus	Direct labor %	0.9994 [pWU/%]			10 %	9.99
Ford F150	Direct labor %	0.8303 [pWU/%]			5 %	4.15
		Total:		**1,520**		**166**

In Table 15, the consumption of resources by activities is shown in terms of energy by reversing the rows and columns of Table 13. In each row of Table 15, the amount of energy in MJ traced to the particular activity by each resource driver is given. The driver levels are not listed like in Table 13, but can be obtained by dividing the resource driver value by the consumption intensity. For example, dividing 585,084 MJ by 585.08 will give a value for the resource driver 'Area' of 1,000 square feet for activity A121. In Table 16, the resource consumption by the activities in terms of waste generation is presented. For brevity, we have left out the cost consumption table, but it is the same in principle.

Table 15 - Consumption of Resources by Activity in terms of Energy.

	Resources										
		Building	Production equipment						Cars		
	Material	Atlanta Techn. Center	Conveyer System	Kitting Equip.	Lathes	Milling Machines	Assembly Equip.	Inspection Equip.	Ford Taurus	Ford F150	**Total [MJ]**
Total [MJ]	796,295	9,683,144	15,810	410	8,990	9,500	1,000	500	61,540	51,280	*9,832,174*
Res. Driver	direct	Area [sq.ft]	direct	direct	direct	direct	direct	direct	Direct labor %	Direct labor %	
Cons. Intens.		585.08 MJ/sq.ft							615.4 MJ/%	512.8 MJ/%	
A111		234,034									*234,034*
A112		5,850,842									*5,850,842*
A113		234,034									*234,034*
A121	320,590	585,084	15,810	410							*906,084*
A122		585,084				9,500					*594,584*
A123		585,084			8,990						*594,074*
A124		585,084					1,000				*586,084*
A125		585,084						500			*585,584*
A21		58,508							6,154	2,564	*67,226*
A221		58,508							24,616	5,128	*88,252*
A222		29,254								5,128	*34,382*
A31		58,508									*58,508*
A41		29,254							24,616		*53,870*
A42		87,763							6,154	38,460	*132,377*
A43		58,508									*58,508*
A44		58,508									*58,508*

Table 16 - Consumption of Resources by Activity in terms of Waste.

	Resources			
	Building	Cars		
	Atlanta Techn. Center	Ford Taurus	Ford F150	**Total [pWU]**
Total [pWU]	25,182.53	99.94	83.03	*25,366*
Res. Driver	Area [sq.ft]	Direct labor %	Direct labor %	
Cons. Intens.	1.52 pWU/ sq.ft	0.9994 pWU/%	0.8303 pWU/%	
A111	609			*609*
A112	15,200			*15,200*
A113	609			*609*
A121	1,502			*1,502*
A122	1,502			*1,502*
A123	1,502			*1,502*
A124	1,502			*1,502*
A125	1,502			*1,502*
A21	152	9.99	4.15	*166*
A221	152	40	8.3	*200*
A222	76		8.3	*84*
A31	152			*152*
A41	76	40		*116*
A42	228	9.99	62.3	*300*
A43	152			*152*
A44	152			*152*

2.8 Step 8 - Find/Compute the Cost, Energy Consumption and Waste Generation of the Objects

Given that we have the actual 'cost' of the activities in terms of monetary cost, energy consumption and waste generation, we can calculate how much each product consumes a particular activity. This is presented in Table 17,

Table 18 and Table 19. In Table 17 and Table 18, it is shown how the activities' monetary cost and energy consumption is traced to the three products. In Table 19, the waste generation is traced to the products.

Note that several different activity drivers are used. For example, 'Annual Production' is used to trace the building related resources to product level. To use 'Annual Production' as activity driver may impose some errors because there is no direct causal relationship between the floor area that an activity occupies and the number of product units being processed. Nevertheless, this is the least inaccurate way of doing so without investing heavily in information gathering. Similarly, it is assumed that direct labor in activity A111 is directly proportional to the number of components used. Operational energy consumption (EC) is assumed to be proportional to direct labor spent.

Table 17 - Cost Consumption (CC) in [$] by Product.

Activities:					Products					
					CW1000		CW4000		CW7000	
	CC Elements		Activity Driver	Consum. Intensity	Driver level	CC [$]	Driver level	CC [$]	Driver level	CC [$]
A111	Direct Labor	19,060	Comp. Use [comp]	0.10[$/comp]	130,000	13,041	36,000	3,611	24,000	2,408
	Area Use	6,308	Production [unit]	0.63 [$/unit]	5,000	3,154	3,000	1,892	2,000	1,262
A112	Area Use	19,093	Production [unit]	1.91 [$/unit]	5,000	9,547	3,000	5,728	2,000	3,819
	Planning	157,704	Comp. Use [comp]	0.83[$/comp]	130,000	107,903	36,000	29,881	24,000	19,920
A113	Direct Labor	19,365	Products Sold [product]	1.91 [$/product]	5,000	9,634	2,950	5,684	2,100	4,046
	Area Use	6,308	Production [unit]	0.63 [$/unit]	5,000	3,154	3,000	1,892	2,000	1,262
A121	Material	215,350	Material CC [$]	1.0 [$/$]		132,800		42,810		39,740
	Operation CC	155,237	Direct Labor [h]	11.09 [$/h]	8,500	94,251	3,300	36,592	2,200	24,394
	Area Use	15,770	Production [unit]	1.58 [$/unit]	5,000	7,885	3,000	4,731	2,000	3,154
A122	Operation CC	297,254	Direct Labor [h]	13.60 [$/h]	13,000	176,856	5,310	72,239	3,540	48,159
	Area Use	15,770	Production [unit]	1.58 [$/unit]	5,000	7,885	3,000	4,731	2,000	3,154
A123	Operation CC	49,356	Direct Labor [h]	14.10 [$/h]	3,500	49,356	0	0	0	0
	Area Use	15,770	Production [unit]	1.58 [$/unit]	5,000	7,885	3,000	4,731	2,000	3,154
A124	Operation CC	136,715	Direct Labor [h]	8.04 [$/h]	12,000	96,504	2,700	21,714	2,300	18,497
	Area Use	15,770	Production [unit]	1.58 [$/unit]	5,000	7,885	3,000	4,731	2,000	3,154
A125	Operation CC	141,860	Direct Labor [h]	47.29[$/h]	1,500	70,930	900	42,558	600	28,372
	Area Use	15,770	Production [unit]	1.58 [$/unit]	5,000	7,885	3,000	4,731	2,000	3,154
A21	Operation CC	115,604	Labor hours [h]	110.10 [$/h]	500	55,050	300	33,030	250	27,525
	Area Use	1,577	Production [unit]	0.16 [$/unit]	5,000	789	3,000	473	2,000	315
A221	Operation CC	117,632	Number of Batches [batch	45.24 [$/batch]	1,000	45,243	900	40,719	700	31,670
	Area Use	1,577	Production [unit]	0. 16 [$/unit]	5,000	789	3,000	473	2,000	315
A222	Operation CC	127,135	Number of Inquiries [inq.]	23.54 [$/Inquiry]	2,000	47,087	1,900	44,733	1,500	35,315
	Area Use	789	Production [unit]	0.08 [$/unit]	5,000	394	3,000	237	2,000	158
A31	Operation CC	17,534	Production [unit]	1.75 [$/unit]	5,000	8,767	3,000	5,260	2,000	3,507
A41	Operation CC	59,711	Number of Comp. [comp]	1,194.23 [$/comp.]	26	31,050	12	14,331	12	14,331
	Area Use	789	Production [unit]	0.08 [$/unit]	5,000	394	3,000	237	2,000	158
A42	Operation CC	198,179	Number of Batches [batch]	39.64 [$/batch]	2,500	99,089	1,500	59,454	1,000	39,636
	Area Use	2,366	Production [unit]	0.24 [$/unit]	5,000	1,183	3,000	710	2,000	473
A43	Operation CC	74,833	Products Sold [product]	7.45 [$/unit sold]	5,000	37,230	2,950	21,966	2,100	15,637
	Area Use	1,577	Production [unit]	0.16 [$/unit]	5,000	789	3,000	473	2,000	315
A44	Operation CC	144,729	Total Direct Labor [h]	2.44 [$/h]	38,500	93,885	12,210	29,775	8,640	21,069
	Area Use	1,577	Production [unit]	0.16 [$/unit]	5,000	789	3,000	473	2,000	315
			Total:			1,229,083		540,598		398,389
			Unit Cost:			**245.82**		**180.20**		**199.19**

Table 18 - Energy Consumption (EC) in [MJ] by Product.

Activities:	EC Elements [MJ]		Activity Driver	Consum. Intensity	Products CW1000 Driver level	CW1000 EC [MJ]	CW4000 Driver level	CW4000 EC [MJ]	CW7000 Driver level	CW7000 EC [MJ]
A111	Direct Labor	0	Comp. Use [comp.]	0.00 [MJ/ comp]	130,000	0	36,000	0	24,000	0
	Area Use	234,034	Production [unit]	23.40 [MJ/unit]	5,000	117,017	3,000	70,210	2,000	46,807
A112	Direct Labor	0	Production [unit]	0.00 [MJ/unit]	5,000	0	3,000	0	2,000	0
	Area Use	5,850,842	Comp. Use [comp.]	30.79 [MJ/comp]	130,000	4,003,208	36,000	1,108,581	24,000	739,054
A113	Direct Labor	0	Products Sold [product]	0.00 [MJ/ product]	5,000	0	2,950	0	2,100	0
	Area Use	234,034	Production [unit]	23.40 [MJ/unit]	5,000	117,017	3,000	70,210	2,000	46,807
A121	Material	320,590	Material EC [MJ]			179,550		83,040		58,000
	Operation EC	16,220	Direct Labor [h]	1.61 [MJ/h]	8,500	9,848	3,300	3,823	2,200	2,549
	Area Use	585,084	Production [unit]	58.51 [MJ/unit]	5,000	292,542	3,000	175,525	2,000	117,017
A122	Operation EC	9,500	Direct Labor [h]	0.43 [MJ/h]	13,000	5,652	5,310	2,309	3,540	1,539
	Area Use	585,084	Production [unit]	58.51 [MJ/unit]	5,000	292,542	3,000	175,525	2,000	117,017
A123	Operation EC	8,990	Direct Labor [h]	2.57 [MJ/h]	3,500	8,990	0	0	0	0
	Area Use	585,084	Production [unit]	58.51 [MJ/unit]	5,000	292,542	3,000	175,525	2,000	117,017
A124	Operation EC	1,000	Direct Labor [h]	0.06 [MJ/h]	12,000	706	2,700	159	2,300	135
	Area Use	585,084	Production [unit]	58.51 [MJ/unit]	5,000	292,542	3,000	175,525	2,000	117,017
A125	Operation EC	500	Direct Labor [h]	0.17 [MJ/h]	1,500	250	900	150	600	100
	Area Use	585,084	Production [unit]	58.51 [MJ/unit]	5,000	292,542	3,000	175,525	2,000	117,017
A21	Operation EC	8,718	Labor hours [h]	8.30 [MJ/h]	500	4,151	300	2,491	250	2,076
	Area Use	58,508	Production [unit]	5.85 [MJ/unit]	5,000	29,254	3,000	17,553	2,000	11,702
A221	Operation EC	29,744	Number of Batches [batch	11.44 [MJ/batch]	1,000	11,440	900	10,296	700	8,008
	Area Use	58,508	Production [unit]	5.85 [MJ/unit]	5,000	29,254	3,000	17,553	2,000	11,702
A222	Operation EC	5,128	Number of Inquiries [inq.]	0.95 [MJ/inq.]	2,000	1,899	1,900	1,804	1,500	1,424
	Area Use	29,254	Production [unit]	2.93 [MJ/unit]	5,000	14,627	3,000	8,776	2,000	5,851
A31	Operation EC	58,508	Production [unit]	5.85 [MJ/unit]	5,000	29,254	3,000	17,553	2,000	11,702
A41	Operation EC	24,616	Number of Comp. [comp]	492.32 [MJ/comp.]	26	12,800	12	5,908	12	5,908
	Area Use	29,254	Production [unit]	2.93 [MJ/unit]	5,000	14,627	3,000	8,776	2,000	5,851
A42	Operation EC	44,614	Number of Batches [batch]	8.92 [MJ/batch]	2,500	22,307	1,500	13,384	1,000	8,923
	Area Use	87,763	Production [unit]	8.78 [MJ/unit]	5,000	43,881	3,000	26,329	2,000	17,553
A43	Operation EC	0	Products Sold [product]	0.00 [MJ/unit sold]	5,000	0	2,950	0	2,100	0
	Area Use	58,508	Production [unit]	5.85 [MJ/unit]	5,000	29,254	3,000	17,553	2,000	11,702
A44	Operation EC	0	Total Direct Labor [h]	0.00 [MJ/h]	38,500	0	12,210	0	8,640	0
	Area Use	58,508	Production [unit]	5.85 [MJ/unit]	5,000	29,254	3,000	17,553	2,000	11,702
			Total:			6,176,953		2,381,635		1,594,176
			Unit Energy Consumption:			**1,235.39**		**793.88**		**797.09**

We can write many pages about the numbers in the three tables, but the most important point to note is the similarity between all three tables. They are basically exactly the same. Observe that the driver levels are also identical. The consumption and generation intensities are obviously different.

Table 19- Waste Generation (WG) in [pWU] by Product.

Activities:					Products					
					CW1000		CW4000		CW7000	
	WG Elements [pWU]		Activity Driver	Generating Intensity	Driver level	WG	Driver level	WG	Driver level	WG
A111	Area Use	609	Production [unit]	0.06 [pWU/unit]	5,000	304	3,000	183	2,000	122
A112	Area Use	15,216	Production [unit]	0.08 [pWU/unit]	130,000	10,411	36,000	2,883	24,000	1,922
A113	Area Use	609	Production [unit]	0.06 [pWU/unit]	5,000	304	3,000	183	2,000	122
A121	Area Use	1,522	Production [unit]	0.15 [pWU/unit]	5,000	761	3,000	456	2,000	304
A122	Area Use	1,522	Production [unit]	0.15 [pWU/unit]	5,000	761	3,000	456	2,000	304
A123	Area Use	1,522	Production [unit]	0.15 [pWU/unit]	5,000	761	3,000	456	2,000	304
A124	Area Use	1,522	Production [unit]	0.15 [pWU/unit]	5,000	761	3,000	456	2,000	304
A125	Area Use	1,522	Production [unit]	0.15 [pWU/unit]	5,000	761	3,000	456	2,000	304
A21	Area Use	152	Production [unit]	0.02 [pWU/unit]	5,000	76	3,000	46	2,000	30
A221	Operation WG	48	Number of Batches [batch]	0.02 [pWU/unit]	1,000	19	900	17	700	13
	Area Use	152	Production [unit]	0.02 [pWU/unit]	5,000	76	3,000	46	2,000	30
A222	Area Use	76	Production [unit]	0.01 [pWU/unit]	5,000	38	3,000	23	2,000	15
A31	Operation WG	152	Production [unit]	0.02 [pWU/unit]	5,000	76	3,000	46	2,000	30
A41	Operation WG	40	Number of Comp. [comp]	0.80 [pWU/unit]	26	21	12	10	12	10
	Area Use	76	Production [unit]	0.01 [pWU/unit]	5,000	38	3,000	23	2,000	15
A42	Operation WG	72	Number of Batches [batch]	0.01 [pWU/unit]	2,500	36	1,500	22	1,000	14
	Area Use	228	Production [unit]	0.02 [pWU/unit]	5,000	114	3,000	68	2,000	46
A43	Operation WG	0	Annual Products Sold [product]	0.00 [pWU/unit]	5,000	0	3,000	0	2,000	0
	Area Use	152	Production [unit]	0.02 [pWU/unit]	5,000	76	3,000	46	2,000	30
A44	Operation WG	0	Total Direct Labor [h]	0.00 [pWU/unit]	38,500	0	12,210	0	8,640	0
	Area Use	152	Production [unit]	0.02 [pWU/unit]	5,000	76	3,000	46	2,000	30
			Total:			15,470		5,921		3,952
			Unit Waste Generation:			**3.09**		**1.97**		**1.98**

2.9 Step 9 - Perform Sensitivity and Other Numerical Analyses

We implement the model using commercially available software - MS Excel® and Crystal Ball®. The Crystal Ball software is used to handle the uncertainty in the model and more importantly, to trace the critical success factors. The Crystal Ball software performs the Monte Carlo simulations that we use. In this model, we chose all uncertainty distributions to be triangular with ±10% variation to facilitate sensitivity analyses. A sensitivity analysis is performed by measuring statistically how the uncertainty of the different assumption cells affects the uncertainty of the forecast cells by rank correlation. The assumption cells with the largest absolute magnitude of the correlation coefficient are the most critical success factors.

However, before we turn to the numerical and sensitivity analyses, let's review the summary of the model, which is given in Figure 5. As can be seen, there are four assessment objects; three products and the company itself. Also, there are 16 activities at the lowest level that are equally split in production activities and support activities. When it comes to the resource elements, there are actually more than shown; too much to fit nicely in a figure. We have used this model extensively for simulation purposes and even modeled the individual workers. The model presented in this chapter is a simpler version for the purpose of illustrating the development of these activity-based cost, energy and waste models. However, even though this is a small model, we see that the number of drivers is quite large; about two activity drivers per activity on average. This gives good and accurate tracing. The number of resource drivers is much lower, and that is not a problem since the production and organization is very simple and almost every production activity is consumed by all products. There are however, a couple of feed-back loops between activities and resources. This is to ensure that when the resource consumption changes all the activities will be correctly affected.

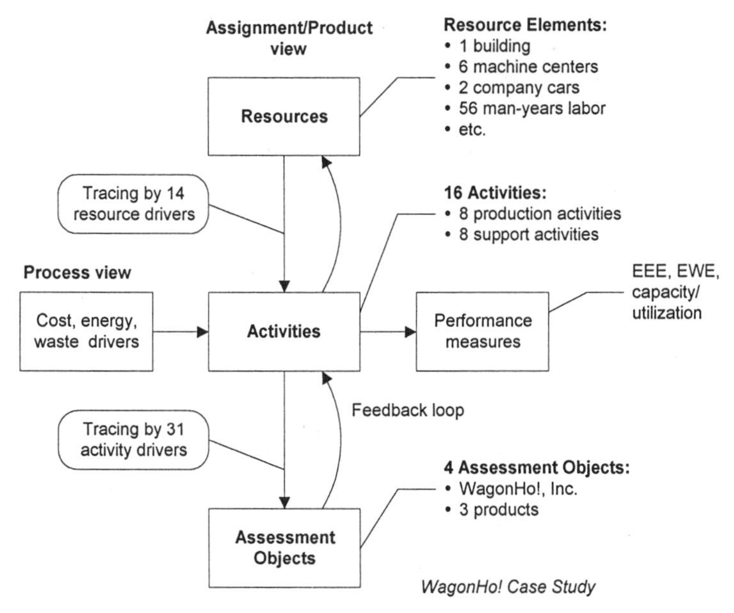

Figure 5 - Working Principle for WagonHo! Activity-Based Cost, Energy and Waste Model.

3. USING THE MODEL

Every name for a lake or river, for mountain or meadow, has its peculiar significance and to tell the Indian title of such things is generally to tell the nature of them also.

Capt. W.F. Butler, F.R.G.S.
In "The Great Lone Land"

Allright, so we created this Activity-Based Cost, Energy, and Waste model and implemented it in a spreadsheet. What can we do with it? That's what we will show in this section. Specifically, we will try to take you through a typical scenario of a company that is in economic trouble and show you how our activity-based approach not only helps management identify an improvement strategy, but also helps in identifying the environmental win-win situations and/or trade-offs. With that in mind, the following issues are presented.

First of all, a key thing to remember is that we added uncertainty into the model in Step 6, but all results shown in the previous chapter were deterministic. So the first thing we are going to do is to run a Monte Carlo simulation and see how much effect the uncertainty has on our results.

Secondly, we already know that the economic results are horrible and we will follow a conventional management approach focused on saving the company first before any environmental improvements are sought. So a critical issue is to develop a strategy that will help WagonHo! to increase its profitability. We will show how identify possible avenues for improvement by using sensitivity charts that are generated from the Monte Carlo simulation.

After having identified an improvement strategy, we will implement some changes and rerun the model again to see what the improvements are. A key question is whether an economic driven improvement strategy can also have environmental benefits. So we will not only look at improvements in terms of monetary cost, but also look at the strategy's effects on energy consumption, and waste generation. We also show how two environmental performance measures, namely the Energy Economic Efficiency (EEE) and Economic Waste Efficiency (EWE), are used.

Because our activity-based model uses three performance measurement dimensions - economic, energy consumption and waste generation - we organize the results accordingly. We also present our results in this order. Within these three dimensions we present the results in this order:

1. WagonHo!, Inc. overall results
2. The results and analyses for the products (CW1000, CW4000, CW7000)
3. Strategy for product improvements based on the analyses.
4. Results from implementation.

3.1 Basic Economic Results and a New Strategy

The economic results are the most reliable in our analyses because the environmental analyses are subject to some data shortages. The point that we like to get across, though, is that the principles and even interpretation of the data and subsequent management actions are the same for the different assessment dimensions.

3.1.1 Economic Results for WagonHo!

From the profitability distribution in Figure 6 which is generated by the Crystal Ball software, we see that the result for WagonHo!, Inc. is very bleak. The mean is -$1,272,831, which is a horrendous result. To find out what contributes to this result most significantly we use sensitivity analysis in Figure 7 (also generated by Crystal Ball).

Due to the possibility of random effects, we cannot trust model variables (to the left) with a smaller coefficient of correlation than about 0.05. We see that the sales price of the CW1000 is the most important success factor, followed by production volume and labor cost for activity A122, which can be explained by the fact that there is no automation in WagonHo!. The improvement efforts that we (and management) can observe from Figure 7 are the following:

1. Increase sales price. An obvious way to improve the result. However, we will assume that this is not a likely option.
2. Increase production. This will lead to economies of scale, assuming that the overhead costs can be kept under control.
3. Cut labor costs. This is the traditional way of improving the profit of a company, but it may not be very feasible due to production requirements, union contracts, and ethics.

In the next section we investigate the performance of the three products.

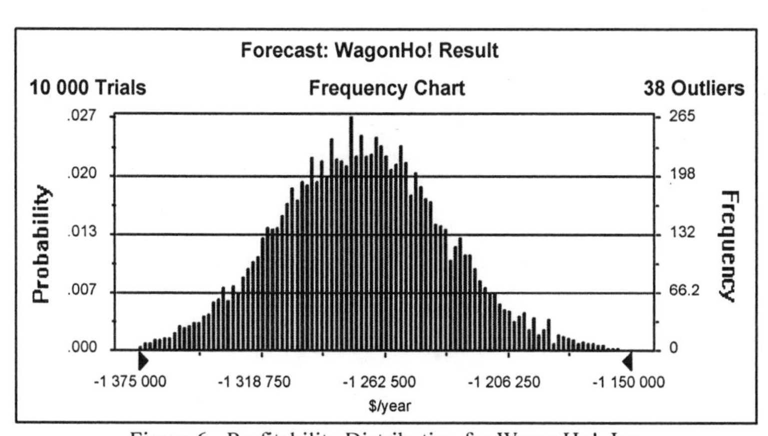

Figure 6 - Profitability Distribution for WagonHo!, Inc.

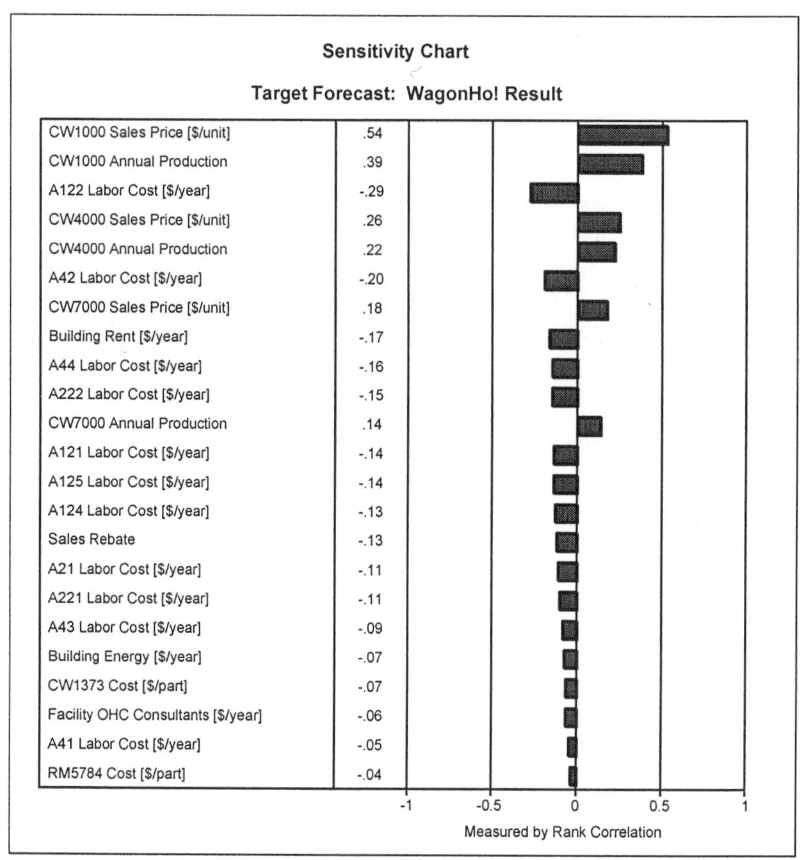

Figure 7 - Sensitivity Chart for WagonHo!, Inc.

3.1.2 Economic Performance of the Products

In Figure 8 the results from the analysis of the CW1000 Wagon is shown. Figure 8 is basically a spreadsheet cell that subtracts the unit cost of the CW1000 from its sales price. We clearly see that the CW1000 is not profitably as the mean is -$143 per unit and it is within a -$120 to -$165 interval with a 10% variation in all variables.

From Figure 9, we see that this product alone contributes over 50% of the WagonHo!, Inc. economic result. It is therefore a crucial product to improve, which brings up the next issue: how to improve the CW1000 Wagon?

Sensitivity charts, like the one in Figure 10, are our primary tools for continuous improvement efforts. There are several interesting findings in the CW1000 sensitivity chart. The most important ones are:

1. WagonHo! must increase sales, as noted before.

2. The labor costs and the amount direct labor [min] are too high. This basically tells us that the CW1000 consumes too much labor time. In fact, the CW1000 has a cycle time of more than 419 minutes. This is far too much and will therefore be one of our prime targets for improvements.
3. We also see that the rate at which they sell and produce the other products also impacts the profitability of CW1000. This is evident from the products' overhead costs. Keeping overhead costs under control is therefore essential.

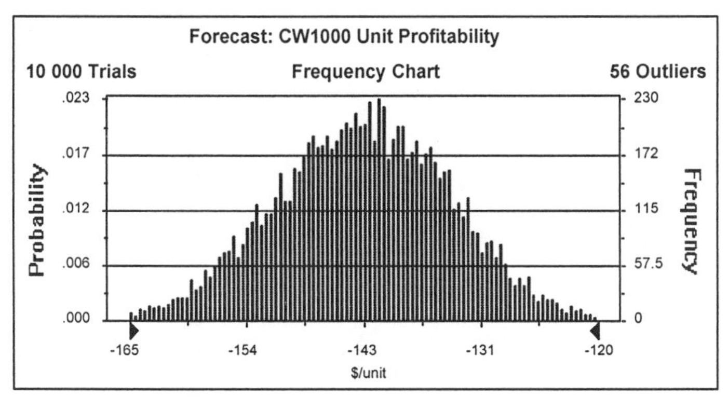

Figure 8 - CW1000 Wagon Unit-Profitability Distribution.

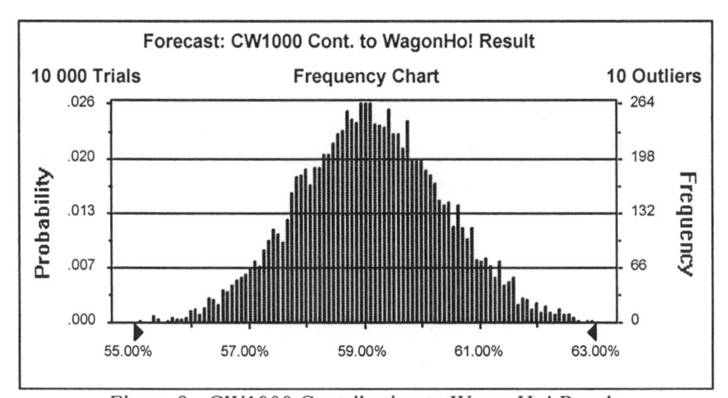

Figure 9 - CW1000 Contribution to WagonHo! Result.

It is very interesting to notice how little the annual production of CW1000 affects the profitability. This is a typical sign of very high amount of material costs and direct labor costs compared to the sales price. If we compare the $106 net sales price to the costs of material and direct labor of $86.29 we see that there is little room for economies of scale. This is a very serious problem for the CW1000 that can only be solved in three ways:

a) Increase sales price. We know that this may lead to reduced sales.

b) Harder bargaining with materials vendors and workers. We assume that the material costs are already at the lowest possible levels.
c) Reduction of cycle time, as already mentioned.

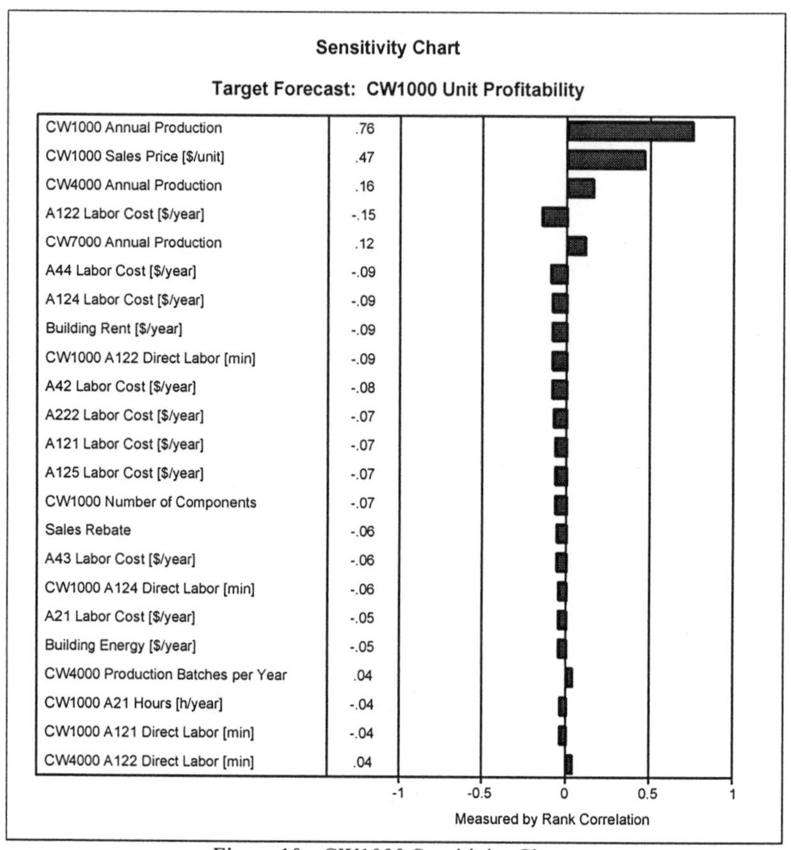

Figure 10 - CW1000 Sensitivity Chart.

So what about the CW4000 and CW7000? We can generate similar distributions as shown in Figure 8 for the CW4000 and CW7000 (not shown here for brevity). It turns out that the CW4000 is also not profitably with a mean of -$95 per unit with a range of -$75 to -$115 given a ±10% variation in all variables. The reasons for this bad performance are basically the same as for the CW1000 Wagon. This can be seen from Figure 11. Interestingly enough, we see that increased production of CW1000 will increase the profitability of the CW4000. This is due to cross-consumption of overhead resources by the CW10000. Essentially, by increasing the CW1000 production the amount of overhead resources traced to CW1000 will increase and the amount traced to the CW4000 will reduce, and hence the profitability of the CW4000 will improve.

Similar analyses show that the CW7000 is in between CW1000 and CW4000 with a mean profitability of approximately -$105 per unit and a profitability range of -$85 to -$125. Again, the reasons for this bad economic performance are similar to the ones for the CW1000 and CW4000. Improvement efforts will therefore also be similar to CW1000 and CW4000.

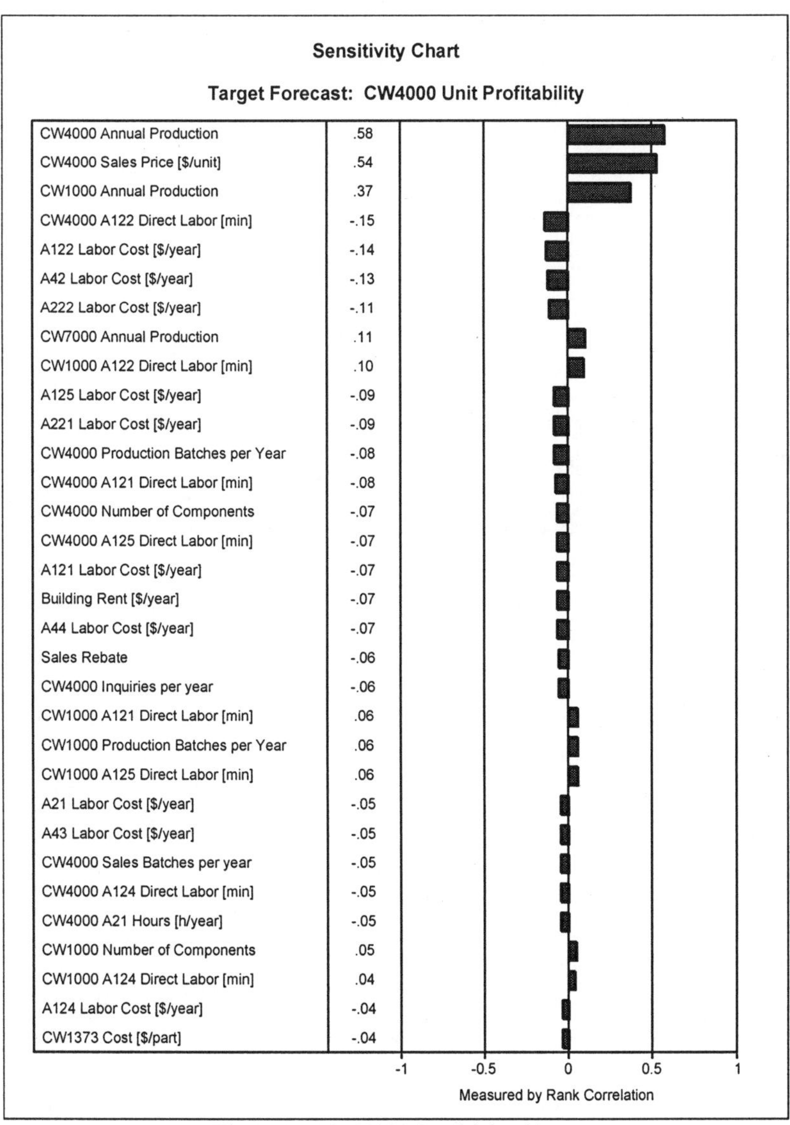

Figure 11 - CW4000 Sensitivity Chart.

3.1.3 Strategy for Economic Improvement

Clearly, WagonHo! is in trouble. In Figure 12 we compare the product profitability distributions. We see that CW1000 is clearly the worst, but it is also the product produced in highest numbers today. The lack of potential economies of scale seems to be a problem for all the products, which means that we have to focus not just on increased sales but also better resource utilization in general. The sensitivity charts in the previous sections clearly indicate that the production labor costs and the amount of direct labor per unit are too high. This can only be interpreted as bad resource utilization at the shop floor, because it takes too long to make a product and it costs too much.

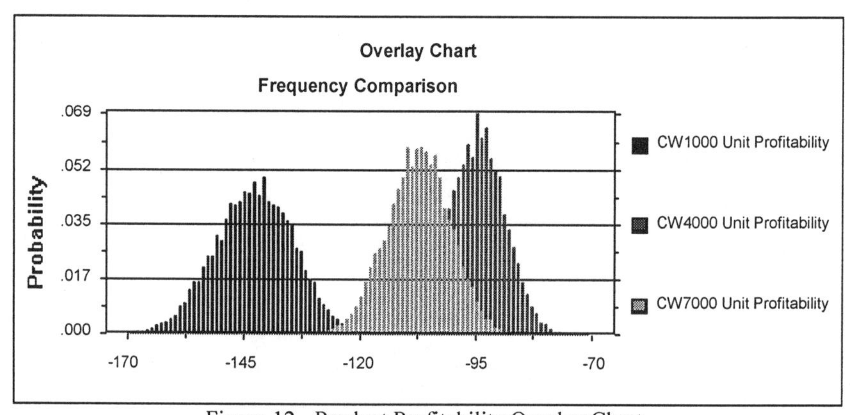

Figure 12 - Product Profitability Overlay Chart.

By adding a capacity versus resource utilization performance measurement in our spreadsheets, we found a mismatch of resources with 1) a bottleneck at the milling machines (activity A122) and assembly (activity A124) and 2) huge resources are wasted at the lathe machines (activity A123) and final inspection (activity A125). To remedy this situation we can suggest the following:

1. Re-measure all the standard times.
2. Sell five (out of six) lathes.
3. Buy two more milling machines.
4. Buy a saw to cut the simple parts, rather than milling them. This will increase the milling capacities and also increase the cutting speed yielding further time reductions.
5. Move people around accordingly. Then we need no layoffs in principle. In reality, the new production level and time savings make it possible to layoff employees. However, before laying people off we recommend to wait and see how the new market strategy works. If it works very well, those employees might be needed.

6. To reduce the cycle times further we also suggest rearranging the shop floor, as shown in Figure 13. This gives savings in transportation times and better process flow in general.

After implementing these suggestions the cycle time was reduced with over 50% for all products as shown in Table 20. Because CW1000 consumes much more resources than the two other products, we also changed the market strategy for WagonHo!. We increased the price from $120 to $225 for the CW1000 and reduced the sales by 50%. The reason for increasing the price that much is to create a feeling around that product that this is the top of the line. This frees up substantial resources enabling WagonHo! to boost sales and production of their other two products by roughly an average factor of five. On one hand, CW1000 is the only product that uses the lathe and it would, from a production and costing point of view, be smart to eliminate that product because then we would eliminate an activity and simplify others. On the other hand, WagonHo! believes that CW1000 is their flagship and it aids the whole product specter. We also suggest going from cost-plus pricing to target pricing for all products. The targets are determined by benchmarking. This gave however, no change in the prices for CW4000 Wheel Barrow and CW7000 Garden Cart.

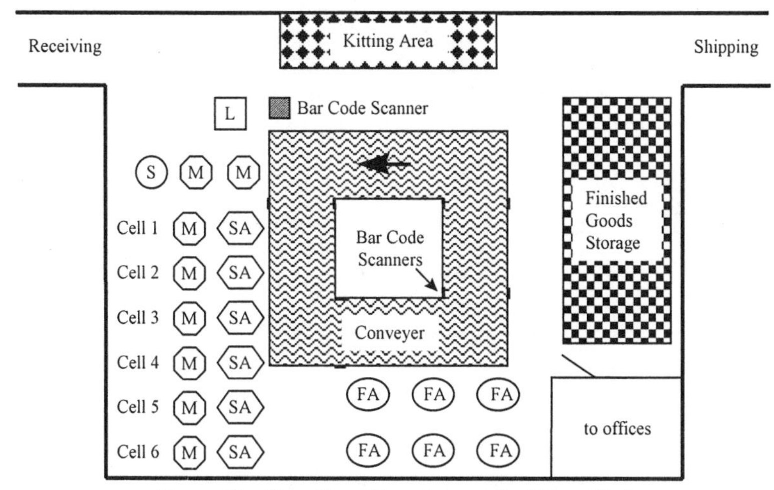

Figure 13 - New Shop Floor Layout for WagonHo!.

Table 20 - Cycle Time Reductions at WagonHo!, Inc.

Product	Old Cycle Time [min]	New Cycle Time [min]	Improvement
CW1000 Wagon	462	192	58.4%
CW4000 Wheel Barrow	244	103	57.8%
CW7000 Garden Cart	259	113	56.4%

3.1.4 Economic Results after Implementation

In Figure 14 the overall profitability of WagonHo! after the implementation of the new strategy is presented. We see that the new strategy could give a mean result for WagonHo! of $225,000, ranging from -$50,000 to $500,000. When it comes to the product performances, we see in Figure 15 that the CW1000 most likely will be negative next year as well, but the CW4000 and CW7000 will be profitable. However, none of them are very strong, which emphasizes the need to continuously improve WagonHo!. The next stage in the improvement efforts at WagonHo!, Inc. is to improve the designs of the toys as well. Maybe new materials can be chosen so that the machining time can be reduced further... In the next section we study the energy consumption results.

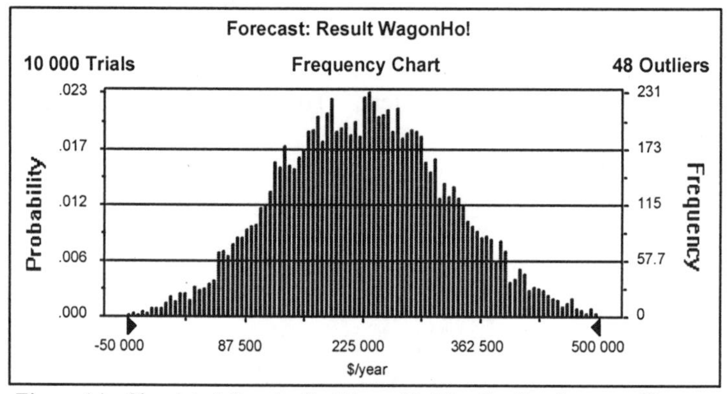

Figure 14 - Simulated Results for WagonHo! Profit after Strategy Change.

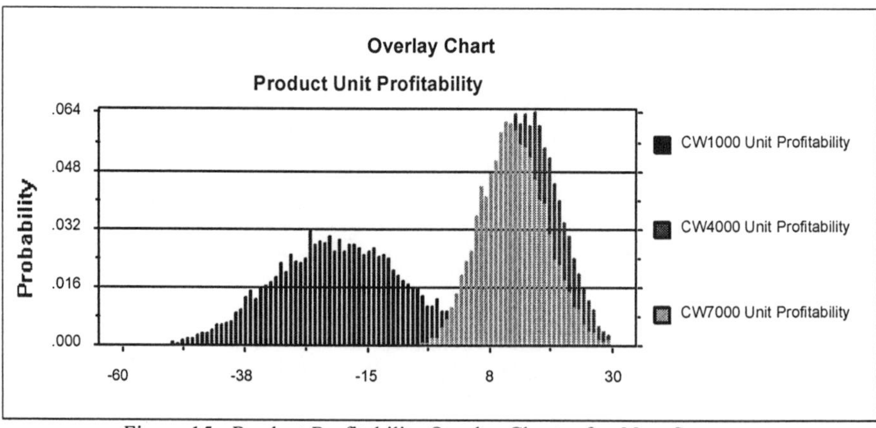

Figure 15 - Product Profitability Overlay Charts after New Strategy.

3.2 Energy Consumption Results

The results prior to the new strategy are not presented here (for brevity), because the WagonHo! management initially was only focusing on costs due to the large losses for WagonHo!. With the new strategy in place, management is also interested in the environmental aspects to find win-win situations between economic and environmental performance can be found. First, we present the environmental results starting with energy consumption.

3.2.1 Energy Consumption Results for WagonHo!

From Figure 16 we see that the annual energy consumption for WagonHo! is roughly 10,125,000 MJ/year. Unfortunately, we cannot really say whether the results presented in Figure 16 are good or bad with respect to some external reference point, but as we conduct more and more case studies we can compare. However, the results can serve as a comparison from year to year and also within WagonHo!.

An interesting number to compute is the total energy consumption divided by total cost, which we call Economic Energy Efficiency (EEE). The EEE essentially measures the amount of energy consumed by the various assessment objects in relation to their overall resource consumption as measured monetarily (adjusted for inventory movements). For WagonHo! as a whole the EEE is 4.54 MJ/$. Every day, week, month, year another EEE can be found and compared; the lower EEE the better performance.

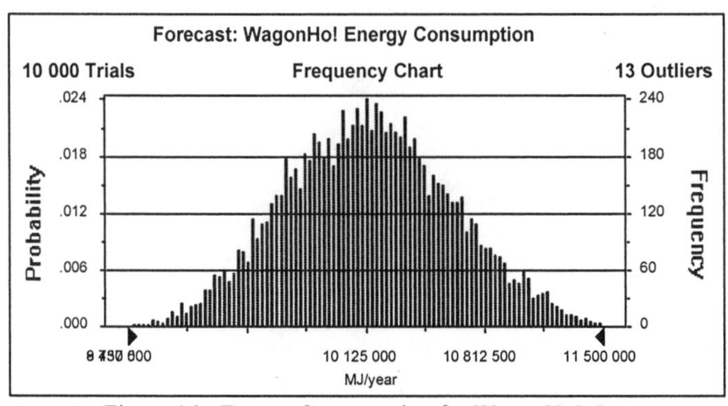

Figure 16 - Energy Consumption for WagonHo!, Inc.

To find out how WagonHo! can improve we again look at the sensitivity chart, which we is shown in Figure 17. We see that the largest contributors are the electricity consumption and the energy content of this electricity. We basically see that increased activity will cause higher energy consumption, unless we do something about the efficiency the energy is consumed with, or the type of electricity consumed. On one hand, in the economic situation of WagonHo!, we do not recommend doing anything drastic, but on the other hand, it is clear that more energy efficient heating, lighting and equipment will yield both economic and energy savings. Note that the electricity consumption is so dominating that not even the energy content of purchased materials shows up in Figure 17.

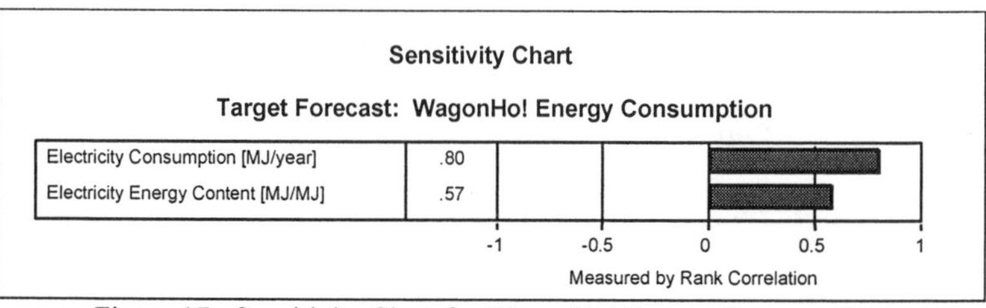

Figure 17 - Sensitivity Chart for WagonHo! Energy Consumption.

3.2.1 Energy Consumption Results for Products

The performance of the CW1000 Wagon is shown in Figure 18. Its EEE is 5.03 MJ/$, 0.49 MJ/USD worse than the company average. From this it follows that the CW1000 is not utilizing the energy efficiently compared to the company average. This is a similar indication to what we found regarding costs: the CW1000 consumes much more resources than average.

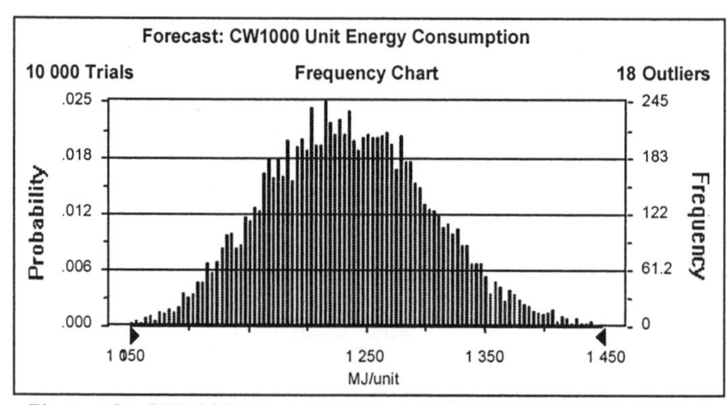

Figure 18 - CW1000 Wagon Unit Energy Consumption Distribution.

If we investigate the sensitivity chart in Figure 19 closely, a paradox appears. For WagonHo!, Inc. as a whole, increased production of units leads to higher energy consumption (due to the energy content of the purchased materials), while according to Figure 19 increased production leads to lower unit energy consumption. The reason is that the overhead energy consumption (like electricity) is so large compared to the direct product energy consumption (like energy content of materials) that the economies-of-scale effect is dominating. This seems to have no solution, except improving the energy efficiency. Then increased production will give better overall performance *and* product performance.

The performance of the CW4000 is shown in Figure 20. It is clear that the CW4000 consumes far less energy per unit than the CW1000. The EEE for the CW4000 is 4.41 MJ/USD, which is substantially better than the CW1000 Wagon and also better than the company average. The CW7000 performance is around 800 MJ/unit, with a range of 675 to 925 MJ/unit. The EEE is an about 4.00 MJ/$. This is the best EEE for all the products even though the CW4000 has slightly better unit energy consumption. However, due to economic inefficiencies, the CW7000 is only second best concerning the EEE. The sensitivity charts for the CW4000 and CW7000 are very similar to the sensitivity chart for the CW1000 Wagon and are therefore omitted here.

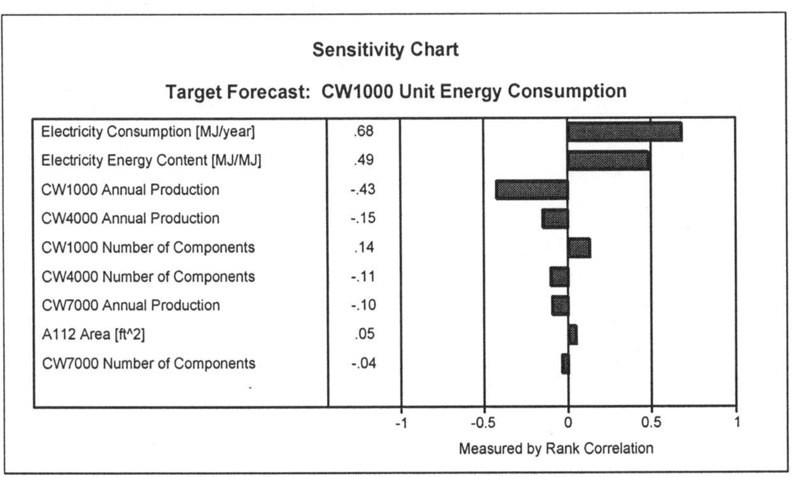

Figure 19 - CW1000 Wagon Energy Consumption Sensitivity Chart.

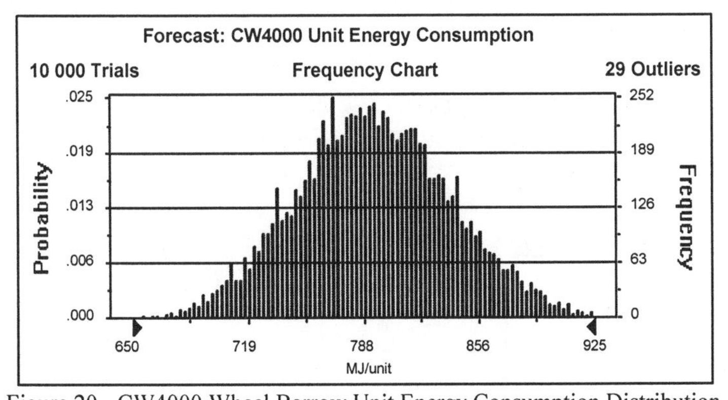

Figure 20 - CW4000 Wheel Barrow Unit Energy Consumption Distribution.

3.2.3 Identifying an Energy Strategy

The question comes to mind how the new economic strategy has affected the energy consumption. Did it result in a win-win situation, or did it cause an increase in energy consumption? To answer this, we look at the energy consumption results after implementing the new economic strategy, as shown in Table 21. The percentages show improvement in the new strategy. We see that the EEE developed positively for WagonHo! as a whole and for CW1000, while negatively for CW4000 and CW7000. This is because no energy efficiency improvements were implemented, while the economic activity grew. However, if we look at the mean values we see that the overall energy consumption increased only by 4.70% while the number of units produced grew from 10,000 per year to 27,500 per year (175% increase). The products improved by roughly 50%. This means that with the implementation of the new strategy outlined in Section 4.1.3, a unit today is produced more environmentally friendly (from an energy perspective) than before, because it uses the resources much better. But, WagonHo! as a whole has become less environmentally friendly in absolute terms, although the EEE improved. Basically, WagonHo! is using more energy than before, but the energy is utilized more efficiently.

So, WagonHo! has made progress in increasing the unit energy efficiency, but has also increased its overall energy consumption. The question is now: *how can we develop an energy strategy for WagonHo so that it can also reduce its overall energy consumption?* We are especially interested in finding win-win situations when the worst economic situation is over. If we look at Figure 17 again, we see that there are two major areas for energy reduction:

1. The overall electricity consumption must be reduced. This can be done by simply finding more energy efficient equipment.
2. The CW1000 is very energy inefficient and should not be produced in such high numbers as before. This is also in line with the economic strategy, hence a win-win situation is already found.

Potential win-win opportunities exist. So now when WagonHo! comes out of its economical ditch, it can start addressing energy consumption related issues as well as economic issues.

Table 21 - Excerpts from Energy Consumption Results after Implementing New Strategy.

Key Numbers from Results	WagonHo!, Inc.	CW1000	CW4000	CW7000
Mean [MJ/unit]	10,628,491 (-4.7%)	615 (+50.2%)	364 (+54.2%)	364 (+54.4%)
Standard Deviation [MJ/unit]	485,146	31	19	19
Range Minimum [MJ/unit]	9,210,620	516	303	304
Range Maximum [MJ/unit]	12,187,166	731	428	429
Mean Standard Error [MJ/unit]	4,851.46	0.31	0.19	0.19
EEE [MJ/USD]	4.27 (+6.3%)	2.86 (+43.1%)	5.08 (-15.2%)	4.58(-14.5%)

3.4 Waste Generation Results

The waste part of our analysis is the least reliable. As with the energy results, it is difficult to relate the waste units to something familiar. Similarly to the EEE, we use an EWE (Economic Waste Efficiency). The EWE is total waste generation divided by total costs.

3.4.1 Waste Generation Results for WagonHo!

The results of the waste generation analysis of WagonHo! are shown in Figure 21. The mean is about 25,500 pWU/year with a range of 21,000 to 30,000 pWU/year. The calculated EWE is 0.0113 pWU/USD.

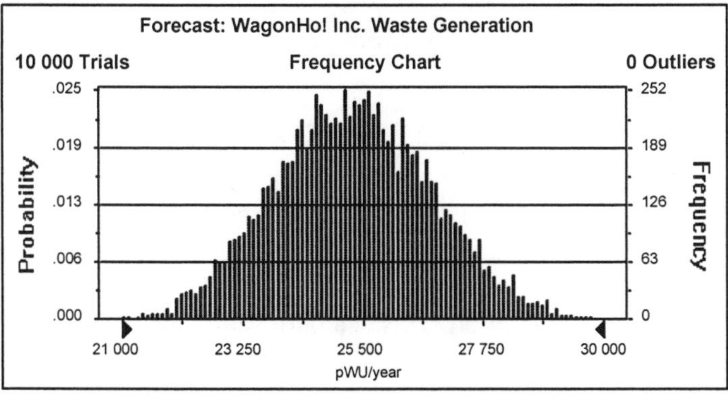

Figure 21 - Waste Generation for WagonHo!, Inc.

From the sensitivity chart in Figure 22, we see that the annual electricity consumption is overshadowing all other waste generation due to the relatively high waste content of the electricity. Hence, using less electricity and trying purchase electricity with lower waste content, such as hydroelectric energy, are the two most important critical success factors for reducing the waste generation.

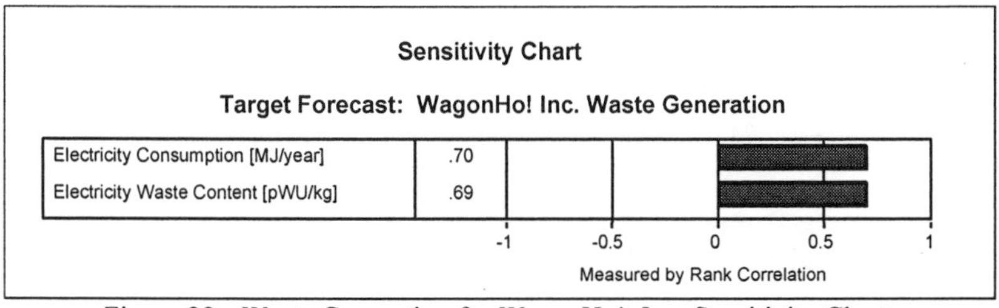

Figure 22 - Waste Generation for WagonHo!, Inc. Sensitivity Chart.

3.4.2 Waste Generation Performance of the Products

As an example, the Unit Waste generation results for CW1000 Wagon is shown in Figure 23. The mean is 3.096 pWU/unit, which gives an EWE of 0.0126 pWU/USD, which is much worse than the company average.

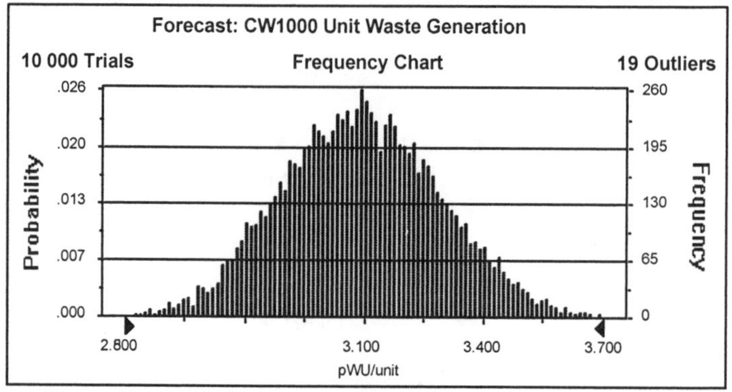

Figure 23 - CW1000 Wagon Unit Waste Generation Distribution.

From the sensitivity chart in Figure 24, we see that, after the electricity related factors, the annual production of CW1000 has the highest effect on the unit waste generation. In fact, Figure 24 shows that increasing production will reduce the waste generation. This is a pure economies-of-scale effect.

Sensitivity Chart

Target Forecast: CW1000 Unit Waste Generation

Electricity Waste Content [pWU/kg]	.61
Electricity Consumption [MJ/year]	.60
CW1000 Annual Production	-.39
CW4000 Annual Production	-.14
CW1000 Number of Components	.12
CW4000 Number of Components	-.10
CW7000 Annual Production	-.08
CW7000 Number of Components	-.04

-1 -0.5 0 0.5 1

Measured by Rank Correlation

Figure 24 - CW1000 Waste Generation Sensitivity Chart.

It is interesting to note that the 'Number of Components' in the products also plays an important part. That is because 'Number of Components' is a cost driver for general management, hence, we have made the implicit assumption that the more components a product contains, the more things can go wrong and thus the more management is needed. Note, however, that the reason 'Number of Components' shows up as so important is that virtually everything in the company (in terms of waste generation) is overhead and volume related. Hence, all the factors that show up in Figure 24, except the two top factors, are various drivers that either are favorable for the CW1000 or not. That is why none of these factors shows up in Figure 22.

The CW4000 unit waste generation is lower than the CW1000 with an estimated mean of 1.976 pWU/unit and an EWE of 0.0110 pWU/USD. It turns out that the results for the CW7000 Garden Cart are quite similar, except that the CW7000 EWE is 0.0099 pWU/USD.

Next, we will evaluate how the changed strategy turns out from a waste generation perspective. It should be noted, however, that the critical success factors for energy and waste are the same. Hence, strategies for reducing energy consumption will also yield waste generation reductions.

3.4.4 Waste Generation Results after Implementation of New Strategy

The waste generation related results are shown in Table 22. If we compare the results in Table 22 with the results in the previous section, then the changes shown in Table 23 can be found. Negative numbers mean that the situation worsened after implementing the new strategy. We see that the EWE developed positively for WagonHo! as a whole and for CW1000, while the situation is opposite for CW4000 and CW7000. The mean improvement of WagonHo! as a whole cannot be real, because the overhead waste generation has been constant, and there is no other waste generation. This is therefore due to random errors: the accuracy of the numbers in Table 23 is therefore roughly ±1%. The improvement of the EWE is, however, real because the waste generation has been constant while the costs decreased.

Table 22 - Excerpts from Waste Generation Results after Implementing New Strategy.

Key Numbers from Results	WagonHo!, Inc.	CW1000	CW4000	CW7000
Mean [pWU/unit]	25,366.27	1.478	0.868	0.867
Standard Deviation [pWU/unit]	1,454.54	0.091	0.054	0.054
Range Minimum [pWU/unit]	21,077.25	1.192	0.698	0.697
Range Maximum [pWU/unit]	30,411.11	1.813	1.061	1.061
Mean Standard Error [pWU/unit]	14.55	0.001	0.001	0.001
EWE [pWU/USD]	0.0102	0.0068	0.0121	0.0109

Table 23 - Waste Generation Improvements due to New Strategy.

Key Numbers from Results	WagonHo!, Inc.	CW1000	CW4000	CW7000
Mean [pWU/unit]	0.9%	52.3%	56.1%	56.2%
EWE [pWU/USD]	9.7%	46.0%	-10.0%	-10.1%

The fact that the waste generation is constant even though the economic activity increases is unlikely and the numbers in Table 23 are therefore extremely optimistically estimated. Nevertheless, the results give guidelines concerning overhead waste generation. With this in mind, we see that CW1000 has improved its EWE due to less activity and decreased costs. CW4000 and CW7000 have also improved significantly economically, but because there were no efforts in improving the waste generation efficiency, the EWE has dropped. The reason for the mean improvements for CW4000 and CW7000 is the common scaling effect (similar to economies-of-scales).

3.5 Critical Evaluation of the Activity-Based Approach

In Table 24 we see how economic and environmental performance changed after implementing the new strategy at WagonHo!. It is apparent that Activity-Based Cost, Energy, and Waste models helped both economic and environmental management. Of course, a result of $226,852 is nothing to be satisfied with, but it does represent a $1.5 million improvement (from -$1.3 million)! And by using the models repetitively we believe that WagonHo! can identify how the products should be designed to reduce for example material costs and cycle time further. WagonHo! should also start looking at energy usage because of the direct savings and the reduced risk in case of energy shortage. Our analysis also shows that increased energy efficiency also will reduce the waste generation substantially. On one hand, this seems obvious, but on the other hand, it is often very hard how to go about solving this in a systematic and guided way. But a systematic and guided way is exactly what the Activity-Based Cost, Energy and Waste models provide, in our opinion.

Table 24 - Improvement Strategy Results.

Improvement Strategy Results	Activity-Based Cost, Energy, Waste Model
Economic Performance	
WagonHo!, Inc. Profitability	226,852 [$] (117.8%)
CW1000 Profitability	-20 [$/unit] (86.0%)
CW4000 Profitability	15 [$/unit] (115.8%)
CW7000 Profitability	12 [$/unit] (112.2%)
Environmental Performance	
WagonHo!, Inc. Energy Consumption	10,628,491 [MJ] (-4.7%)
CW1000 Energy Consumption	615 [MJ/unit] (50.2%)
CW4000 Energy Consumption	364 [MJ/unit] (54.2%)
CW7000 Energy Consumption	364 [MJ/unit] (54.4%)
WagonHo!, Inc. Waste Generation	25,366.27[pWU] (0.9%)
CW1000 Waste Generation	1.478 [pWU/unit] (52.3%)
CW4000 Waste Generation	0.868 [pWU/unit] (56.1%)
CW7000 Waste Generation	0.867 [pWU/unit] (56.2%)

Finally, there is only one thing left to emphasize; this is not a *real* case study. The purpose of it is not to prove that WagonHo! indeed went from loss to profits, but to merely illustrate possible scenarios using our assessment approach.

4. IN CLOSING

We are born weak, we need strength; helpless, we need aid; foolish, we need reason. All that we lack at birth, all that we need when we come to man's estate, is the gift of education.

Jean Jacques Rousseau

This case study is special since it concerns a simulated company and not a real company. This has bearings on the reliability of both the input information and the way we interpret the output from the models. The main purpose of this case study is to serve as learning case study for readers not familiar with Activity-Based Cost, Energy, and Waste models and how to use them for Activity-Based Cost and Environmental Management. The other case studies, except the Farstad Shipping case study in Chapter 5, are much more complex so that a little 'warm up' case study is very useful.

When it comes to the development of Activity-Based Cost, Energy, and Waste models, including the interpretation of results, there is not much to say except it is repetitive and consistent. That will be clear when one reads the other case studies presented in this book. The repetitiveness and consistency illustrate a major point with Activity-Based Cost, Energy, and Waste models that can be lacking in conventional environmental assessment methods: our Activity-Based Cost, Energy, and Waste modeling approach always works, and it always work the same way.

5. ACKNOWLEDGMENTS

We thank Research Engineer Greg Wiles at Center for Manufacturing Information Technology in Atlanta for his cooperation and the data he provided, which made this case study possible. We would also like to thank all those who helped gathering data for the costing part of the Activity-Based Cost, Energy, and Waste model.

Chapter 6

THE FARSTAD SHIPPING CASE STUDY

To me there is only one ocean, with its constant currents rotating around islands we call continents. To some the Northern Sea is, therefore, a global problem - otherwise it will be like removing poison out of one side of a cup, hoping it is safe to drink from the other.

Thor Heyerdahl

Imagine you are a European shipping company managing a fleet of ships and you want to be proactive; you want to assess the impact your ships have on the environment, and you want to reduce that impact in a manner good for business as well as the environment. This was the initial motivation for the case study presented in this chapter. In contrast to the other case studies, which primarily deal with manufacturing, this case study deals with how to develop an Activity-Based Cost, Energy and Waste model for the use (or operational) phase, of a product, namely a UT 705 Platform Supply Vessel (PSV) owned by Farstad Shipping ASA (Farstad in short) in Ålesund, Norway. The case study is based on a real product, from a real company with real data.

1. HISTORY OF THIS CASE STUDY

God set a limit to man's locomotive ambition in the construction of his body. Man immediately proceeded to discover means of over-riding this limit.

Mahatma Gandhi

The case study presented in this chapter is based upon a project that took place in 1996 between Farstad Shipping ASA and Møre Research on assessing the environmental impact of a PSV operating in the North Sea, presented in (Fet, Emblemsvåg *et al.* 1996). Farstad Shipping ASA is one of the leading supply shipping companies in Norway with roughly 25 supply vessels in their fleet at the time. A platform supply vessel (PSV) brings raw materials, food and supplies to oil platforms and takes back to shore waste, equipment for repair, etc. The UT 705 PSV being investigated in this study was designed by Ulstein International AS in Ulsteinvik, Norway, and the particular vessel studied is called FAR Scandia, owned by Farstad. The two basic questions of interested were:

- *What is the environmental impact?*
- *What are the critical economic and environmental success factors?*

The case study presented in this chapter is a rework of the 1996 project because in the 1996 project the environmental impact analysis, which was carried out along the lines of an ISO 14000 LCA, was inconclusive and incapable of producing results due to the following problems:

- The functional unit definitions were problematic.
- The unit processes of the vessel did not fit the unit processes in the SimaPro 3.5 software that was used.
- The impact assessment was impossible.

The impact categorization was the main reason why the environmental analysis failed in 1996 and the project team was incapable of getting beyond the inventory analysis: As stated in (Fet, Emblemsvåg *et al.* 1996):

> *The key findings are summarized in Table 16. However, there has not been drawn any conclusions concerning which outlets gives the most serious environmental impacts.*

This Table 16 from (Fet, Emblemsvåg *et al.* 1996) is presented in Table 1[1].

As part of the 1996 project, an Activity-Based Costing (ABC) model was developed to identify the critical economic success factors. Comparing to the problems with the environmental assessment, the ABC method, the 1996 ABC model handled the problems set out to solve very well by providing:

1. A detailed and useful economic analysis of the maintenance part of the vessel by allowing the detailed modeling of the maintenance jobs.
2. An accurate and useful economic analysis of the PSV operational pattern.
3. An detailed and useful economic analysis of fuel alternatives, where two fuel options were thoroughly assessed; Marine Gas Oil (MGO) versus IF 40 Heavy Fuel Oil (HFO). MGO has higher fuel costs but lower maintenance costs than HFO. Under what conditions is one preferable to the other? This question was resolved and recommendations regarding when to use MGO and when to use HFO were provided.

In Table 1, some activity references, e.g. A222, can be found that were used in the 1996 economic assessment/ABC model. This activity notation must in no circumstance be mixed with the activity notation used in this chapter because the economic assessment in the 1996 project different (more detailed) than the models presented in this chapter

In the rest of this chapter, we present the rework of the 1996 project, but now using our approach to the same problem - analyzing the environmental performance of an UT 705 PSV. We focus on the ship's hull and main engines, and we use our activity-based method to identify the critical economic and environmental success factors for the vessel's two main components.

[1] Please note that British spelling was used in Table 1.

Table 1 - Inventory Table; Presentation of Material and Energy Consumption, Releases to Air and Sea and Solid Waste generated during the 10 Years Operation of a Platform Supply Vessel (Fet and Oltedal 1994). Table 16 in (Fet, Emblemsvåg *et al.* 1996).

Activity (Activity Number)	Impact	Component	Amount per 10 years	
			MGO (tonnes)	HFO (tonnes)
Main engine emissions (A11)	Releases to air	SO_2	300	855
		CO	240	230
		HC	75	70
		NO_x	1,650	1,570
		CO_2	97,500	98,300
Release of toxic substances from the hull coating (A11)	Releases to sea	Tributyltin	1.5 tonnes	
		Copper Oxide	1.7 tonnes	
		Coal tar	173 kg	
Polishing of anodes (A11)	Releases to sea	Zinc	1,000 kg	
Dock mobilization (A21)	Energy use		560 kWh	
Sand-blasting of hull (A221)	Material consumption	Sand	22 tonnes	
		Cured paint	274 kg	
(360 m² per 10 years)	Energy consumption		Not specified	
	Releases to air	Flying dust	227 kg	
	Releases to sea	Sand	4.0 tonnes	
	Solid waste	Collected sand => storage	15.5 tonnes	
		Paper => recycled	4.0 tonnes	
		Pallets => storage	184 kg	
		Removed paint => storage	274 kg	
		Heavy dust => surroundings	2.2 tonnes	
Recoating of hull (A222)	Material consumption	JVA 002/003 Primer	342 kg	
		GTA 853 Thinner	51 kg	
(3,500 m² per 10 years)		HISOL BFA 954/956 AF	7,840 kg	
		GTA 007 Thinner	380 kg	
	Energy consumption		Not specified	
	Releases to air	Trimetylbenzene	1.4 litres	
		Xylene	919 litres	
		4-Metyl 2-Pentanon	84 litres	
		Metylisobutylene	22 litres	
		5-Metyl 2-Hexanan	224 litres	
	Releases to sea	Tributyltin (combinations)	77.0 kg	
		Tar Epoxy	10.5 kg	
		Copper Oxide (combinations)	77.0 kg	
	Solid waste	Xylene	42 litres	
		White Spirit	28 litres	
		Pallets	3.5 m³	
Cleaning of hull (A223)	Material consumption	Freshwater 280 bar	28,800 litres	
	Energy consumption		Not specified	
	Releases to air		None	
	Releases to sea	Tributyltin (combinations)	187.2 kg	
		Tar epoxy	86.4 kg	
		Copper Oxide (combinations)	216 kg	
	Solid waste		None	
Lubrication oil change (A232)		Lubrication oil	23,200 litres	
Anodes replacement (A24)		Zinc (50% replacement rate)	1 tonne	
Sand blasting of tanks (A251)	Material consumption	Sand	164 tonnes	
		Cured paint	1.1 tonnes	
	Energy consumption	Heat	30 MWh	
	Releases to air		None	
	Releases to sea		None	
	Solid waste	Sand before blasting => yard	2.4 tonnes	
		Sand after blasting => yard	5.7 tonnes	
		Used sand => storage	154 tonnes	
		Paper => recycling	0.9 tonnes	
		Pallets => storage	1.4 tonnes	
		Removed paint => storage	1.1 tonnes	

It should be noted that the problems we address in this case study are simplified compared to the problems (Fet, Emblemsvåg *et al.* 1996) set out to solve. This has been done to make it easier to follow, as well as to remove confidential data. Especially the cost model is simpler here than in (Fet, Emblemsvåg *et al.* 1996). In (Emblemsvåg and Bras 1997a) an even simpler version of this case study is found that did not include accurate data on the waste generation. *The main focus in this chapter is to show how an integrated Activity-Based Cost, Energy and Waste model can be obtained for the operational phase of a product.* First, however, we need to present some terminology used in the offshore industry and the problems of investigation.

2. SUPPLY VESSEL TERMINOLOGY

I have always thought that the substitution of the internal combustion machine for the horse marked a very gloomy milestone in the progress of mankind.

Winston Churchill
In "Churchill Reader, A Self Portrait"

In general, operating a PSV is not an easy task. PSVs face heavy competition and high risks. Risks stem from the combination of uncertain costs and revenues from long-term contracts. Typically, the longer the time-charter contract, the harder the competition. Costs are mostly related to the maintenance, service and repair activities, which are defined as follows:

- Maintenance is defined as all maintenance activities that can be done while the vessel is in service.
- Repair is defined as unplanned service.
- Service is defined as planned maintenance activities that require the vessel being out of service. This happens every time the vessel is docking. How often the vessel docks depends on the policy of Farstad, but the vessel has to be docked every at least once every five years to renew class certifications done by Det Norske Veritas (DNV).

Two more important definitions are needed for this case study:

- Lay day: The number of days specified in the contract between ship owners and charterer when necessary repair, service and maintenance can be done.
- Off-hire: The vessel is incapable of fulfilling the contract. In the contract between the ship owners and charterer this is specified in detail. Planned services on dock (kept within the lay days) are not considered off-hire.

Depending on the contract, off-hire will come into effect in different situations. For FAR Scandia, which is the particular UT 705 PSV we got data for, the contract is specified as follows:

- Planned service on dock is not considered off-hire.
- Unplanned repairs are considered off-hire.

- The ship owners are given one lay day per month, but the maximum annual aggregated number of lay days is six per year. If the time for annual service exceeds six days, the additional time is considered off-hire.

This is the context, in which the question was raised, *how can we reduce the cost, energy consumption, and waste generation of the PSV?* In the following, we show how an Activity-Based Cost, Energy and Waste model was developed to help answer that question. Although we assess all three dimensions: costs, energy consumption and waste generation in this case study, the two latter will receive most attention.

3. DEVELOPING THE ACTIVITY-BASED COST, ENERGY AND WASTE MODEL FOR FARSTAD

Resource efficiency is the rising tide that will float all the boats higher.

Ray C. Anderson
In "Mid-Course Correction"

We will again follow the steps discussed in Chapter 4 for developing an Activity-Based Cost, Energy and Waste model for assessing and managing the economic and environmental impact of the operational phase of Farstad's PSV.

3.1 Step 1 - Identify the Assessment Objects.

The absolute first thing to be done is to ask "why are we doing this assessments?" and "what do we want to find out?" The answers to these two basic questions will define the system boundaries and scope of the model.

In this case, the focus of the assessment (i.e., the assessment object) is the UT 705 Platform Supply Vessel and we want to know what its economic and environmental impact is and how to influence this impact. Therefore, the ultimate system boundaries are everything that the assessment object (the UT 705 PSV) directly impacts or is impacted by either economically or environmentally. The scope of the model is to assess the costs and environmental impact of operating the UT 705. Furthermore, we want to trace all critical success factors. In accordance with Farstad, only the major ship sections are included in the model, i.e. the hull and the main machinery, see Figure 1. Hence, the assessment underestimates the 'true' PSV impact. Figure 1 uses the SFI (Skipsforskningsinstituttet, i.e., Ship Research Institute) grouping system of ship components, which is used in the shipping industry to categorize vessel parts. It is distributed by Norwegian Shipping and Offshore Services AS, and it is the most frequently used system in Norway.

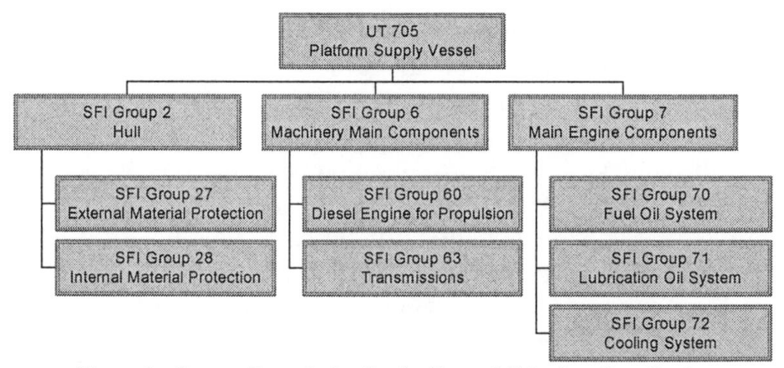

Figure 1 - System Boundaries for the Farstad Shipping Case Study.

3.2 Step 2 - Create an Activity Hierarchy and Network.

First, an activity hierarchy is formed followed by an activity network. In Table 2, the activity hierarchy for the model is presented (this is not the only way of breaking down the use phase, but it is useful for this model). Always choose activities about which good information can be found, provided that the chosen activities *do* capture costs, energy consumption and waste generation well. As shown in Table 2, there are two different levels of activities. For example, activity A3 ('Service Vessel') consists of three level 2 activities - 'Maintain Vessel', 'Service Vessel on Dock' and 'Repair Vessel'. In the activity network (see Figure 2), we use the lowest level activities from the activity hierarchy (the shaded cells in Table 2).

Table 2 - Activity Hierarchy for the UT 705 PSV.

Level 1		Level 2	
Use Vessel	A1	Load Vessel	A11
		Be in Service	A12
		Stand By	A13
		Service Platform	A14
		Be out of Service	A15
Certify Class	A2		
Service Vessel	A3	Maintain Vessel	A31
		Service Vessel on Dock	A32
		Repair Vessel	A33

A11 → A12 → A13 → A14 → A15 → A2 → A31 → A32 → A33

Figure 2 - UT 705 PSV Activity Network.

3.3 Step 3 - Identify and Quantify the Resources

Resource elements are defined as 'economic, energy related or waste related elements that are consumed or generated by the performance of activities'. First, you should define what the resource elements are. Then, you should quantify them.

The resources were identified in consultation with employees of Farstad and the main elements are listed in Table 3. The amounts (i.e., quantification) of the resource elements are also given with the totals presented in bold. The costs are given in Norwegian Kroner (NOK) where NOK 7 ≈ $1. Five main types of resources can be identified:

1. Classification - the vessels must be certified and classified periodically.
2. Lubrication Oil - the lubricant of the machinery.
3. MGO Fuel - FAR Scandia ran on MGO (Marine Gas Oil).
4. Overhead resources - the costs related to running the Farstad organization.
5. Shipyard Work - the labor performed in the shipyard when the supply vessel is being serviced, maintained and repaired.

Table 3 - UT 705 PSV Annual Resource Consumption.

Resource Element	Economic Cost [NOK/year]	Energy Consumption [MJ/year]	Waste Generation [pWU/year]
Classification:	**27,105**		
Annual EO Class	11,067		
EO Class	8,824		
Annual IOPP Certificate	4,774		
Intermediate IOPP Certificate	1,061		
IOPP Certificate	1,379		
Lubrication Oil	**132,597**	**914,103**	**884**
MGO Fuel	**22,029,954**	**147,362,603**	**3,359,274**
Overhead Resources:	**14,318,000**		
Finance	7,000,000		
Crew	6,698,000		
Insurance	620,000		
Shipyard Work	**500,000**	**110,016**	**144**
Total	**37,007,657**	**148,386,722**	**3,360,302**

3.3.1 Quantifying the Cost

To quantify the resources, we typically recommend using annual, quarterly, or monthly values that correspond to the general ledger or other accounting sources. Using resource time-scales, values, and units that match the company's accounting practices will reduce the information gathering and conversion significantly. The values for the economic cost were obtained from inspecting Farstad's books, where it was documented exactly what it spent in 1995 on classification, lubrication oil, MGO fuel, shipyard work and overhead.

3.3.2 Quantifying the Energy Consumption

Data from accounting also is used to calculate the energy consumption. For example, in 1995 the amount of MGO fuel consumed was 3,067,000 kg (or 3,067 ton), a number that was easy to obtain from the company's billing records. In order to get a true energy unit (i.e., MJ/year in this case), we multiply the mass by the unit energy content of MGO. Obtaining the unit energy content is tricky because not many people or companies keep track how much energy they spent in creating a product. They do keep track of how much money they spend in creating a product.

So how do you obtain the unit energy content of MGO? Like before, we used the IDEMAT material database that includes numbers for waste created and energy spent on the production of materials. Other Life-Cycle Inventory or Assessment databases could be used as well. Worst case, one could simply use numbers for embedded energy listed in chemical handbooks as a reference number, but these (typically) do not include the energy spent upstream in the value-chain. By using the IDEMAT material database and by assuming that MGO fuel's unit energy content would be roughly equivalent to the unit energy content of North-Sea crude oil, we assumed the unit energy content to be 48.05 MJ/kg. The unit energy content (48.05 MJ/kg) times the total amount of MGO fuel consumed (3,067,000 kg) gives the energy consumption of 147 million MJ/year found in Table 3 (deviations are due to round-off).

Similarly, we calculated the energy consumption for the lubrication oil by obtaining the amount consumed in a year (19,024 kg.) by the assumed unit energy content of 48.05 MJ/kg, resulting in a corresponding energy consumption of 914×10^3 MJ/year.

The energy consumption from shipyard work was obtained from Table 1 (the original data), specifically, activities A21 - dock mobilization and A251 - sand blasting of tanks, resulting in a total of 30,560 kWh which equals 110,016 MJ (1 kWh = 3.6 MJ).

3.3.3 Quantifying the Waste Generation

We used the Waste Index (WI) to quantify the waste generation, like in the WagonHo! case study. In Table 4 and Table 5, the information needed to compute the WI for the resources' waste generation is shown. In Table 4, the resources' waste elements and their amounts are shown, as well as their upstream waste content (in *pico* Waste Units (pWU) per kilogram) and the waste releases from using the resources in the operation of the PSV.

Table 4 - Resource Waste and Waste Content.

Resource	Annual Amount		Upstream Waste Content	Waste from Use	Annual Waste Generation [pWU]
Lubrication Oil		19,024.0 kg	0.0465 pWU/kg	N/A	**884.6**
MGO Fuel					**3,362,966**
	MGO	3,067,000 kg	0.0465 pWU/kg	1.05 pWU/kg	3,362,966
	Tributyltin	150.0 kg	N/A	1.48E-08 pWU/kg	0.00000221
Shipyard Work					**144.41**
	Coat	454.4 kg	N/A	0.32 pWU/kg	144.41
	Tributyltin	26.4 kg	N/A	1.48E-08	0.000000390

Table 5 - Unit Waste from Resource Use.

Resource Element	Waste Release	Degrades into	Unit Release	$V.T_N/2A_N$ Ratio (Appendix A)	Unit Waste from Use [pWu/kg]
MGO					**1.05E+00**
	CO2	Appears naturally	3.18E+00	2.15E-02	6.85E-02
	CO	Appears naturally	7.83E-03	2.21E-01	1.73E-03
	SO2	Appears naturally	9.78E-03	6.28E-01	6.14E-03
	CH4	Appears naturally	2.45E-03	1.72E+00	4.22E-03
	NOx	Appears naturally	5.38E-02	1.80E+01	9.71E-01
Coat					**3.17E-01**
	Trimetylbenzene	CH4	2.76E-04	1.76E+00	4.86E-04
	p-Xylene	CH4	1.74E-01	1.82E+00	3.17E-01
Tributyltin					
Shipyard	Tributyltin	Chlorine and NaCl	1.00E+00	1.48E-08	1.48E-08
Open Sea	Tributyltin	Chlorine and NaCl	1.00E+00	1.48E-08	1.48E-08

The annual waste generation for each resource waste element is found by multiplying the waste element amount by the sum of the upstream waste content and the use waste (from the release). For example, to calculate the MGO annual waste generation, we take the 'waste from use' in Table 4, which is 1.05 pWU/kg, and add it to the upstream waste content of MGO (0.0465 pWU/kg). Next, we multiply this sum (1.0965 pWU/kg) by the waste element amount (3,067,000 kg), which yields the 3,362,966 pWU listed in the right most column of Table 4. The total annual waste generation for the MGO Fuel resource is aggregated with the annual waste from using tributyltin (TBT). TBT is an acrylic copolymer used in coatings as anti-fouling to prevent juvenile fouling organisms settling on the hull, thus reducing friction. Seawater reacts with the TBT copolymer, removing TBT from the copolymer, which is released as anti-fouling biocide. The reaction is confined to the top few nanometers of the paint surface (Anderson 1993). Such coatings are often referred to as self-polishing. The TBT's unit waste from release is 1.48×10^{-8} and its total waste index value is 2.21×10^{-6} (see Table 4), thus negligible. MGO and TBT are aggregated because the resource driver is identical, as we will see, and aggregation simplifies the modeling effort.

The annual waste generation value for Shipyard Work is obtained in similar fashion, see Table 4. We assumed no waste releases from using lubrication oil, because this oil is recycled and not released into Nature.

The upstream waste content numbers are from Table 3 in Appendix A. The information needed to compute the unit waste release from use is given in Table 5. This value is calculated by multiplying each waste element's unit release (measured as released chemical compound in kg per unit waste element, e.g. kg CO released per kg MGO consumed) by the $V.T_N/2A_N$-ratio for the release. The $V.T_N/2A_N$-ratios for all releases are from Table 2 in Appendix A.

3.3.4 Information Assumptions

The data needed for the Waste Index calculations were obtained from the following sources (see also Appendix A):

- Data for atmospheric and oceanic control volume calculations are from (Lang 1992; Seidelmann 1992), (COESA 1976) and the U.S. 1967 Geological Survey. These are needed to scale and weight releases to the ocean versus releases to the atmosphere.
- The North Sea control volume is 53,730 km^3 according to (The Encyclopædia Britannica 1911). We assumed that the air control volume above the North Sea is equal to the sea control volume, i.e. 53,730 km^3. Furthermore, we estimated the harbor control volume to be roughly 0.025 km^3 by assuming a depth of 25 meter over a 1 km by 1 km area, which is a large shipyard area or a smaller one with significant sea currents.
- Data for computing the gas $T_N/2A_N$-ratios were from (IPCC 1993) and (Mackay, Shiu *et al.* 1992a). We assumed that 1,2,3 trimethylbenzene and *p*-Xylene would essentially behave as methane after a rapid decomposition into methane. Halftimes were used to estimate how fast they would become methane. The T_N and A_N values were adjusted accordingly. We assumed normal temperature and pressure, i.e. $25^{\circ}C$ and 1 atmosphere.
- We used data for crude oil to approximate the data for MGO and lubrication oil. The waste content of crude oil was calculated using material production release data obtained from the IDEMAT'96 database.
- By using coastal dogwhelk (a marine mollusk) as a biological indicator, we assumed that the T_N is roughly 6 years for TBT. Determining the A_N value was much more difficult because the TBT biocide apparently goes into biological entities and stays there until they die. What happens next is hard to say, but the biocide consists of tin (Sn^+) and chloride (Cl^-) ions and some less harmful ions. Since tin is a metal, it would probably stay in the dead tissue while the Cl^- ion is released and becomes chlorine (Cl) and sea salt (NaCl) as it reacts with the Na^+ ion in the seawater from the soluble part of the reaction between TBT and seawater. We therefore let the A_N in this case be the amount of NaCl in the seawater plus the amount of chlorine. These numbers were found in (Encyclopedia of Climate and Weather 1996).

For some chemical compounds listen in the inventory table, Table 1, it was simply impossible to find data for and they were therefore discarded from the analysis. The effect of this is minimal because these compounds decay rapidly and only constitute a tiny fraction of the PSV's total release.

3.4 Step 4 - Identify and Quantify the Resource Drivers and Activity Drivers and their Intensities.

In this example, there is no need to define activity drivers because the model only contains a single product (assessment object) that consumes all activities. But we do have multiple activities and therefore need to define resource drivers. A resource driver is defined as a measure of the quantity of resources consumed by an activity. In the UT 705 PSV Activity-Based Cost, Energy and Waste model we utilize the following resource drivers:

- Running Hours: This is a cost driver used to determine the use pattern of the vessel, and it plays a key role in determining when the vessel is off-hire. Furthermore, 'running hours' is the cost driver by which the overhead and lubrication oil is distributed.
- Fuel Consumption: This is a cost, energy and waste driver that simply keeps track of fuel costs, energy consumption, and waste generation for the vessel.
- Labor Hours: This is a cost, energy and waste driver that keeps track of costs, energy consumption and waste generation associated with servicing the vessel, e.g. sand-blasting and painting.
- Activity Frequency: This is a cost driver only as its sole purpose is to keep track of classification and certification costs.

To find the levels and intensities of these cost, energy and waste resource drivers, historical data was obtained by:

- Asking the crew on FAR Scandia and two technical managers from Farstad Shipping ASA, Ålesund, Norway, as well as a technical manager from Farstad Shipping Ltd., Aberdeen, Scotland, to fill out survey forms.
- Using invoices up to four years old from different shipyards.

In Table 6, the obtained historical data used to quantify the fuel consumption resource driver is shown. Note that 'nm' means 'nautical miles', which is 1,852 meters. One item to explain is the 4 ton/day nominal fuel consumption in activity A13 - 'Stand By'. This is only nominal fuel consumption whereas the other number (running hours) is real. Hence, the PSV does not spend any time in the A13 - 'Stand By' mode/activity, but if it did, it would consume 4.0 ton/day. We could have removed activity A13 from the model, but kept it because management was interested in exploring what-if scenarios in which the amount of time spent in stand-by was increased.

To quantify the impacts and costs over a full year, we scaled the fuel consumption for a typical mission (which lasts about 48 hours) to a full year, which is roughly 6,500 running hours per year. The resulting resource driver values for each activity are given in Table 7, 8, and 9. These driver values were provided by Farstad or calculated based on numbers from Farstad. The consumption- and generation intensity values for the drivers are also given in Table 7, 8, and 9.

Table 6 - Typical FAR Scandia Mission in 1995.

Activity	Mode of Use	Running Hours [h]	Speed [nm/h]	Nominal Fuel Consumption [ton/day]
A11	In port	9.3	0.0	1.0
A13	Stand By	0.0	0.0	4.0
A12	Economic Speed	2.1	10.0	14.3
A12	Full Speed	15.3	14.0	21.4
A14	Service Platform	21.5	0.0	4.8

3.5 Step 5 - Identify the Relationships between Activity Drivers and Design Changes.

Because no design efforts are defined, this step is not applicable here. The 1996 ABC model, however, did solve some design and operational issues related to fuel choices, as noted in Section 1.

3.6 Step 6 - Model the Uncertainty

In this particular model, uncertainty is modeled solely for tracing purposes, see Section 3.9. We did not attempt to model the actual uncertainty.

3.7 Step 7 - Find/Compute the Cost, Energy Consumption and Waste Generation of the Activities

In Table 7, 8, and 9, the cost consumption, energy consumption and waste generation of the activities is given. As explained in Chapter 4, to determine the cost, energy and waste of an activity, simply multiply each resource driver in the activity by its respective consumption intensity and sum the totals for the activity. For example, for calculating the cost of activity A11, multiply 113 ton/year (the resource driver) times 7,183 NOK/ton (the corresponding consumption intensity), repeat the same for the consumption of lubrication oil and overhead resources, and a total cost of 5,304,679 NOK/year for activity A11 is found. The same procedure is followed for each activity. The energy consumption and waste generation is calculated in the same way. Note that the driver levels are the same in all three tables, but the intensities are (obviously) different.

Table 7 - Activity Cost Consumption (CC).

Resource	[NOK/ year]	Resource Driver	Consumption Intensity	Activities																	
				A11		A12		A13		A14		A15		A2		A31		A32		A33	
				Driver level	CC	Driver level	CC	Driver level	CC	Driver level	CC	Driver level	CC	Driver level	CC	Driver level	CC	Driver level	CC	Driver level	CC
Annual EO Class	11,067	Activity Frequency	11,067 [NOK/instance]											1.0	11,067						
EO Class	8,824	Activity Frequency	44,121 [NOK/ instance]											0.2	8,824						
Annual IOPP Certificate	4,774	Activity Frequency	4,774 [NOK/ instance]											1.0	4,774						
Intermediate IOPP Certificate	1,061	Activity Frequency	5,306 [NOK/ instance]											0.2	1,061						
IOPP Certificate	1,379	Activity Frequency	6,895 [NOK/ instance]											0.2	1,379						
Lubrication Oil	132,597	Running Hours	15.26 [NOK/h]	2,705	41,265	2,678	40,853	0	0	3,309	50,479										
MGO Fuel	22,029,954	Fuel Consumption	7,183 [NOK/ton]	113	811,668	2,292	16,463,207	0	0	662	4,755,080										
Overhead Resources	14,318,000	Running Hours	1,646 [NOK/h]	2,705	4,451,746	2,678	4,407,311	0	0	3,309	5,445,777	8	13,166								
Shipyard Work	500,000	Labor Hours	71.23 [NOK/h]													3,500	249,288	3,500	249,288	20	1,425
Total	37,007,657				5,304,679		20,911,371		0		10,251,336		13,166		27,105		249,288		249,288		1,425

Table 8 - Activity Energy Consumption (EC).

Resource	[MJ/ year]	Resource Driver	Consumption Intensity	Activities																	
				A11		A12		A13		A14		A15		A2		A31		A32		A33	
				Driver level	EC	Driver level	EC	Driver level	EC	Driver level	EC	Driver level	EC	Driver level	EC	Driver level	EC	Driver level	EC	Driver level	EC
Annual EO Class	0	Activity Frequency	0 [MJ/instance]											1.0	0						
EO Class	0	Activity Frequency	0 [MJ/ instance]											0.2	0						
Annual IOPP Certificate	0	Activity Frequency	0 [MJ/ instance]											1.0	0						
Intermediate IOPP Certificate	0	Activity Frequency	0 [MJ/ instance]											0.2	0						
IOPP Certificate	0	Activity Frequency	0 [MJ/ instance]											0.2	0						
Lubrication Oil	914,103	Running Hours	105.17 [MJ/h]	2,705	284,474	2,678	281,635	0	0	3,309	347,994										
MGO Fuel	147,369,350	Fuel Consumption	48,050 [MJ/ton]	113	5,429,650	2,292	110,130,600	0	0	662	31,809,100										
Overhead Resources	0	Running Hours	0 [MJ/h]	2,705	0	2,678	0	0	0	3,309	0	8	0								
Shipyard Work	110,016	Labor Hours	15.67 [MJ/h]													3,500	54,851	3,500	54,851	20	313
Total	148,393,469				5,714,124		110,412,235		0		32,157,094		0		0		249,288		54,851		313

Table 9 - Activity Waste Generation (WG).

Resource	[pWU/ year]	Resource Driver	Generation Intensity	Activities: A11 Driver level	A11 WG	A12 Driver level	A12 WG	A13 Driver level	A13 WG	A14 Driver level	A14 WG	A15 Driver level	A15 WG	A2 Driver level	A2 WG	A31 Driver level	A31 WG	A32 Driver level	A32 WG	A33 Driver level	A33 WG
Annual EO Class	0	Activity Frequency	0 [pWU/ instance]											1.0	0						
EO Class	0	Activity Frequency	0 [pWU/ instance]											0.2	0						
Annual IOPP Certificate	0	Activity Frequency	0 [pWU/ instance]											1.0	0						
Intermediate IOPP Certificate	0	Activity Frequency	0 [pWU/ instance]											0.2	0						
IOPP Certificate	0	Activity Frequency	0 [pWU/ instance]											0.2	0						
Lubrication Oil	884	Running Hours	0.10 [pWU/h]	2,705	275	2,678	272	0	0	3,309	337										
MGO Fuel	3,359,274	Fuel Consumption	1,095.30 [pWU/ton]	113	123,768	2,292	2,510,419	0	0	662	725,086										
Overhead Resources	0	Running Hours	0 [pWU/h]	2,705	0	2,678	0	0	0	3,309	0	8	0								
Shipyard Work	144	Labor Hours	0.02 [pWU/h]													3,500	72	3,500	72	20	0.41
Total	3,360,302				124,044		2,510,692		0		725,423		0		0		72		72		0.41

3.8 Step 8 - Find/Compute the Cost, Energy Consumption and Waste Generation of the Objects

We only have one assessment object, namely the UT 705 PSV, and (implicitly) assume that the resources and activities are all attributable to this single assessment object. Thus, the total cost, energy consumption, and waste generation of the UT 705 PSV equals the total cost, energy consumption, and waste generation of the resources: 37,007,657 NOK/year, 148,386,722 MJ/year and 3,360,602 pWU/year, respectively.

3.9 Step 9 - Perform Sensitivity and Other Numerical Analyses

Like the other case studies, we implemented the model using commercially available software - MS Excel® and Crystal Ball®. The latter is used to model the uncertainty and more importantly to trace the critical success factors. A high number (10,000) of trials is used to ensure a high accuracy of the mean and low mean standard error. As stated in Section 3.6, we chose all uncertainty distributions to be triangular with ±10% variation. This works well enough for sensitivity analyses that measure statistically how the uncertainty of an assumption cell affects the (resulting) distribution of the forecast cells by rank correlation. The assumption cells with the largest absolute magnitude of the correlation coefficient are the most critical success factors. In Figure 3, the model is illustrated schematically in terms of the working principles and model infrastructure.

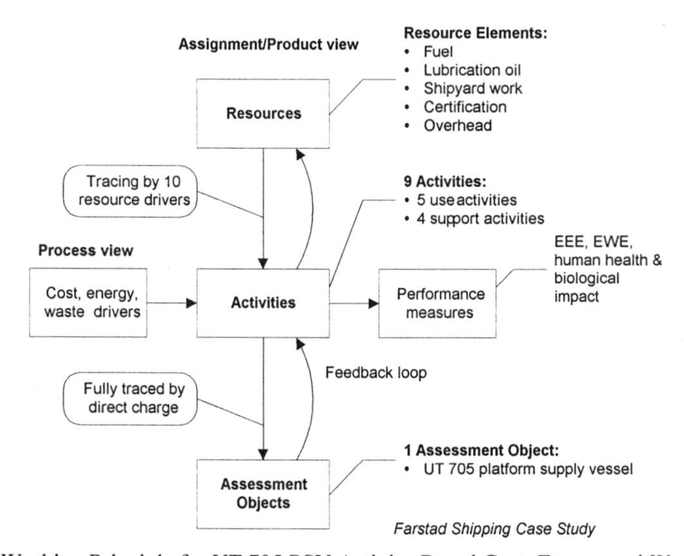

Figure 3 - Working Principle for UT 705 PSV Activity-Based Cost, Energy and Waste Model.

Clearly, the Farstad model is a small model compared to the Westnofa Industrier case study in Chapter 8. The total number of assumption cells is only 45 while the number of forecast cells is just 8; a total of 54 variables.

3.10 Step 10 - Interpret the Results and Iterate if necessary

We will interpret and discuss the results in detail in Section 4, but before we go to the results, let us briefly talk about the quality of the model. The design of the model was done in several stages because it was hard to get information about TBT and a couple of other chemical compounds. All in all, the model is fairly satisfactory, even though it includes some numbers that are not directly measured, such as the balance time of TBT.

4. INTERPRETATION & DISCUSSION OF RESULTS

It is an important and popular fact that things are not always what they seem. For instance, on planet Earth, man had always assumed that he was more intelligent than the dolphins because he had achieved so much - the wheel, New York, wars and so on - while all the dolphins had ever done was muck about in the water having a good time. But conversely, the dolphins had always believed that they were far more intelligent - for precisely the same reasons.

Douglas Adams
In "The Hitchhiker's Guide to the Galaxy"

In essence, Table 7, 8, and 9 represent the results. These numbers, however, are purely deterministic. Part of our approach is to use Monte Carlo simulations to identify and trace critical success factors. In this section, we present and discuss the results obtained from the simulation, starting with costs; then energy consumption and waste generation followed by the efficiency measures.

4.1 Cost Results

In Figure 4, the forecasted distribution for the total annual operating cost ('Annual Life-Span Costs') is presented. Figure 4 is a frequency chart that is a numerical approximation of the exact uncertainty distribution. We see that the mean expected Annual Life-Span Cost is roughly 37,000,000 NOK/year, and that it can range between about 33,000,000 to 41,000,000 NOK/year given 10% uncertainty in all model assumption cells. Uncertainty in the annual costs comes from the MGO fuel price, future certification costs, etc. Having an uncertainty range around an estimate is an advantage of our approach because no costing model is 100% accurate, hence knowing what the potential range can be provides valuable extra information for management.

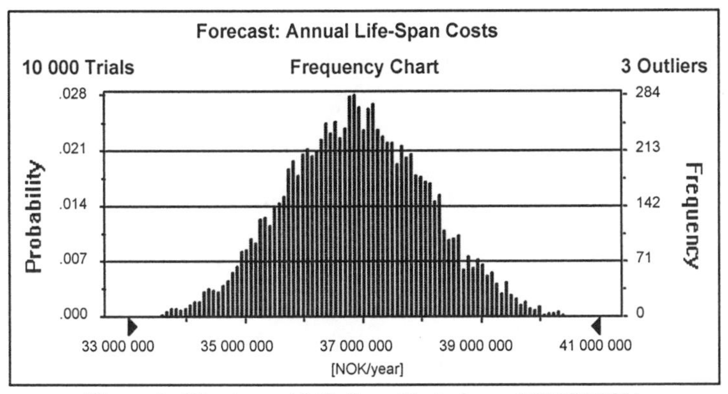

Figure 4 - The Annual Life-Span Costs for an UT 705 PSV.

The sensitivity chart in Figure 5 can be (and was) used by a ship owner to identify where and how the operational costs can be reduced. As can be seen, the MGO fuel price and MGO consumption when the PSV is in service (activity A12) overshadow all other costs. Some options for Farstad to reduce the risk of high fuel cost are 1) make the charterer pay for the fuel, and 2) improve the tuning of the machinery. In the project between Møre Research and Farstad different fuel options were also investigated (again, the cost model was more comprehensive), but it was found that the MGO provides greater savings for the ship owners than the competing IF 40 HFO in most cases.

In Figure 5, we also see that the finance costs and crew costs are heavy burdens. Thus, refinancing is an option for increasing overall profitability, and Farstad has indeed done so.

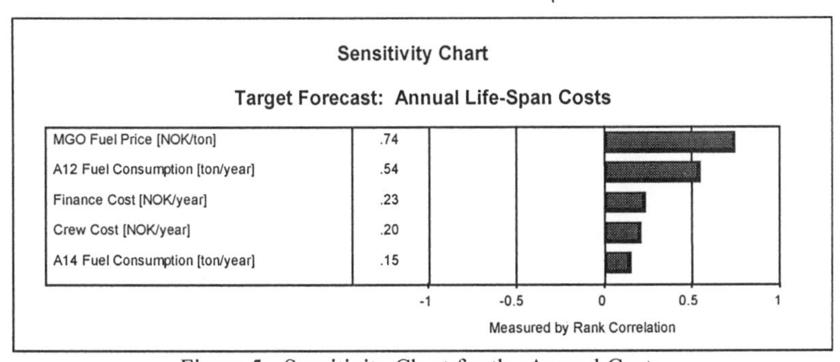

Figure 5 - Sensitivity Chart for the Annual Costs.

The factors in Figure 5 overshadowed any other cost factors in importance. Suppose Farstad wanted to see what other factors could be changed, how would they know? To handle that question, we simply *remove the uncertainty in the assumption cells* listed in Figure 5 and rerun the model.

Doing that gave the sensitivity chart shown in Figure 6. Thus, now factors insignificant in the first run become detectable. This is the way of performing tracing in multiple steps. It is a useful approach in uncertainty analyses, because then all cells that contribute to the total uncertainty in the assessments can be found.

Even now, in Figure 6, fuel related costs are at the top, but now for the loading activity A11. The insurance and shipyard costs are next. By running the model over and over again in this fashion, every assumption can be ordered and its importance on the cost detected, giving good and useful foundations for Critical Assumption Planning (CAP), see also (Emblemsvåg and Bras 1998b).

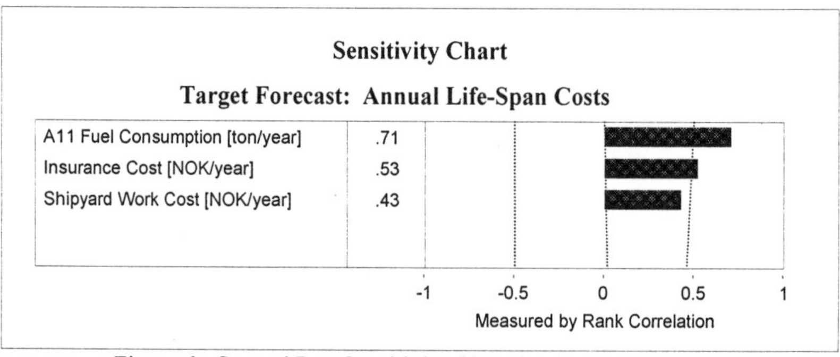

Figure 6 - Second Run Sensitivity Chart for the Annual Costs.

4.2 Energy Results

4.2.1 Annual Energy Consumption and Critical Success Factors

The energy part of the analysis is treated in a similar fashion as the monetary cost, and in Figure 7 the frequency chart for the energy consumption in 1995 is shown. The estimated mean is 147,500,000 MJ/year. Unfortunately, this number does not tell us much on its own because we have no good frame of reference.

However, as more case studies are undertaken these numbers will be just as meaningful as the cost estimates. We can, of course, compare it to, e.g., Westnofa Industrier, see Chapter 8, which consumed 246,450,000 MJ in the same year (1995). Thus, one PSV is capable of consuming close to 60% of an entire medium sized company's energy consumption (!). When one realizes that the sole purpose of a PSV is to ensure a steady supply of oil, we can just start imagining the amount of resources spent and energy consumed in ensuring a steady energy supply. And if this is not enough, both estimates are probably very low due to the complete absence of energy accounting, see also (Emblemsvåg and Bras 1998a). As mentioned before, (Lovins and Lovins 1997) estimate that in the US alone $300 billion is wasted in poor energy management annually. We cannot let this comparison between a PSV and Westnofa pass unnoticed. It is a major motivating factor for our integrated Activity-Based Cost and Environmental Management approach, which would be facilitated energy accounting, analogous to monetary accounting, was established. A comparison like this also provides useful insight in the whole energy debate.

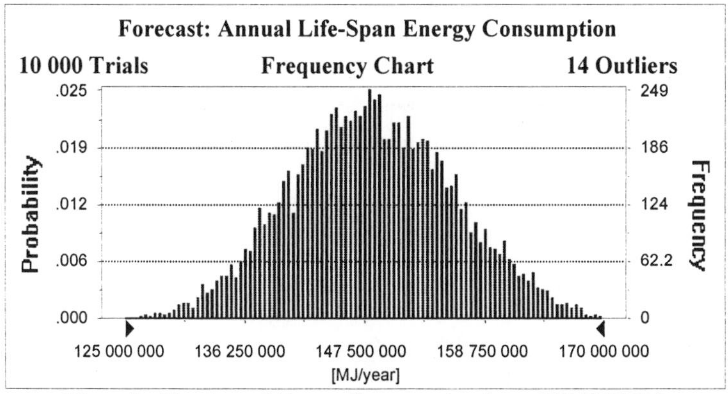

Figure 7 - The Annual Energy Consumption for an UT 705 PSV.

Just as the sensitivity chart in Figure 5 can be used to trace cost factors, the sensitivity charts in Figure 8 can be used to trace energy factors. Here we see that there are three large energy drivers that are basically identical to the cost drivers in Figure 5:

1. MGO Fuel Energy Content - the energy spent in making the fuel from extraction through oil refining and sale. This is equivalent to the MGO price in a cost perspective. Unfortunately, Farstad Shipping cannot affect this driver at all, because they have no information on which manufacturer uses the least amount of energy in production of fuel.
2. The use of the vessel to and from the platform (being in service - activity A12) is the single most energy demanding activity.
3. The use of the vessel at the platform (activity A14) is third on the list.

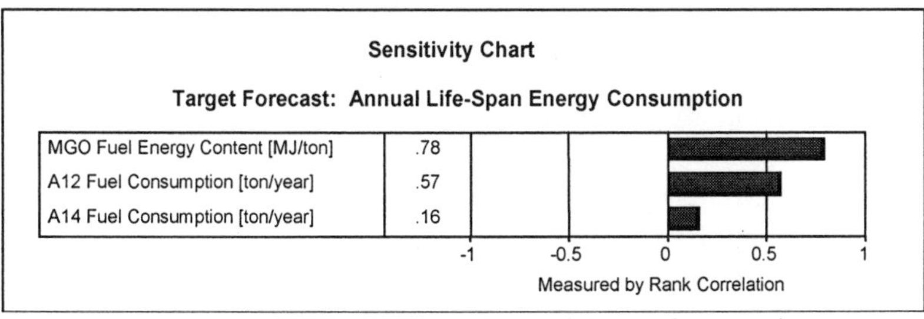

Figure 8 - Sensitivity Chart for the Annual Energy Consumption.

There are other energy drivers, but these are so relatively small that we have to run the model without the three previously mentioned energy drivers to detect them. The sensitivity chart for the second run is given in Figure 9.

As we can see in Figure 9, the energy drivers related to the fuel consumption totally dominate the second run sensitivity chart, although here it is activity A11 (Load Vessel). However, something not related to the fuel consumption, i.e. lubrication oil energy content and lubrication oil consumption, also appeared.

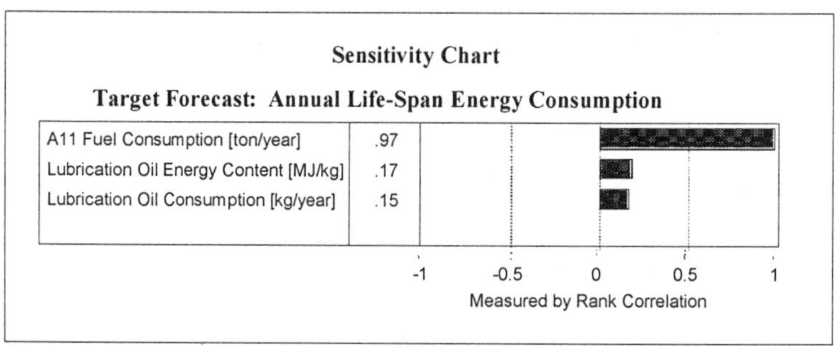

Figure 9 - Second Run Sensitivity Chart for the Annual Energy Consumption.

If we trace all the resource assumption cells in the energy model, then the complete ranking in order of importance would be as follows:

1. MGO Fuel Energy Content [MJ/ton].
2. A12 Fuel Consumption [ton/year].
3. A14 Fuel Consumption [ton/year].
4. A11 Fuel Consumption [ton/year].
5. Lubrication Oil Energy Content [MJ/kg].
6. Lubrication Oil Consumption [kg/year]
7. Shipyard Work Energy Consumption [MJ/year].

This list is the list of critical success factors for the energy consumption of an UT 705 PSV running under conditions that FAR Scandia does. Hence, if Farstad Shipping wants to reduce the energy consumption they should start at the top of the list and work downwards, as the factors on top are the most important ones. In principle, this can be done in any model regardless of size; it is just a matter of software capabilities and budget.

4.2.2 Economic Energy Efficiency

To find the efficiency of energy consumption, we utilize the Economic Energy Efficiency (EEE) performance measure that should be as low as possible. For the UT 705 the EEE is 3.998 MJ/NOK. From the other case studies we will realize that this is very high. In other words, the UT 705 energy utilization per NOK spent is poor. The reason can be found in Figure

10, namely, the high energy content in the MGO fuel, which the UT 705 consumes in large quantities (over 3,000 ton/year).

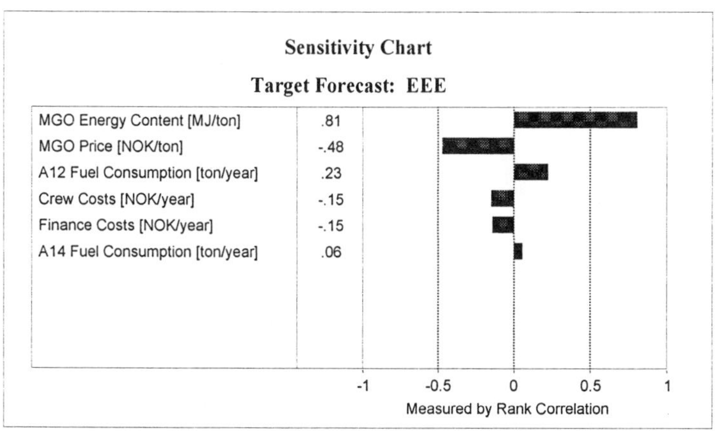

Figure 10 - UT 705 Platform Supply Vessel EEE Sensitivity Chart.

In Figure 10, the bars that show negative correlation (i.e., to the left) are cost-related assumption cells, so if the costs increase the EEE will go down. Clearly, the EEE cannot be used by itself, but it has to be used in conjunction with cost (or profitability) estimates <u>and</u> energy estimates. Consequently, increasing the MGO price and consumption will increase the costs, and therefore will not be an option in reducing the EEE. *So why not just use energy estimates?* The reason is that energy estimates by themselves (in MJ) do not have a direct relation to costs and therefore do not show where (monetary) resources are used in an energy inefficient manner. The purpose of the EEE is to bridge this gap and make such tracing possible.

4.3 Waste Results

4.3.1 Annual Waste Generation and Critical Success Factors

In Figure 11, we see that the mean expected waste generation is roughly 3,350,000 pWU/year as measured by WI. By comparing this waste generation estimate to the waste generation estimate of another vessel, we could identify which vessel was more environmentally friendly and relatively by how much. In Figure 12, the sensitivity chart for the waste generation is presented. We see that every factor showing up is related to fuel consumption. Thus, the consumption of fuel is the most important waste driver. It is also the most significant energy and cost driver. This means that by reducing the fuel consumption the overall environmental impact of the vessel can be reduced while reducing the costs. <u>A win-win-win situation</u>.

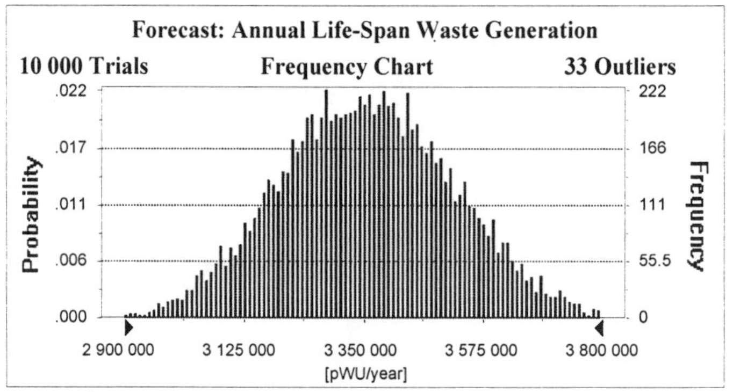

Figure 11 - The Annual Waste Generation for an UT 705 Platform Supply Vessel.

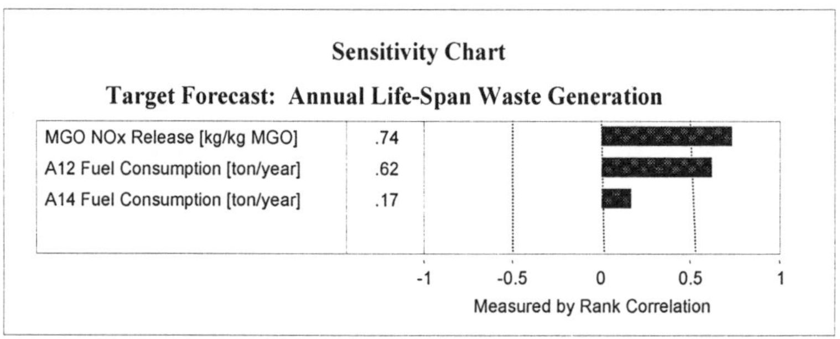

Figure 12 - Sensitivity Chart for the Annual Waste Generation.

When it comes to the actual releases, we see that the NO_x releases have the highest environmental impact. The reason is that the natural amount of NO_x is small, but the release is (relatively) large, which means that a relatively large imbalance is created. How this imbalance affects the environment we can only guess, but according to the WI the NO_x is the worst effect associated with MGO waste releases.

Because the waste drivers in Figure 12 are so dominant, the model is run once more to find out which waste drivers are next important. The results are given in Figure 13 and we see that fuel related factors are again on the top; the CO_2 release, the waste content of the fuel, then the A11 consumption of it, the SO_2 and CH_4 releases associated with fuel consumption. Thus, fuel consumption is *the* dominant source of waste. The CO_2 is dominant due to its long balance time in Nature, that is, it takes a long time to degrade.

By running the model a third time new waste factors can be detected, see Figure 14, and the last fuel-related effects show up; the CH_4 and the CO releases, followed by the lubrication oil waste content and consumption.

So far nothing has indicated any importance of the use of self-polishing coat. This is because the effects associated with TBT are the least important on a holistic scale.

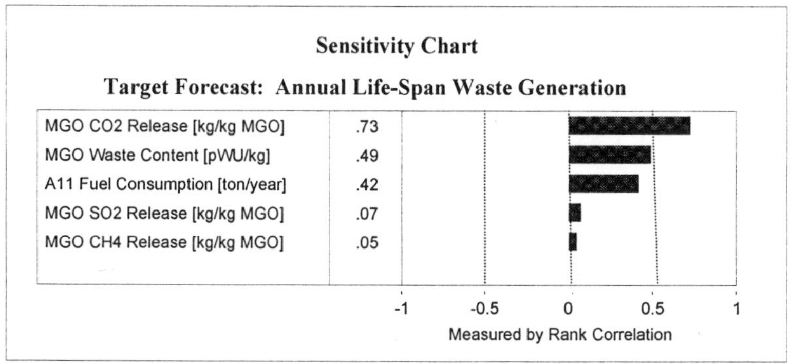

Figure 13 - Second Run Sensitivity Chart for the Annual Waste Creation.

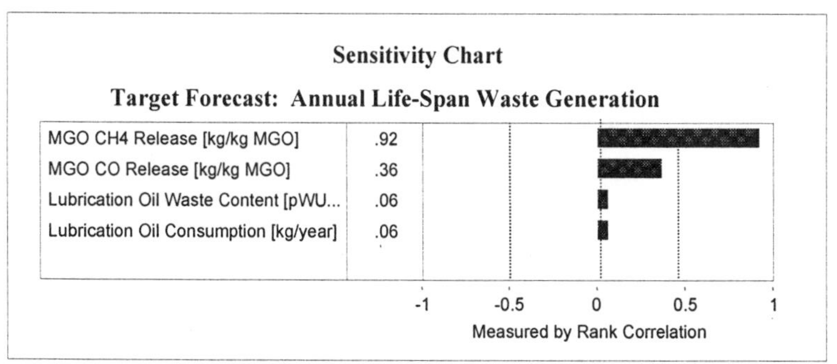

Figure 14 - Third Run Sensitivity Chart for the Annual Waste Generation.

4.3.2 Economic Waste Efficiency

The mean for the Economic Waste Efficiency (EWE) for the UT 705 PSV is 0.091 pWU/NOK according to our model. As will be evident, this is also very high. The reasons are the same as for the EEE. The sensitivity chart for the EWE is given in Figure 17.

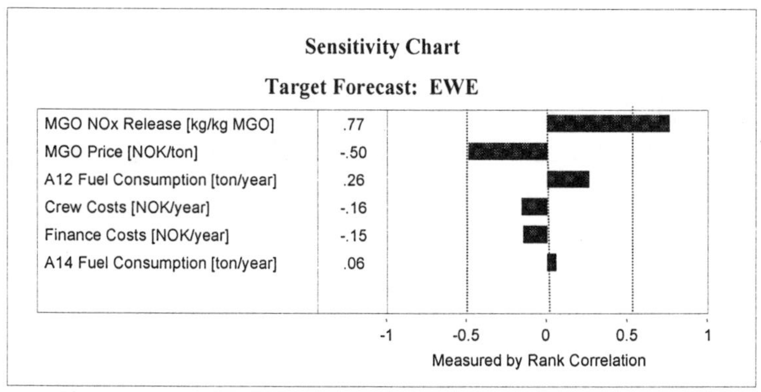

Figure 17 - UT 705 Platform Supply Vessel EWE Sensitivity Chart.

4.4 Identifying Win-Win Situations

Recall the various sensitivity charts for the various runs. It turns out that after seven runs all assumption cells are traced. A complete list of all the critical success factors in order of importance and how they contribute to the three assessment dimensions (cost, energy and waste) is given in Table 10.

We see that the energy dimension is the simplest, in terms of number of factors to model and keep track, while the costing perspective is the most comprehensive. This is mostly due to the amount of data available, assumptions made, and comprehensiveness of the modeling effort.

Management can use the information in Table 10 to identify economic and environmental win-win situations. Clearly, reducing MGO fuel consumption is the biggest win-win factor to focus on.

Table 10 - List of the Critical Success Factors for the UT 705 PSV.

	Cost Factors	Energy Consumption Factors	Waste Generation Factors
1	MGO Price [NOK/ton]	MGO Energy Content [MJ/ton]	MGO NO_X Release [kg/kg MGO]
2	A12 Fuel Consumption [ton/year]	A12 Fuel Consumption [ton/year]	A12 Fuel Consumption [ton/year]
3	Finance Cost [NOK/year]	A14 Fuel Consumption [ton/year]	A14 Fuel Consumption [ton/year]
4	Crew Cost [NOK/year]	A11 Fuel Consumption [ton/year]	MGO CO_2 Release [kg/kg MGO]
5	A14 Fuel Consumption [ton/year]	Lubrication Oil EC [MJ/kg]	MGO Waste Content [pWU/kg]
6	A11 Fuel Consumption [ton/year]	Lubrication Oil Consum.. [kg/year]	A11 Fuel Consumption [ton/year]
7	Insurance Cost [NOK/year]	Shipyard Work EC [MJ/year]	MGO SO_2 Release [kg/kg MGO]
8	Shipyard Work Cost [NOK/year]		MGO CH_4 Release [kg/kg MGO]
9	Lubrication Oil Cons. [kg/year]		MGO CO Release [kg/kg MGO]
10	Lubrication Oil Price [NOK/kg]		Lub. Oil Waste Content [pWU/kg]
11	Daily Revenue Loss [NOK/year]		Lubrication Oil Cons. [kg/year]
12	A15 Running Hours [h/year]		Shipyard Coat Use [kg/year]
13	Annual EO Class [NOK]		Coat p-Xyl. Release [kg/kg Coat]
14	Annual EO Class No. of Times		Coat Trimet.ben. Rel. [kg/kg Coat]
15	EO Class [NOK]		Operational TBT Release [kg/year]
16	EO Class No. of Times		Shipyard TBT Release [kg/year]
17	Annual IOPP Cert. No. of Times		
18	Annual IOPP Certificate [NOK]		
19	IOPP Certificate No. of Times		
20	IOPP Certificate [NOK]		
21	Intermediate IOPP Cert. [NOK]		
22	Interm. IOPP Cert. No. of Times		

5. ANALYSIS OF BIOLOGICAL DIVERSITY DEGRADATION AND HUMAN HEALTH RISKS

Chance favors only the prepared mind.

Louis Pasteur

Once we have gone through the effort of gathering all the information and creating an Activity-Based Cost, Energy and Waste model, we can also do some other analyses with the same data set. In this section, we show how a simple analysis can be done whether the releases from the PSV cause risk with respect to biological degradation and human health. There are much more sophisticated analyses available for determining risks with respect to biological degradation (BD) and human health (HH), but we wanted to show that simple analyses are not out of your reach.

In Table 11, we list some threshold values (in kg/m^3) above which releases can cause risk. Let's assume that we want to check whether any risk occurs in either the North Sea or in the shipyard harbor. For this, we assumed the North Sea control volume to be 53,730 km^3 and the harbor control volume to be 0.025 km^3. Given these control volumes and the threshold levels (in kg/m^3) listed in Table 11, we can calculate the total allowable threshold levels in kilograms for these volumes. For example, to pose any human health risks releases of trimethylbenzene must exceed 6.60E+09kg in the air of the North Sea basin.

Table 11 - Biological Diversity (BD) and Human Health (HH) Risk Indicator Calculations.

Waste Element	Threshold Amount [kg/m^3] (Appendix A)	Total Allowable Amount [kg]	Actual Release [kg]
MGO (North Sea)			
CO_2	9.00E-03 (HH)	4.84E+11	9.75E+03
CO	2.86E-05 (HH)	7.16E+02	2.40E+01
SO_2	2.82E+08 (HH)	2.82E+08	3.00E+01
NO_X	4.70E-05 (HH)	2.53E+09	1.65E+02
Coat (North Sea)			
Trimetylbenzene	1.23E-04 (HH)	6.60E+09	1.25E-01
p-Xylene	4.34E-04 (HH)	2.33E+10	7.92E+01
Tributyltin (TBT)			
Shipyard	8.00E-10 (BD)	***2.00E-02***	***2.64E+01***
Open Sea	8.00E-10 (BD)	4.30E+04	2.64E+01

If we compare these maximum allowable levels with the actual releases, then we see that only the TBT release in the shipyard can potentially cause a risk. The reason for distinguishing TBT releases at shipyards from the ones at open sea is the enormous amount of TBT releases during maintenance. Thus, even though TBT has no significant large-scale effects, it is the only release that has potential local effects. Therefore, if shipyards operate in areas with stronger sea currents and more water, the bad local effects of maintenance in the shipyards will be reduced.

With respect to TBT, in 1992 the San Francisco Bay Regional Water Quality Control Board set the biological diversity reduction concentration level for San Francisco Bay Area for Tributyltin (TBT; C_3H_9ClSn) to be 5.0ppt (Department of Pesticide Regulations 1996). In the UK, a 0.8 ng Sn/liter limit is set (Matthiessen, Waldock *et al.* 1995). It is suspected that damage of oyster colonies and imposex (i.e, induction of male characteristics in the female) in coastal dogwhelk (*Nucella lapillus*), and (in extreme cases) suppression in breeding is attributable to TBT releases. In 1989, TBT was banned in coating used on vessels smaller than 25 meters (Council Directive 89/677/EEC 1989). Roughly six years later (Evans 1995; Waldock 1995) and (CEFIC 1996) reported that biological recovery was observed in many coastal dogwhelk and oyster populations. This is why we assume that the natural degradation time for TBT is roughly 6 years.

We have used the UK limit for TBT because it is the most conservative. According to (Davies, Bailey *et al.* 1998), 68 tons of TBT is annually released in the North Sea by shipping activities. Given the volume of the North Sea, this gives an average of 1.15 ng TBT/liter, or 0.61 ng Sn/liter, which is just slightly lower than the UK limit. Hence, one can argue that the North Sea, certainly regions of it, is becoming dangerously contaminated.

6. ACTIVITY-BASED MASS ASSESSMENTS

Every disadvantage has its advantage.

Johan Cruyff

In this case study we made a straightforward Activity-Based Cost, Energy and Waste model without any further discussions on what additional (to the three dimensions we always use; cost, energy consumption and waste generation) performance measures could have been employed. Most commonly, such discussions will revolve around two issues; 1) environmental indicators and 2) using mass.

6.1 Using Mass as an Environmental Indictor

Using mass as an indicator, as discussed in earlier chapters, has a major disadvantage; the impact of the various materials on the environment is ignored and one could say that we are comparing apples and pears. This disadvantage is coupled by an equally major advantage; mass data are easy to obtain. So, we are left with the issues of correctness versus data availability.

We realize that many of our readers are novices in the area of environmental assessments (who isn't?) and may only have mass data available. So, in order to take away any excuse you may have for NOT doing an environmental analysis, we will show you how easy you can use mass as a measure of environmental impact in our Activity-Based Cost and Environmental Management method.

When it comes to using mass as indicator, however, we must distinguish between input mass and output mass, and two distinct situations are possible:

- *Use input mass*, either with or without (upstream) waste content included.
- *Use output mass*, either with or without (upstream) waste content included.

"Wait a minute", you may think, "Isn't mass 'in' supposed to equal mass 'out'?" Yes, but many people in business tend to forget a whole bunch of mass coming in, such as the oxygen needed to burn fuel. The main reason why these input mass-streams are 'forgotten' is because you don't have to pay for it directly (where is oxygen on the general ledger or Bill of Materials?) Nevertheless, even when we have the mass balance correct, still there will be a difference between the type of input and output mass-streams because chemical reactions occur. Thus, the composition of the input-mass is different from the output mass, although the total mass amounts are equal. In the following two sections, we will use both approaches to illustrate a) the flexibility and consistency of the Activity-Based Cost and Environmental Management, and b) the difference in results from using the different mass measures.

6.2 Using Input Mass

Using input mass as a measure is very simple. In Table 13, we provide an activity mass consumption table that shows how the resources in terms of mass are consumed by the activities. Note that compared to Table 7, 8, and 9, only the resource quantification and consumption intensities have changed. This is exactly what we already described in Chapter 4, Section 1. This view on input mass is focusing on tangible items that are also associated with monetary cost. This activity-based 'input mass' assessment is therefore more suitable for traditional waste management that focuses on 'trash' reduction. For example, in an office you may need to reduce the use of paper but you do not really know what causes the paper consumption. An analysis similar to the one in Table 13 would enable you to find out what activities cause the most input mass consumption, which allows you to target the most significant activities.

6.3 Using Output Mass

If we use output mass as a measure of waste, we should include all releases. In Table 5, we find the output elements and their mass that a particular input mass resource is responsible for. Using this information, we can compute the total amount of the various output releases for the PSV resources and the results are given in Table 12. Simply looking at this table, we can conclude that based on mass as environmental indicator, the CO_2 output from MGO consumption is overshadowing all other environmental factors. NO_x from MGO is second. Table 12 represents yet another way to quantify the resources and the consumption of the resources can be traced to the activities in the usual fashion. We have done this in Table 14 and we can see that Activity A12 consumes the most mass by far and more specifically that MGO is the biggest mass driver. It is clear from Table 13 and Table 14 that input mass $\neq$ output mass in this example.

Table 12 - Output Mass from Operating the PSV.

Waste Element	Unit Release [kg/kg]		Total Release [kg/year]
Lubrication Oil (19,024 kg)	CO_2	3.18E+00	60,496.3
	CO	7.83E-03	149.0
	SO_2	9.78E-03	186.1
	CH_4	2.45E-03	46.6
	NOx	5.38E-02	1,023.5
Total			**61,901.4**
MGO (3,067,000 kg)	CO_2	3.18E+00	9,753,060.0
	CO	7.83E-03	24,014.6
	SO_2	9.78E-03	29,995.3
	CH_4	2.45E-03	7,514.2
	NOx	5.38E-02	165,004.6
Total			**9,979,588.6**
Shipyard work			
Coat (454.4 kg)	Trimetylbenzene	2.76E-04	0.1
	p-Xylene	1.74E-01	79.1
Tributyltin (26.4 kg)	Tributyltin	1.00E+00	26.4
Total			**105.6**
Total			**10,041,595.6**

Table 13 - Activity Mass Consumption (MC).

				Activities																	
				A11		A12		A13		A14		A15		A2		A31		A32		A33	
Resource	[kg/ year]	Resource Driver	Consumption Intensity	Driver level	MC	Driver level	MC	Driver level	MC	Driver level	MC	Driver level	MC	Driver level	MC	Driver level	MC	Driver level	MC	Driver level	MC
Annual EO Class	0	Activity Frequency	0 [kg/ instance]											1.0	0						
EO Class	0	Activity Frequency	0 [kg/ instance]											0.2	0						
Annual IOPP Certificate	0	Activity Frequency	0 [kg/ instance]											1.0	0						
Intermediate IOPP Certificate	0	Activity Frequency	0 [k/ instance]											0.2	0						
IOPP Certificate	0	Activity Frequency	0 [kg/ instance]											0.2	0						
Lubrication Oil	19,024	Running Hours	2.19 [kg/h]	2,705	5,920	2,678	5,861	0	0	3,309	7,242										
MGO Fuel	3,067,000	Fuel Consumption	1,000 [kg/ton]	113	113,000	2,292	2,292,000	0	0	662	662,000										
Overhead Resources	0	Running Hours	0 [kg/h]	2,705	0	2,678	0	0	0	3,309	0	8	0								
Shipyard Work	481	Labor Hours	0.068 [kg/h]													3,500	240	3,500	240	20	1.4
Total	3,086,505				118,920		2,297,861		0		669,242		0		0		240		240		1.4

Table 14 - Activity Mass Waste Generation (WG).

				Activities																	
				A11		A12		A13		A14		A15		A2		A31		A32		A33	
Resource	[kg/ year]	Resource Driver	Generation Intensity	Driver level	WG	Driver level	WG	Driver level	WG	Driver level	WG	Driver level	WG	Driver level	WG	Driver level	WG	Driver level	WG	Driver level	WG
Annual EO Class	0	Activity Frequency	0 [kg/ instance]											1.0	0						
EO Class	0	Activity Frequency	0 [kg/ instance]											0.2	0						
Annual IOPP Certificate	0	Activity Frequency	0 [kg/ instance]											1.0	0						
Intermediate IOPP Certificate	0	Activity Frequency	0 [k/ instance]											0.2	0						
IOPP Certificate	0	Activity Frequency	0 [kg/ instance]											0.2	0						
Lubrication Oil	61,901	Running Hours	7.12 [kg/h]	2,705	19,264	2,678	19,072	0	0	3,309	23,565										
MGO Fuel	9,979,589	Fuel Consumption	3,254 [kg/ton]	113	367,686	2,292	7,457,847	0	0	662	2,154,055										
Overhead Resources	0	Running Hours	0 [kg/h]	2,705	0	2,678	0	0	0	3,309	0	8	0								
Shipyard Work	106	Labor Hours	0.015 [kg/h]													3,500	53	3,500	53	20	0.3
Total	10,041,596				386,950		7,476,919		0		2,177,621		0		0		53		53		0.3

6.4 Mass as a Measure - Pros and Cons Revisited

"Great", you may think, "I can use mass without a problem!" Careful! *It depends*. Look back at the results we showed in Sections 3 and 4. Using mass as environmental indicator can lead to somewhat 'wrong' conclusions:

- Compared to the results from using the WI, we see mass does not capture the *time* aspect of environmental impact. It also does not look at how much the natural balance is disturbed. In Figure 12, the MGO NO_x release was identified as the main waste factor contributing to the waste, not CO_2 identified with the output mass-based assessment.
- We also 'conveniently' forgot to include one very important aspect; the upstream waste content. This can be a significant factor.
- Also look at the difference in values between simply using input mass versus total output mass. Basically, our input mass balance is incorrect.

But, it is clear that using mass as an indicator did not lead to completely different results. Hence, mass can serve as an environmental indicator in this case. The reason is that some mass drivers/releases dominate others by magnitudes and that this domination makes the 'fine-tuning' provided by bringing the time aspect into the equation *almost* redundant. The generation of mass by activities is, however, not far off the mark, but that is because MGO is the largest cost, energy, waste and mass driver.

In the general case, however, we cannot expect some mass drivers to overshadow others to the extent shown in this case study. Using mass as an environmental indicator should therefore be preceded by a more thorough WI assessment up-front to estimate the distortion of using mass as indicator. Our opinions are in keeping with conventional ISO 14000 LCA guidelines that also state that mass-based LCI provide good indications of environmental impact, but are to be treated with care.

The problem of incomparability will always be present when using mass as an environmental indicator because different chemical compounds are incomparable from an environmental impact perspective, and comparability is an indispensable property for those that want to prioritize and make decisions accordingly. Therefore, a very important issue to consider is what the purpose of the assessment is. In cases where companies can identify significant areas of improvements by simple means and common sense, a full-blown Activity-Based Cost, Energy and Waste model implementation does not make much sense. But companies that want to systematically improve their products, processes, and their management of business processes, will inevitably need an integrated and systematic approach that can capture all the crucial economic and environmental aspects of their company. ABCEM is in our opinion a method that can most effectively and efficiently do this whether mass is used as indicator or not.

6.5 Using Environmental Indicators as Measures

We hope that by now it is clear to you that our ABCEM method does not restrict you in your choice of assessment dimensions and/or units of measurement. If desired, you could use conventional environmental indicators such as the Dutch Eco-Indicator'95 (or now Eco-Indicator'99) or the Swedish Environmental Priority System (EPS). However, be warned; the Eco-Indicator '95 failed in this case study in 1996 because (then) the PSV activities and materials were not contained in SimaPro's Eco-Indicator database, and the complexity of composing an Eco-Indicator value for a new material or process is daunting. Other similar indicators face the same problems. That is why we choose a step in between by using the Waste Index, which contains information that is in our opinion easier obtainable and always comparable.

7. CRITICAL EVALUATION AND CLOSURE

Man, from the activity of his mental faculties, cannot avoid reflection; past impressions and images are incessantly and clearly passing through his mind.

Charles Darwin
In "The Descent of Man"

Clearly, we did create a model in this case study that produced some results; hence Activity-Based Cost and Environmental Management can be used. However, it is probably more interesting to show how *well* it works. To do that, we need to look at several criteria that can indicate the quality of the model. We chose the following:

Is the model logically constructed? First we checked the logic by examining the various links and computations in the model. There were no errors. Then, we investigated the sensitivity charts closely, and from all the sensitivity charts presented in this case study we were unable to identify any logical problems. Also the results seem logic. Consequently, we can conclude that the model is indeed logically set up.

How does the model respond to changes? The model also responds to changes in a logical fashion. That is, if a variable in the model is changed, the model must behave in a reasonable way. This was also tested using the sensitivity charts.

Did the customers, i.e. Farstad, like the model? Technical managers at Farstad *did* find the cost model satisfactory. The environmental part of the Activity-Based Cost, Energy and Waste model presented in this chapter was never shown to them because the project was long finished by the time the model presented here was completed. But, their initial approval is indeed an indication of quality. For the costing part, Farstad's technical managers were involved in supplying data and checking the model logically and they approved the model in 1996 when the comprehensive version in (Fet, Emblemsvåg *et al.* 1996) was designed.

How much did it cost? The final issue is that the model has to be set up in a reasonable amount of time within budget. In the project between Farstad and Møre Research, the budget was only $15,000. About 50% of this was allocated to the environmental LCA which was done using the conventional (ISO 14000) LCA method. In both the environmental and cost analysis, only the hull and main machinery were investigated, in accordance with Farstad. Taken into account that the conventional environmental model failed to produce useful results, while the initial ABC model produced satisfactory results, we can conclude, at least for this case, that the conventional LCA is not only more expensive to conduct, but also produces inferior results.

8. WHAT IS NEXT?

Computers are useless. They can only give you answers.

Pablo Picasso

Although this was a real case study was a real, the model is still simple enough that you can actually implement in a spreadsheet yourself and 'play' with it. We also showed you that:

- You can use the ABCEM approach for assessing the cost and environmental aspects of the use-phase of products. This can be particular useful for those readers who are in the business of fleet-management.
- You can use input mass and output mass as measurement dimensions too, but you have to be careful with interpreting the results. Environmental indicators can also be used, provided you can find the data.

Next, we will go back to a manufacturing context; to a case study is where the ABCEM implementation focused on energy and waste assessments of a real US company that wants to be leading what they call the 'Second Industrial Revolution' towards sustainable development; Interface Flooring Systems, Inc.

9. ACKNOWLEDGMENTS

We would like to thank Farstad Shipping ASA and Møre Research for allowing the use of information and results from the project carried out between the two in 1996.

Chapter 7

THE INTERFACE CASE STUDY

Whatever you can do, or dream you can, begin it.
Boldness has genius, power and magic in it.

Goethe

Many readers may not know Interface Flooring Systems, a carpet manufacturer located in Troup County, Georgia, USA. But those who are familiar with the Sustainable Development movement in the United States, may recognize the name Ray C. Anderson. He is the founder of Interface, Inc., and a visionary who seeks to lead his company into what he calls the Second Industrial Revolution. In late 1998, we were given the opportunity to create an Activity-Based Energy and Waste model for Interface Flooring Systems' plants in Troup County. In this chapter, we show you what we did for Interface Flooring Systems, Inc.

1. PURPOSE OF THIS CASE STUDY

What's the gas mileage of your company?

Dave Gustashaw

What is unique about this case study? First of all, this case study demonstrates that you do not have to use a cost dimension, but one could include only energy and waste in the assessments.

In this case study, we also illustrate how you can depreciate building and equipment not only financially, but also from an energy and waste content perspective. That is, you can use the same principles from finance for depreciating, say, the energy spent in building a factory. This is something that is rarely done because traditional environmental management and assessment typically only focus on the direct impacts attributed to the products.

For those interested in quality and ISO 9000, we will also show how the Activity-Based Energy and Waste models can be used to assess the environmental impact from quality loss - a classic management performance metric.

Plus we show how even the best leading environmentalist may sometimes be misguided in their prioritization of the problems. Specifically, we compare our findings to Interface's internal goals and note that our findings are quite different than what Interface thought was important. One objective of the case study was to compare our findings to Interface's own suggestions in the PLETSUS program, see (Anderson 1998), to see if our Activity-Based Environmental Management approach could add value to the management and design efforts towards sustainability.

2. DESCRIPTION OF INTERFACE

All men of sense weigh, consider, and use great circumspection, before they enter upon any private business of momentous consequence.

Erasmus
In "The Complaint of Peace"

We start with a company overview, followed by a presentation of the facilities and the product lines. Up-front we should state that the model shown in this chapter was based on 1997 product and process data. Since 1997, a number of the processes have changed significantly, so the model and results presented here are not representative of current activities.

2.1 Company Overview

Interface, Inc. is a holding company for Interface Flooring Systems, Inc. We use 'Interface,' for simplicity as a denotation for Interface Flooring Systems, Inc. Interface is one of the largest modular carpet tile manufacturers in the world, with roughly 40% market share in the commercial segment. The product lines, which are offered in 110 countries, include carpet tiles, commercial broadloom carpet, textiles, chemicals and architectural products. Interface, Inc. was founded in 1973 by Ray C. Anderson, went public in 1983 on NASDAQ (the technology index) on the New York Stock Exchange and in 1988 was listed as a Fortune 500 company. Today, the company has 29 manufacturing sites in the United States of America, Canada, the United Kingdom, The Netherlands, Australia and Thailand. The annual turnover in 1997 was $1,135,290,000 out of which 70% was generated in the Americas, which gave a profit (before tax) of $61,271,000. So far, Interface seems like just another major corporation. However, in August 1994, Chairman and Chief Executive Officer of Interface - Ray C. Anderson - read Paul Hawken's book titled *The Ecology of Commerce* (Anderson 1998):

> *I read it. And it changed my life. It was an epiphany. I wasn't halfway through it before the vision I sought became clear, along with a powerful sense of urgency to do something. Hawken's message was a spear in my chest that remains to this day.*

Interface has since then sought to become a *sustainable* company by several initiatives such as waste elimination, using renewable energy and basically reengineering the whole company. In (Interface 1998) the strategy is outlined, a strategy based on the Natural Step, see Chapter 2. Interface's manufacturing sites in Troup County are leading in the quest towards sustainability. In this case study, we are examining these manufacturing sites more closely.

2.2 Facilities

There are six facilities that make up Interface in Troup County:

1. The 74,000 ft^2 Vaughan Warehouses at the city of West Point store raw materials for the Ray C. Anderson Plant (until late 1997).
2. The 113,000[1] ft^2 Ray C. Anderson (R.C.A.) Plant in West Point where the actual manufacturing process starts. West Point is roughly 13 miles away from the other facilities, which are located in La Grange.
3. The 147,500 ft^2 Kyle Plant is the main manufacturing plant in La Grange. On a side wing of the Kyle Plant we also find the administrative building.
4. The 54,000 ft^2 Kyle 2 Plant, a.k.a. Building 10, is next door to the Kyle Plant. Until late 1997, this plant was mostly used for printing and beck dying.
5. The 100,000 ft^2 Graham Scott Technical Center (G.S.T.C.) houses the design activities. Located next door to both Kyle and Kyle 2 plants, it was also used as storage in 1997 for rolled goods and for tufting yarn for fusion bonded products. It was in 1997 also partly empty.
6. The 39,000 ft^2 Finished Goods Warehouse (F.G.W.) is where all the finished tile goods are kept before shipping to customers. This facility is also located near to the three other facilities in La Grange.

We should point out that some of these facilities have changed their purpose since 1997. For example, in Building 10 (the Kyle 2 plant), the printing and beck dying have been removed and a new production line is now in operation.

2.3 Product Line Overview

In Troup County, Interface produces 'Modular' and 'Broadloom' products. There are hundreds, if not thousands, of variants of these products (one of Interface's niche market is custom-ordered and custom-made carpet). Hence we need to group them into some product lines. We used the following ten product lines:

1. Tufted 1/10 gage Slat tiles, which we refer to as '<u>1/10 Slat Tiles</u>'. The tiles are mostly 50 cm by 50 cm in dimensions.
2. Tufted 1/10 gage Slat rolls, which are denoted '<u>1/10 Slat Rolls</u>'.

---

[1] The Ray C. Anderson Plant was expanded by roughly 50% on December 13, 1997. Before that expansion the plant was 113,000 ft^2.

3. Tufted 5/64 gage tiles. This product line is referred to as simply '5/64 Tiles'.
4. Tufted 5/64 gage rolls, which we refer to as '5/64 rolls'.
5. Tufted Intersculpt 1/12 gage tiles, which are denoted 'Intersculpt 1/12 Tiles'.
6. Tufted Intersculpt 1/12 gage rolls. This product line is denoted 'Intersculpt 1/12 Rolls'.
7. Fusion I-bonded tiles which we call 'I-Bond Tiles'.
8. Fusion I-bonded rolls referred to as 'I-Bond Rolls'.
9. Fusion U-bonded tiles. This product line is called 'U-Bond Tiles'.
10. Fusion U-bonded rolls, which are denoted 'U-Bond Rolls'.

Clearly, there are two major lines: tufted and fusion bonded. The tufted product lines are dominating sales and can essentially be described as being manufactured by huge two-dimensional sewing machines, aptly called tufting machines. The fusion bonded product lines are the oldest, but in 1997 they only accounted for seven percent of production volume and no fusion bonded rolled goods were manufactured. Fusion bonded products are manufactured by basically gluing short threads of yarn perpendicular to a backing.

We measure production volume in square yards [yd^2] and not in physical discrete units like in all other case studies in this book. This is convenient because carpet is not produced in comparable discrete units, e.g., tiles or rolls. Square yards are also the measurement that Interface uses to manage production and account for costs and product sales. Thus, using a familiar metric makes collecting and verifying data easier for everyone.

3. DEVELOPING AN ACTIVITY-BASED ENERGY AND WASTE MODEL FOR INTERFACE

An expert is someone who knows some of the worst mistakes that can be made in his subject and how to avoid them.

Werner Heisenberg

In this section, we will again follow the steps discussed in Chapter 4, Section 2. As you will see, the Activity-Based Energy and Waste model for this case study is more elaborate than the preceding case studies. For the purpose of brevity, and also because of some confidentiality issues, we do not describe all the details of how we built the model like we did in the preceding chapters.

3.1 Step 1 - Identify the Assessment Objects

The main goals of the case study were to create an activity-based model for assessing the energy consumption and waste generation of Interface in Troup County in 1997, and to identify the environmental critical success factors of the product lines and the manufacturing process. We were also interested in finding out what the impact is of loss of quality in terms of energy and waste, as well as what the aggregated impact of the facility (building) itself is in terms of energy and waste. The impact of facilities is often excluded/ignored in conventional LCAs and other environmental assessments. Given these goals and interests, we identified the following 13 assessment objects for which we want to assess the energy consumption and waste generation:

- Interface Flooring Systems, Inc, in Troup County as a whole
- The ten product lines.
- The facility.
- The quality loss (which can also be seen as a performance measure).

Regarding production data and product line definitions, we assumed that the production volume ratio between tiles and rolls is the same for all tufted product lines. In 1997 the rolls made up 14.60% of total production volume for tufted product lines. The product lines are also distinguished in terms of tufted versus fusion bonded. Furthermore, for the tufted products the product lines can further be organized in terms of the yarn gage while the fusion bonded product lines can be broken into I-bonded and U-bonded.

In this case, the management at Interface chose the system boundaries to be everything that is in their control, except administrational resources. Thus, overhead related to marketing, finance, etc. and overhead related to production is excluded from the modeling. However, overhead resources that are related to facilities *are* included. Hence, this case study is a case study where overhead issues are mostly excluded and it can thereby, to some extent, be called a *variable absorption* model, i.e. only the traditional variable resources - resources that are consumed proportionate to volume - are included. In ABC, we do not use the terms *variable* and *fixed* in the traditional sense. Thus, we prefer to call this a *partial absorption* model (as opposed to *full absorption*). Actually, having a partial instead of a full absorption model allows us to illustrate that Activity-Based Cost, Energy, and Waste models can be scaled up or down in size and complexity to whatever you are interested in.

3.2 Step 2 - Create an Activity Hierarchy and Network

Compared to our case studies, defining the activity network and the activity hierarchy for Interface in Troup County was the most difficult. In Table 1 we present the activity hierarchy.

Table 1 - Interface Troup County 1997 Activity Hierarchy.

Level 1	Level 2	Level 3
Use Vaughan Warehouse **A1**	Receive Tufting Raw Materials **A11**	
	Keep Warehouse **A12**	
	Ship to Ray C. Anderson Plant **A13**	
	Transport to R.C. Anderson Plant **A14**	
Use G.S. Technical Center **A2**	Receive Yarn **A21**	
	Keep Warehouse **A22**	
	Warp Yarn **A23**	
	Ship Yarn to Kyle **A24**	
	Transport Yarn to Kyle **A25**	
Produce at R.C.A. Plant **A3**	Receive Materials **A31**	Receive Yarn **A311**
		Receive Chemicals **A312**
	Produce Products **A32**	Warp Yarn **A321**
		Tuft Yarn **A322**
		Shear 1 Out of 4 Interfaces **A323**
		Coat Back **A324**
	Ship Products **A33**	Ship to La Grange **A331**
		Transport to La Grange **A332**
Produce at La Grange **A4**	Receive Materials **A41**	Receive from R.C.A Plant Kyle **A411**
		Receive Backing Raw Materials **A412**
		Receive from R.C.A Plant Kyle 2 **A413**
		Receive Chemicals Kyle 2 **A414**
	Mix Chemicals **A42**	Mix Backing Chemicals Kyle **A421**
		Mix Dye Chemicals Kyle 2 **A422**
	Keep Work In Progress (WIP) **A43**	Keep Tufted WIP **A431**
		Keep Fusion Bonded WIP **A432**
	Produce Tufted Products **A44**	Produce on Backing Line 1 **A441**
		Inspect/Cut Rolls **A442**
		Produce on Backing Line 4 **A443**
		Cut Off-line **A444**
		Inspect/Sort after Back. Line 4 **A445**
		Package Tile Products Line 4 **A446**
	Produce Fusion Bonded Products **A45**	Make I-Bond (Guillotine) **A451**
		Make U-Bond (Pleater) **A452**
		Split in Two **A453**
		Produce on Backing Line 2 **A454**
		Inspect after Backing Line 2 **A455**
		Package Tile Products Line 2 **A456**
	Print Pattern and Package Products **A46**	
	Produce Some 5/64 **A47**	Use Beck Dye **A471**
		Ship to R.C.A Plant **A472**
		Transport to R.C.A Plant **A473**
	Reinspect and Package Products **A48**	
	Send to Finished Goods Warehouse **A49**	Send from Kyle **A491**
		Send from Kyle 2 **A492**
Handle Finished Goods **A5**	Store Rolls in Warehouse Annex **A51**	
	Store Tiles in F. Goods Warehouse **A52**	
	Ship to Customers **A53**	
Develop Products **A6**	Design Products **A61**	
	Design Custom Products **A62**	
	Test Products **A63**	
	Run Pilot Plant **A64**	

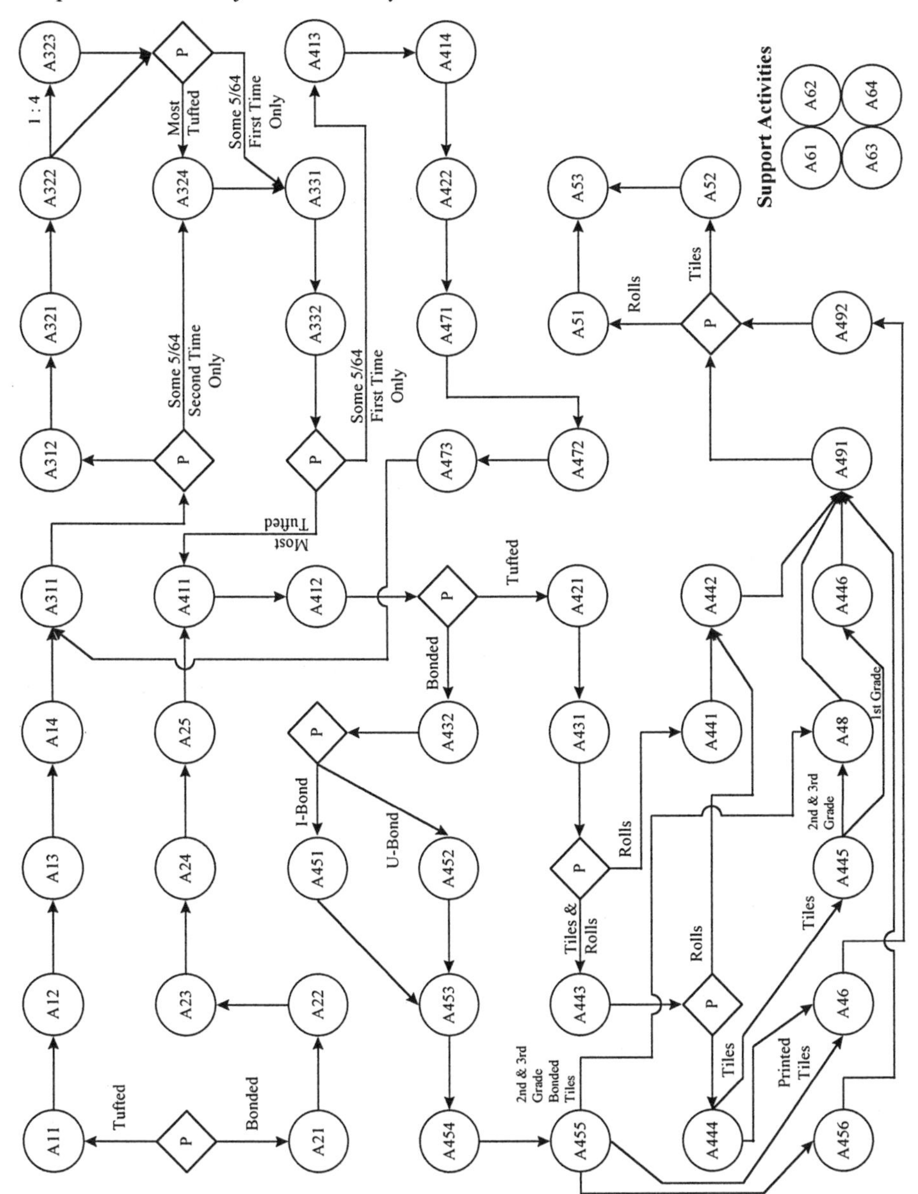

Figure 1 - Interface Troup County 1997 Activity Network.

Usually, the number of activities indicates the complexity of the organization, but the activity network of Interface, Troup County, see Figure 1, is the most complicated in this book. The reason is that the process is fairly complex with many different routing options depending on facility use, production rates, and product mix. There are 49 activities, which is about half the number of activities in the Westnofa Industrier case study in Chapter 8.

The activity network starts in lower left corner and ends in the upper right corner of Figure 1. The routing decisions are typically tufted versus bonded or tiles versus rolls. It is interesting to note the routing with the A471 through A473 activities. Basically, some of the 5/64 tufted products are being sent from one facility to the other, causing a lot of unnecessary material handling and transportation. So just from creating an activity network we can already suggest an improvement, which has already been implemented by Interface in Troup County. Another issue is that such a complicated manufacturing process can cause excessive production planning overhead and general information management becomes more difficult. Because our model (currently) does not include overhead issues, it is hard to assess how much of the overhead resources can be eliminated by reorganizing the manufacturing process. Nevertheless, this could be done quite easily because this is where standard Activity-Based Costing excels in, see Chapter 3.

3.3 Step 3 - Identify and Quantify the Resources

As usual, resource identification should include as much as possible.

3.3.1 Resource Overview

Table 2 contains the main resource categories that we identified based on the production assumptions listed in Step 1. In our model, these categories were further divided hierarchically in more detailed resource elements, but because of confidentiality reasons, we can only list the aggregated resources and their amounts in terms of energy consumption and waste generation. The actual model had over 50 resource elements that were identified and quantified ranging from water, gas and electricity usage to energy and waste content of production chemicals, machines and individual buildings.

Table 2 - Interface Annual Resource Consumption (1997).

Resource Category	Energy Consumption [MJ/year]	Percentage	Waste Generation [pWU/year]	Percentage
Depreciation and Maintenance	2,567,020	0.06 %	1,786	0.01 %
Buildings	344,008,015	8.63 %	700,917	2.86 %
Freight within Troup County	915,212	0.02 %	3,837	0.02 %
Material Usage in Production	3,637,859,399	91.28 %	23,827,025	97.25 %
Total:	**3,985,349,646**	100.00 %	**24,533,565**	100.00 %

With respect to quantities, it is very interesting to see the enormous amount of energy in the resource 'Material Usage in Production'. This is actually the energy content of purchased materials, i.e., the energy spent earlier in the value chain. This energy consumption constitutes 91.43% of the total energy consumption. Hence, we already know that most of the environmental impact due to energy consumption is outside Interface's direct reach. In other words, if Interface is to become sustainable, they must work closely with the suppliers to reduce the energy content of the materials they purchase and recycle the materials effectively.

The second largest category is the 'Buildings' category. This category contains the bulk of the energy spent directly by Interface, Troup County, which include electricity, the energy content of the electricity and the energy content of propane, natural gas and water. Also, estimated depreciation of the buildings are included, however, that contribution does not add much. In fact, the largest contributor (58.0%) is the energy content of the electricity and the electricity itself. This high number is primarily due to various losses on the distribution of electricity and also the energy used in generating the energy.

In Table 2 the resources related to waste generation are also presented. Clearly, as in the case of energy consumption, the largest waste improvement possibilities (97.25% for purchased materials) lie outside Interface's direct reach. Now, remember that in this case study overhead resources are not included. If they *had* been included, we would expect that the amount of waste generated and energy consumed by overhead resources to be around 30% of the total resource consumption because monetary overhead costs are in that range. In fact, the overhead energy consumption and waste generation would possibly be even higher because subsidies make the cost of raw material extraction artificially low (Brown, Flavin et al. 1999) and hence makes a lot of equipment, particularly with rare metals (e.g., computers), cheaper than might otherwise be.

3.3.2 Information Assumptions

Due to the lack of value chain wide environmental accounting systems, we had to recreate some of the value chain of the materials purchased by Interface. The data we have regarding Interface are as good as can be given the time and budget that we had available (two months and zero dollars, respectively, but a lot of help and support). The following assumptions were made for model:

- The energy content data and data used for computing the waste content of materials was taken from (Ashby 1992) and IDEMAT.
- The yarn consists of various types of nylon. We did not have any information regarding production of nylon; hence we approximated the waste content of nylon as the waste content of polyester because polyester and nylon have many similar characteristics.
- LUTRA 83, which is used as the primary backing of the carpet, consists mostly of polyester and therefore is approximated as such. The S/7615 Blue Label substrate, the 7611 Blue Label and the 2211 Bayex Scrm are approximated as E-fiberglass.

- The pre-coat material is BS-211 Latex Precoat. The composition of this material is proprietary and can therefore not be revealed here. However, the main ingredients are Ethylene Vinyl Acetate (EVA), Calcium Carbonate, Phosphate Ester, Sodium Laurel Sulfate (i.e. soap) and Poly Vinyl Chloride (PVC). Due to lack of data for most petrochemical compounds used at Interface, we had to make some approximations. We approximate EVA, Calcium Carbonate, Phosphate Ester, Sodium Laurel Sulfate (i.e. soap) ATH and Phthalate Esters as PVC.
- The chemical mixes/batches T-900, T-902, T-907, T-925 and T-926 are also proprietary, but they mainly consists of PVC, EVA, Calcium Oxide (CaO), Aluminum Trihydrate (ATH), Fumed Silica, Phthalate Esters, Carbon, Calcium/Zinc/Hydrocarbons. The latter was approximated simply as zinc. We use data for coal to approximate the data for Carbon. CaO is mined similar to coal. So we approximated the energy content of CaO as the energy content of coal.
- The gas used at Interface, Troup County, was approximated as natural gas from Birmingham, Alabama, and the data needed to calculate its waste and energy content were found in IDEMAT and (Reed 1978).
- Propane is the most common of two possible gases that are often called Liquefied Petroleum Gas (LPG). We therefore use LPG data in IDEMAT.
- Compared to car emissions, we assumed that emissions from propane combustion are 93% lower for CO and 57% lower for NO_x. These assumptions are based on data from the SIGECO Energy Company found at *www.ecofuels.com*.
- Car fuel information was obtained from David Hart, Centre for Environmental Technology, Imperial College, London, and Günter Hörmandinger, Energy Policy Research, London.
- The energy content of water was approximated by the energy content of gravel, found in (Ashby 1992). This sounds strange, but the resources needed to bring either water or gravel to production is very low, and was deemed equivalent.
- The amount and type of materials embodied in machines and buildings was established in consultation with Interface's Director (now Vice-President) of Engineering Dave Gustashaw. We assumed that machines and buildings were depreciated linearly at 10% and 3% per year, respectively. The energy consumption and waste generation of electricity, gas, propane, and water resources were aggregated into the buildings.

3.4 Step 4 - Identify and Quantify the Resource and Activity Drivers and their Intensities.

Table 3 contains most of the drivers used. The same drivers are associated with energy consumption and waste generation.

Table 3 - Main Resource Drivers and Activity Drivers.

Resource Drivers	Activity Drivers
Area	Direct charge to involved products
Asset effect	Production volume
Direct charge to involved activities	Yarn mass

Although the activity network is very complex and the product variety is high, which would normally lead to non-linearity, the whole manufacturing is remarkably volume based, thus linear. But, the linearity is only approximate, as there are batch differences due to setup times. However, these are so negligible that they are ignored. Also, batch information is not readily available and we therefore omitted it given the trade-off between the cost of the information, its utility, and our budget. Thus, 'production volume' is the single most important activity driver. Another important activity drivers is 'yarn mass', which is indicative of how much yarn is used.

The resource drivers are also few and intuitive. With 'Asset Effect' we refer to the electricity use [kW] of various assets such as machinery, lights and the like. With respect to resource driver values, the A441 ('Produce on Backing Line 1'), A443 ('Produce on Backing Line 4') and A444 ('Cut Off-line') activities required special attention due to their high consumption and generation levels. We relied on Dave Gustashaw and his understanding of these activities to develop adequate approximations. Ideally, sub-metering of electricity, natural gas, water and propane is needed to be able to trace the resources to the A441, A443 and A444 activities accurately, something that is currently being implemented.

There are two categories of consumption intensities and generation intensities:

- The simplest category is when the intensities are independent of the driver magnitudes. This occurs typically for direct resources such as materials.
- The other category is when the intensities are dependent on the driver values. Then we take, for example, the energy consumption attributed to a driver and divide it by the total number of driver units. In this case, the intensities are dependent on the driver magnitudes and they change, therefore, whenever the activity consumption and/or resource consumption changes.

This latter type of intensity allows us to investigate the effects of economies-of-scales and to assign indirect energy consumption (e.g., the energy of heating a building) to products. The temperature needed in a building is a factor of worker comfort and not production volume, but heating a building for producing 1000 yd^2 is more efficient than for only 10 yd^2. Thus, producing 1000 yd^2 results in a lower energy content for the products than if only 10 yd^2 were produced using the same building and the same heating.

3.5 Step 5 - Identify the Relationships between Activity Drivers and Design Changes

Because we are dealing with product lines here, it can be difficult to relate activity drivers to design changes. However, in this case we can look at certain product parameters that determine, e.g., the speed of the tufting machines, the backing lines and the tile lines. We did not do that here, because the overriding effect is (and will be) the energy (and waste) content of the materials. Hence, from a product design perspective, one should focus on reducing materials consumption. From this it intuitively follows that the 5/64 gage is the most environmentally friendly product since it is the least material intensive. More on this will follow in the result section.

3.6 Step 6 - Model the Uncertainty

As in the Farstad case study, we chose all uncertainty distributions to be triangular with ± 10% variation to facilitate sensitivity analyses.

3.7 Steps 7, 8 and 9 - Find/Compute the Cost, Energy Consumption and Waste Generation of Activities and Objects, and Perform Sensitivity and Other Analyses

In Figure 2, the working principle of the Interface model is shown. We have listed the main model characteristics, such as number of activities, number of resource drivers, objects, etc. Because we implemented the model in a spreadsheet, we discuss here steps 7, 8 and 9 of our method together. The results of our analyses are discussed in Section 4.

The energy and waste dimensions of the model are implemented as two separate Microsoft Excel® spreadsheet models. The Crystal Ball® software is again used to perform the Monte Carlo simulations and the sensitivity analyses on both models.

The energy model has 260 assumption cells and 19 forecast cells. The waste model has 271 assumption cells and 22 forecast cells. This gives a total of 531 assumption cells and 41 forecast cells, or 572 variables in total. The fact that there are 13 assessment objects in total, but the number of forecast cells is 41 is because we use extra forecast cells to keep track of several performance measures for every assessment object. For the 1/10 Slat Tile product line, for example, we have the following performance measures:

- 1/10 Slat Tile Unit Energy Consumption [MJ/yd^2]. This indicates how much energy is consumed per square yard 1/10 Slat Tiles. We use this measure to find out if the energy consumption is acceptable and how to reduce the energy consumption for this particular product line.

- 1/10 Slat Tile Contribution to Result (for energy) [%]. This measures the relative energy consumption of the entire 1/10 Slat Tile product line in relation to total energy consumption. Such information is important as a relative, aggregate product line measure.
- 1/10 Slat Tile Unit Waste Generation [MJ/yd^2]. This measure is similar to the unit energy consumption measure except it is for waste generation purposes.
- 1/10 Slat Tile Contribution to Result (for waste) [%]. This is the waste equivalent to the energy version discussed in point 2.

The number and type of performance measures are only limited by the creativity of the implementers. However, it is important not to create too many performance measures as they can become irrelevant, too narrowly focused and thereby fail to aid decision-makers, see also (Keegan, Jones et al. 1991).

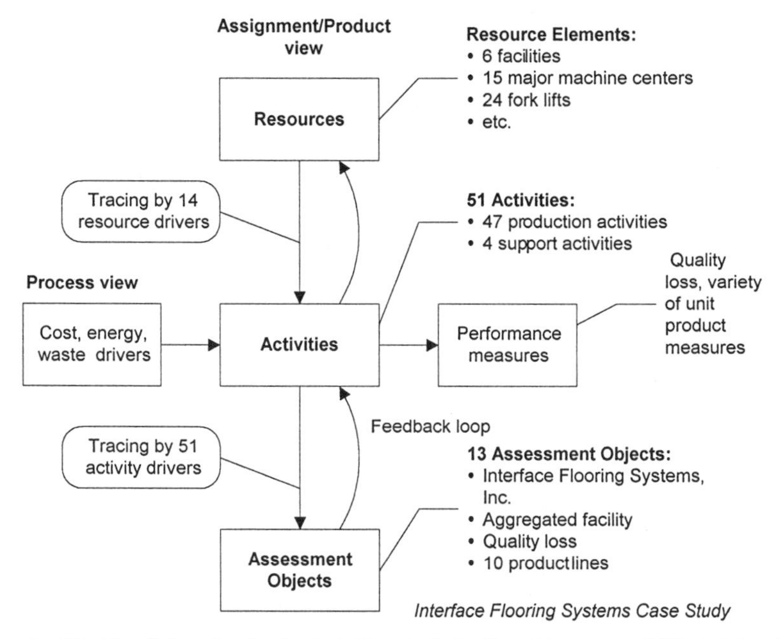

Figure 2 - Working Principles for the Interface Activity-Based Energy and Waste Model.

In Figure 3 an overview of the model infrastructure is shown. The arrows signify how information flows from which spreadsheet to which spreadsheet. The names of the spreadsheets are in bold. AEU is the energy consumption model for the product lines, while the AWU is the waste generation model for the product lines. The information generally flows from the bottom upwards and from left to right. Clearly, it is not the simplest of models.

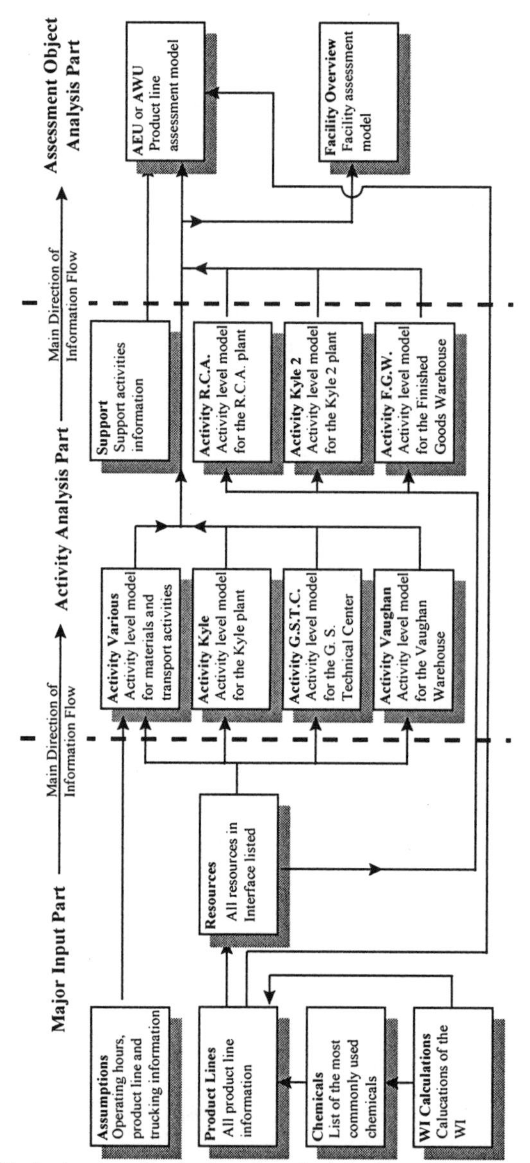

Figure 3 - Model Infrastructure for the Interface Activity-Based Energy and Waste Model.

3.8 Step 10 - Interpret the Results and Iterate if Necessary

The quality assessments are part of the activity analysis part of the model because all the quality assessments are process assessments. Quality can be measured in numerous ways, but at Interface, Troup County, it is measured by:

- *Scrap percentage* - how much of the finished goods were scrapped due to holes in the carpet, incorrect carpet seams and so on.

- *Off-quality percentage* - how much of the finished goods were not first quality.
- *Cut-off waste* - how many square yards of carpet were deliberately cut-off to generate tiles.
- *Working loss* - how much the carpets have stretched/shrunk during manufacturing. This quality measure primarily affects the revenue stream, not energy and waste streams, and was not included in the models.

The model worked satisfactorily, but it is the only model presented in this book where cost and overhead issues were largely ignored; the amount of time we had was not sufficient for establishing a good model for overhead issues. However, this offered us the opportunity to illustrate that Activity-Based Cost, Energy and Waste models can be implemented partially (without the cost and overhead dimensions) and still provide useful results.

4. RESULTS FOR INTERFACE FLOORING SYSTEMS

Intellectuals solve problems; geniuses prevent them.

Albert Einstein

Again, in this case, we investigated energy and waste issues only - costs were excluded. Therefore, we organize this section accordingly, starting with energy consumption analysis. Furthermore, analyses are presented as follows:

1. Interface, Troup County, overall performance.
2. Aggregated production facilities performance.
3. Quality issues.
4. 1/10 Slat Tile product line.
5. 5/64 Tile product line.
6. I-Bond product line.

4.1 Energy Consumption Results

4.1.1 Interface, Troup County, Total Energy Consumption

With ±10% variation in all the 260 assumption cells in the energy consumption model, we get a company wide energy consumption as shown in Figure 4. We see that the energy consumption can vary from 3.6 billion MJ to 4.4 billion MJ. The mean is estimated to be roughly 4 billion MJ for 1997. To relate this energy to something physical, we use a power plant as reference point. In 1997, one of Georgia's largest power plants - the Vogtle Electric Generating Plant near Waynesboro - generated net 18,580,935 MWh, or 66,891,370,000 MJ from its two nuclear reactors. Thus, the total annual energy consumption of Interface is 6% of the annual Vogtle plant production.

This may seem small, but remember that Interface is just one out of many consumers of electricity in Georgia. Note that our calculations include the energy content of purchased materials. If we only look at the electricity consumption at the various Interface manufacturing sites and exclude the energy content of the materials, then only 0.03 Vogtle plants are needed. From this discussion it should be clear that a) not including the entire life-cycle/value chain can lead to very wrong decisions and b) accounting systems are needed to keep track of the energy consumption.

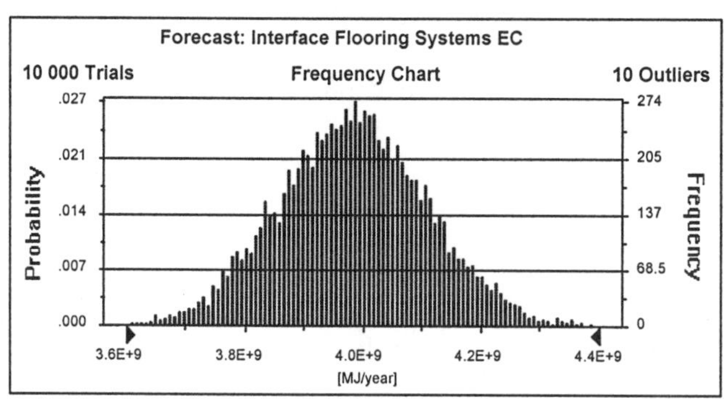

Figure 4 - Interface Flooring Systems, Troup County, Energy Consumption.

To find out how to reduce the overall energy consumption at Interface, we utilize a sensitivity chart as shown in Figure 5 where the numbers are correlation factors between the various assumption cells and the forecast. We see that all the critical success factors are related to two main factors:

1) Unit material consumption factors, such as the amount of yarn per unit [oz./yd^2]. The importance of these factors is evident, and to reduce material consumption is mainly a product design issue.
2) The energy content of the material [MJ/oz.]. This is outside Interface's sphere of control, but there are three ways of dealing with this challenge:
   - It can be treated as a purchasing decision, i.e., Interface should purchase the materials with lowest energy content. However, since their suppliers are not carrying out similar analyses, this is essentially a future vision as of today.
   - Interface can ask their suppliers for this information. A problem with this approach is of course that Interface may be too small to really be capable of influencing large corporations like Exxon, DuPont and BASF, which are their main suppliers.
   - Look into the possibility of using recycled material. Of course, one would have to analyze first the amount of energy it takes to recycle.

To further analyze the energy consumption, we simply remove the assumption cells with correlation factor equal or larger than 0.05, see Figure 5, and run the sensitivity analysis again. Instead of doing that, however, we can use the facility energy consumption to look at factors that are not related materials in any fashion. This is done next.

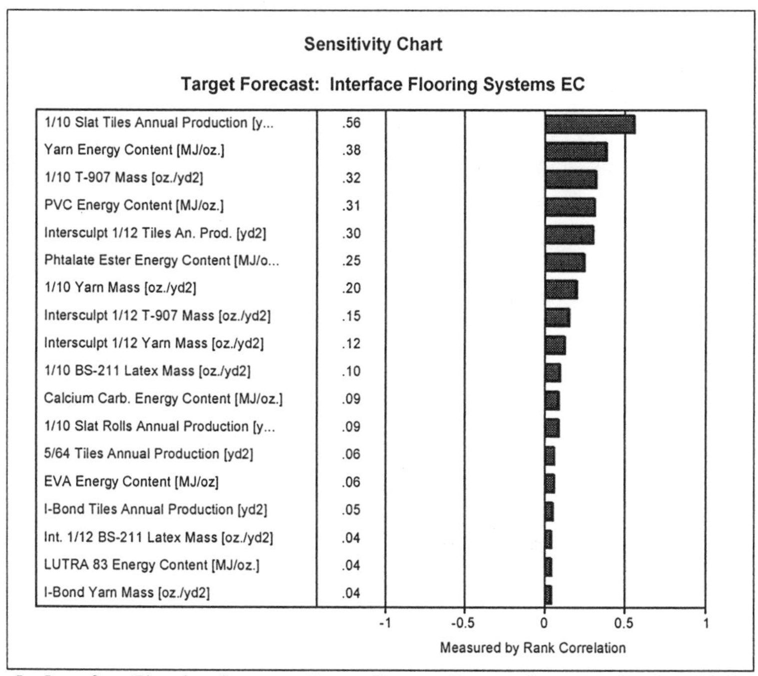

Figure 5 - Interface Flooring Systems, Troup County, Energy Consumption Sensitivity Chart.

4.1.2 Aggregated Facility Energy Consumption

The energy consumption of the production facilities (not including production materials) is shown as an uncertainty distribution in Figure 6. We see that the aggregated facility energy consumption is over ten times smaller than the energy content of purchased materials. Hence, trying to make Interface sustainable by simply trying to reduce the facility energy consumption is not enough. Nonetheless, reducing the energy consumption of the facilities will have positive effects.

To find out how reductions in energy consumption can be achieved we must again investigate the corresponding sensitivity chart. The sensitivity chart for the aggregated energy consumption is shown in Figure 7. We see that the single largest factor is the energy content of the electricity. In fact, according to our data it takes roughly 3.7 MJ to create 1 MJ net. Thus, the losses in energy generation are very large. To avoid this, Interface can do several things:

1. Generate the electricity locally to reduce distribution losses. Decisions must of course be based upon energy consumption assessments of the

alternatives. One should avoid local improvements that lead to overall higher energy consumption.

2. Redesign the processes and products so that less energy is required in the first place. E.g., reducing the 'wetness' of the various carpet layers will be beneficial, because water has a very high specific heat value and (thus) large amounts of energy are needed for carpet drying. Product and process redesign is by no means an easy task, as it will require systematic work over many years. Of course, the alternative(s) should not lead to higher energy consumption elsewhere in the value chain.
3. Reduce quality losses, which are discussed separately in Section 4.1.3.

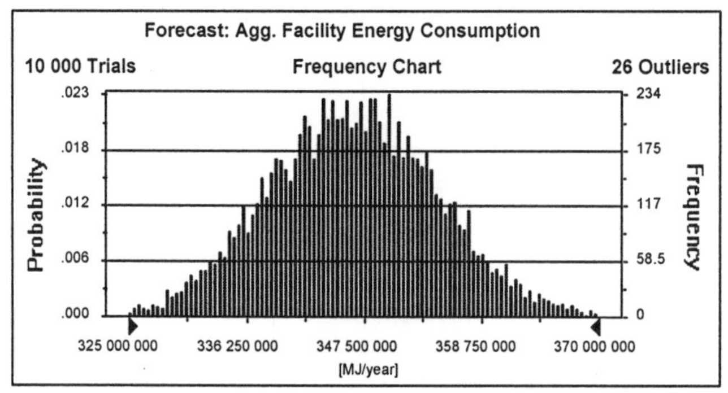

Figure 6 - Aggregated Facility Energy Consumption.

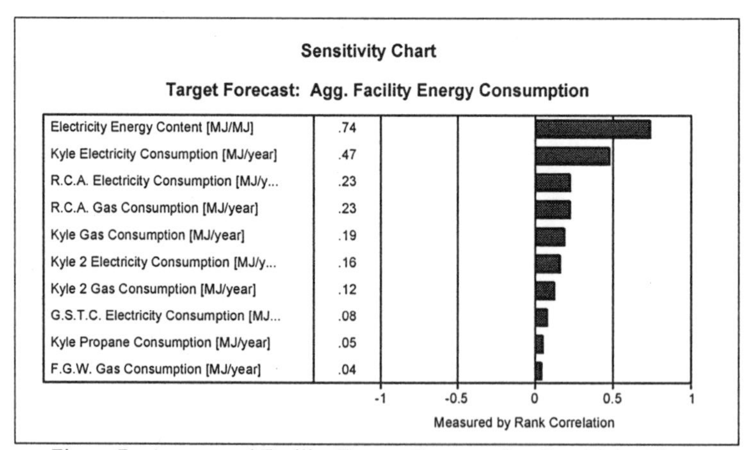

Figure 7 - Aggregated Facility Energy Consumption Sensitivity Chart.

Similar analyses can be conducted for all the critical success factors of the facility energy consumption. Our main point, however, is to illustrate the potential of Activity-Based Cost, Energy and Waste models as an important *attention directing* tool to aid better decision-making. We sincerely hope that those readers who work in manufacturing see how our approach could be applied in their manufacturing operations.

4.1.3 Quality from an Energy Perspective

Avoiding quality loss is an important issue in manufacturing. From Figure 8 we see that the total energy consumption due to quality problems (and subsequent loss of first quality products) is 175,300,000 MJ, or 4.40% of the total energy consumption (including energy content of the materials). Thus, the effect of quality problems is minor. Clearly, an Activity-Based Cost, Energy and Waste model can assess the impact of quality problems of the energy dimension just as an ABC model can assess the costs of quality loss. The interesting question, however, is how quality can be improved. The first step in answering such a question is of course to identify the quality problems. To do that we use the sensitivity chart shown in Figure 9.

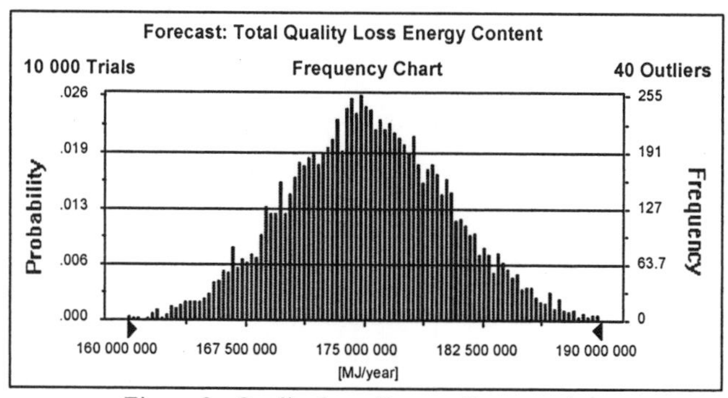

Figure 8 - Quality Loss Energy Consumption.

In Figure 9, we see that the off-quality problems in activity A46 (tile printing) are the single largest source of quality problems. Please note that the yarn energy content and similar factors do not represent quality loss per-se, but factors that are significant for the numerical magnitude of the quality loss. Hence, a reduction of the yarn energy content will result in lower value for quality loss, but the quality itself is not improved. According to Dave Gustashaw, the reason for this poor performance of the A46 activity is that the printing production was to be stopped in 1997 because of declining business which led to declining attention to A46 in general. Similar discussions can be made regarding every critical success factor listed in Figure 9, but that will be omitted here for brevity reasons.

The A46 activity has also the largest numerical off-quality percentage. However, one should not generalize this to saying that activities with the greatest off-quality percentages also have the greatest loss of quality. In this case it just happened to be so, but as we shall see for the waste generation this is not always the case. Thus, looking at scrap and off-quality percentages without linking them to the value of the specific activities yields incorrect results in general. This is particularly important to keep in mind because quality work, as all work, is to be committed under budget constraints.

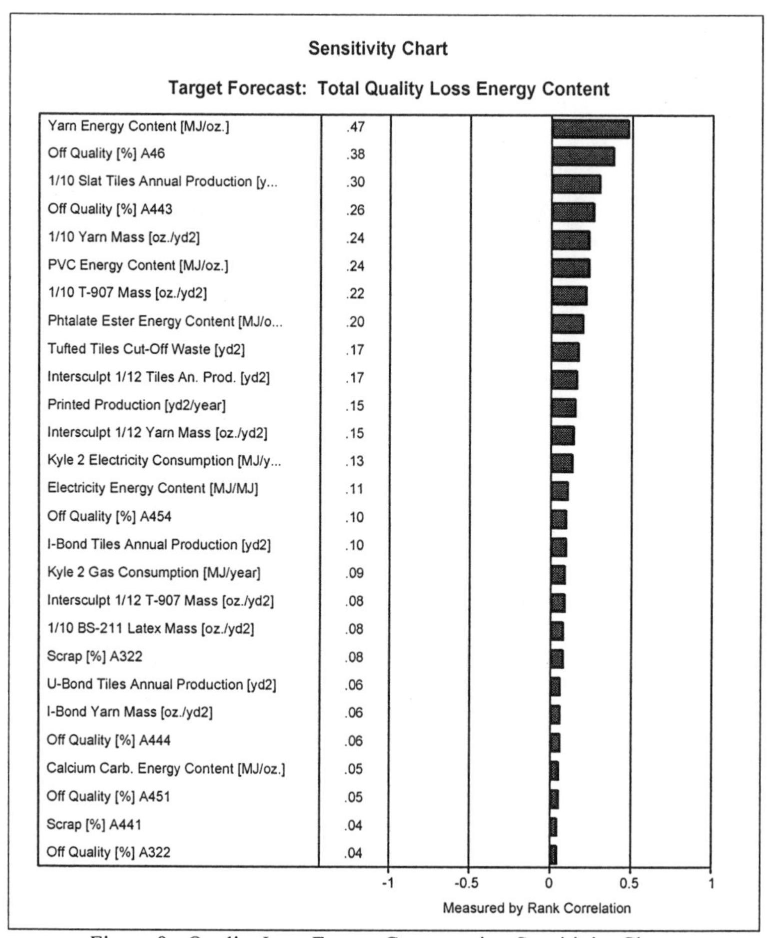

Figure 9 - Quality Loss Energy Consumption Sensitivity Chart.

4.1.4 Product Lines Energy Consumption - Trend Chart

Before we continue to investigate some of the product lines, we present a chart of the unit energy consumption of the product lines in Figure 10. The various bandwidths (100%, 70% etc.) are levels of confidence. That is, the 70% bandwidth indicate that the energy consumption assessment will be within that bandwidth 70% of the time given the assumptions in the assumption cells. This chart is useful for comparison reasons (and trends). We see that the bonded product lines perform worse than the rest. This is primarily due to two reasons:

1. The carpet face mass of the bonded product lines is higher than the rest. The face of a carpet is what we walk upon, i.e. the carpet without the backing.
2. The annual production is significantly lower than the rest, which is unfavorable from an economies-of-scale perspective.

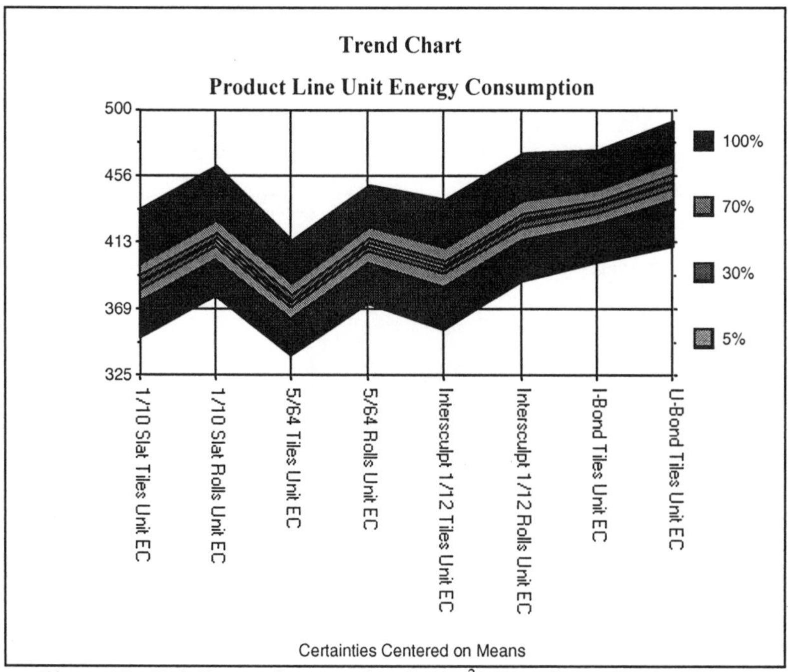

Figure 10 - Unit Energy Consumption [MJ/yd^2] Chart for the Product Lines.

What is strange, after the initial investigation of this chart, is that the tufted tiles perform better than the tufted rolls even though the tufted tiles lead to cut-off waste. This can be explained by at least two factors:

1. The economies-of-scale effect is important because the roll production only accounts for 14.60% of total production.
2. The fact that the rolls have several specific activities that only rolls consume, makes the lack of economies-of-scale even worse.

In the following sections, we discuss each product line in detail.

4.1.5 1/10 Slat Tile Energy Consumption

From the trend chart in Figure 10 we see that the 1/10 Slat Tiles product line is among the product lines that consume the least energy per unit. From the uncertainty distribution in Figure 11 we can get a more detailed picture. We see that the 1/10 Slat Tiles consume about 385 MJ/yd^2, i.e., 107 kWh/yd^2.

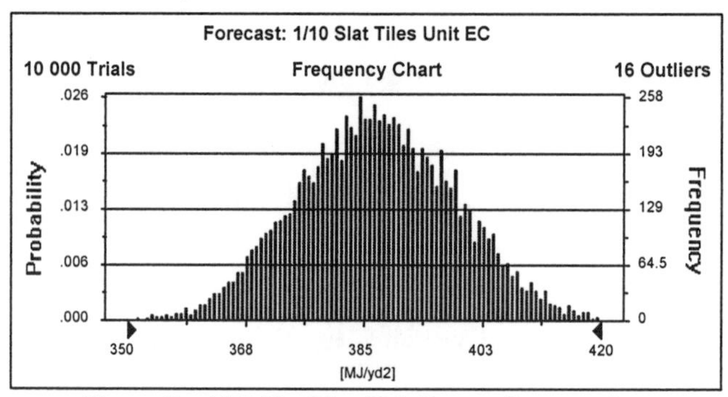

Figure 11 - 1/10 Slat Tiles Unit Energy Consumption.

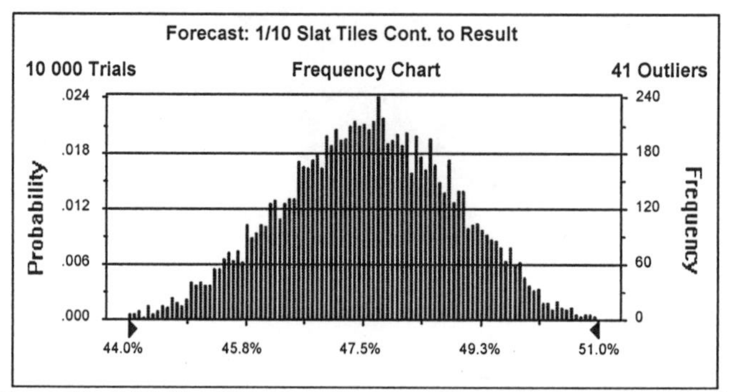

Figure 12 - 1/10 Slat Tiles Contribution to Energy Consumption Result.

By using the contribution forecast in Figure 12, we can assess to which extent the energy consumption at Interface in Troup County is consumed by the 1/10 Slat Tiles product line. We see that the 1/10 Slat Tiles product line most likely contributes 47.5% to the Interface energy consumption (denoted as 'Result' in Figure 12). Hence, this product line is of vital importance for Interface, Troup County, from an energy perspective. To find the critical success factors, and hence how to improve the 1/10 Slat Tiles product line, we turn to the sensitivity chart shown in Figure 13.

For the 1/10 Slat Tiles product line, the amount of the T-907 chemical mix is the largest single energy consumption factor with a correlation factor of 0.62, see Figure 13. We see the increasing the annual production is, in fact, the only factor that, if increased, will yield a lower (negative correlation) unit energy consumption that is significant enough to make it on this sensitivity chart. Note, however, from Figure 5 that an increase in production will *increase* the overall energy consumption at Interface. This is because increased production gives an economies-of-scale effect on *product line level* with respect to production related overhead energy consumption, but unless the production overhead energy consumption is reduced, it will give higher *overall* energy consumption. Basically, what will happen if production of 1/10 Slat Tiles increases (and if everything else remains constant), is that the total energy content of purchased materials increases while the overhead energy consumption is spread out over more units of 1/10 Slat Tiles. Thus, on one hand, the unit energy consumption of 1/10 Slat Tiles is reduced, but on the other hand, because the total production overhead energy consumption remains the same, the overall energy consumption of Interface increases. Hence, *we see here that focusing on product lines in isolation from the big picture can yield significant errors*. With respect to the ISO 14000 LCA standards, this means that focusing on product by product as ISO seems to recommend, can be counterproductive and outright wrong. Also, this tradeoff situation shows the need for being able to handle all products and the entire business unit (Interface, Troup County in this case) in the same model at the same time.

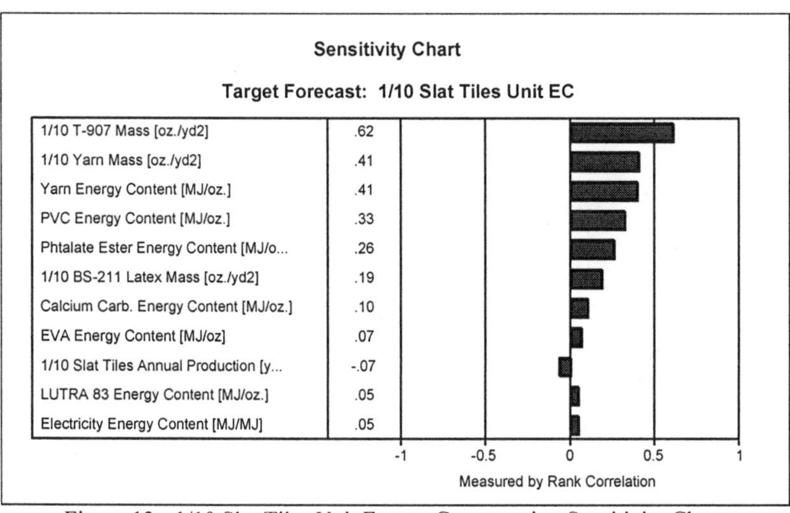

Figure 13 - 1/10 Slat Tiles Unit Energy Consumption Sensitivity Chart.

4.1.6 5/64 Tile Energy Consumption

According to our assessments, the 5/64 Tile product line consumes the least energy per unit, roughly 375 MJ/yd^2, see Figure 14. Hence, from an energy perspective, the 5/64 Tiles can be called the most environmentally friendly product line. The contribution to the Interface energy consumption is only 5.3%. Please note that this might change significantly if all overhead resources were included. The reason is that this product line is produced in one of the lowest volumes, hence economies of scale would become increasingly important as the model became a full absorption model. However, due to the chronic lack of data in general, it is likely that the distortion of the model would be more serious than the uncertainty caused by not including all the overhead resources. To resolve this question is therefore basically a matter of belief. Those that believe that all overhead resources should be included, although the data regarding such issues are very uncertain, would prefer an implementation as shown in the Westnofa Industrier case study. Those that, on the other hand, would like to eliminate the uncertainty while accepting the risk of simplification would prefer an implementation as shown here. When it comes to financial costs there is no reason for *not* making a full absorption model because of the existence of comprehensive accounting systems. Thus, full absorption should be the rule for the economic dimension. We also believe that it should be the rule for the environmental dimension even though the data are not good simply because it makes the whole approach more consistent. In addition, we think it is better to discuss the problem of data explicitly instead of implicitly trying to omit it. An explicit discussion also sheds light over the fact that the lack of energy and waste accounting in the value chain is a major problem.

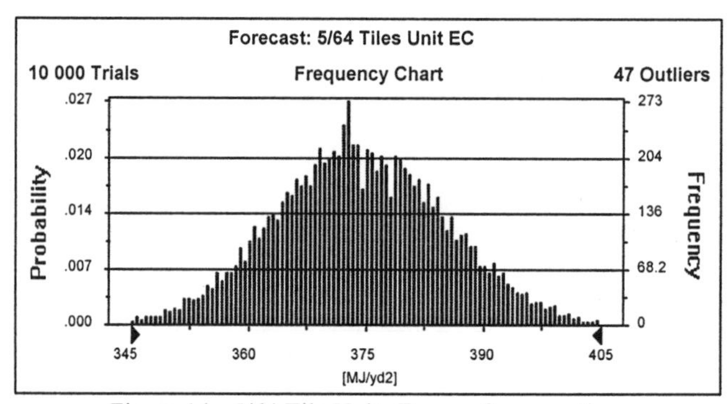

Figure 14 - 5/64 Tile Unite Energy Consumption.

As usual, we identify the critical success factors by using the sensitivity chart as shown in Figure 15. The T-907 chemical mix also shows up as the leading critical factor. The 1/10 annual production also shows up. This means that the cross-consumption of activities by 1/10 Slat Tiles is more important than the consumption of activities by the 5/64 Tile.

The last critical success factor, i.e. the 'A46 Gas Consumption [%]', can be random, since the correlation coefficient is so low, but we believe it is real. This effect occurs as a result of the way that energy consumption is traced in the Kyle 2 facility. Basically, a certain percentage of gas is traced to the A46 activity while the rest is traced to the other activities. Hence, if the A46 consumption increases, there will be less gas for the other activities and hence it can turn out positive for products that consume those other activities. From this we learn that understanding the model and the processes in the company is very important. If the A46 activity were metered separately this problem would not occur.

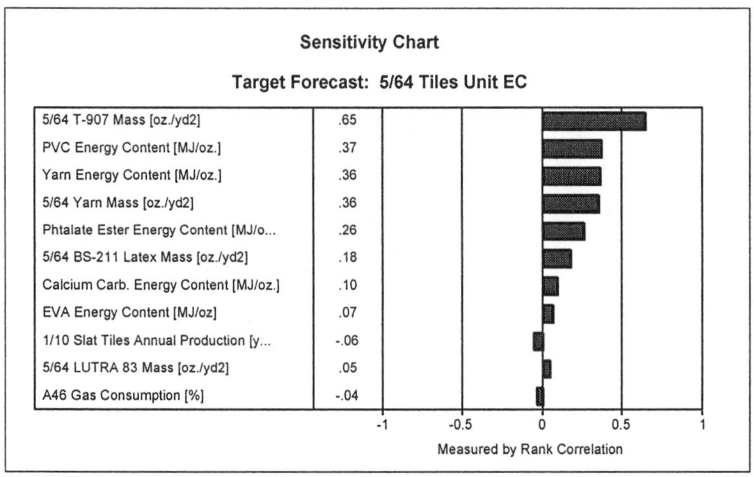

Figure 15 - 5/64 Tile Unit Energy Consumption Sensitivity Chart.

4.1.7 I-Bond Energy Consumption

From the second major product line, the fusion bonded products, we have chosen the I-Bond Tiles as a representative sample as it has the highest sales. The I-Bond Tiles' unit energy consumption uncertainty distribution is shown in Figure 16. The mean is roughly 435 MJ/yd^2, or 16.8% higher than the 5/64 Tiles. Despite this fairly high unit energy consumption, the I-Bond Tiles contribute only 5.2% to the overall result, which is similar to the 5/64 Tiles.

In Figure 17 we can see the critical success factors of the I-Bond Tiles. Again, all the energy content and mass related factors dominate the sensitivity chart. It is, however, interesting to note that we can here spot a significant economies-of-scale potential for the I-Bond Tiles. This is due to the fact that the I-Bond Tiles are produced in very low quantities (470,797 yd^2 in 1997). But again, an increase in production will not yield an overall energy consumption reduction for Interface. Thus, the face mass is the basic problem of the I-Bond Tiles compared to the tufted products. For the U-Bond Tiles it is even worse because the face mass is higher. Fusion bonded product lines are simply resource inefficient compared to the tufted product lines. Hence, from an energy perspective, the elimination of the fusion bonded products, in favor of the tufted products, will yield a net benefit. If all overhead resources were included in the model, this would probably be even clearer to see because overhead resources are rarely volume related and would therefore not be traced necessarily to high volume products.

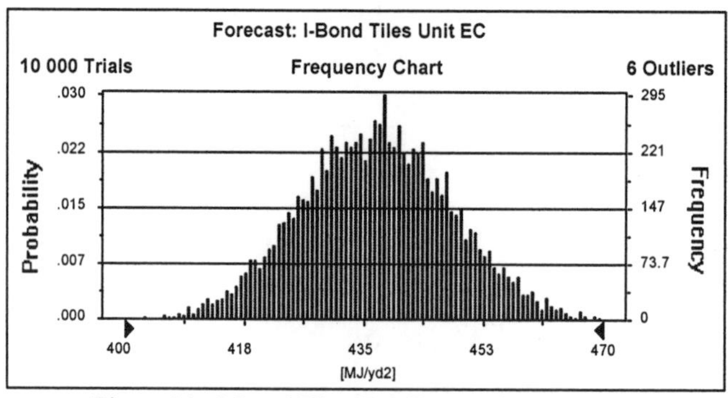

Figure 16 - I-Bond Tiles Unit Energy Consumption.

Sensitivity Chart

Target Forecast: I-Bond Tiles Unit EC

Variable	Rank Correlation
I-Bond Yarn Mass [oz./yd2]	.47
Yarn Energy Content [MJ/oz.]	.46
PVC Energy Content [MJ/oz.]	.38
Phtalate Ester Energy Content [MJ/o...	.27
I-Bond T-900 Mass [oz./yd2]	.23
I-Bond T-902 Mass [oz./yd2]	.23
I-Bond T-926 Mass [oz./yd2]	.22
I-Bond Tiles Annual Production [yd2]	-.18
Kyle Electricity Consumption [MJ/year]	.15
Electricity Energy Content [MJ/MJ]	.14
ATH Energy Content [MJ/oz.]	.12
I-Bond T-925 Mass [oz./yd2]	.10
Kyle Gas Consumption [MJ/year]	.07
U-Bond Tiles Annual Production [yd2]	-.06
A454 Effect [kW]	.06
A443 Effect [kW]	-.05
Intersculpt 1/12 Tiles An. Prod. [yd2]	-.04

-1 -0.5 0 0.5 1

Measured by Rank Correlation

Figure 17 - I-Bond Tiles Unit Energy Consumption Sensitivity Chart.

4.2 Waste Generation Results

The waste generation results are presented in the same fashion and order as the energy consumption results, starting with the total waste generation.

4.2.1 Interface, Troup County, Total Waste Generation

The total waste generation of Interface, Troup County, was 24,533,736 pWU in 1997 (see Figure 18 for the uncertainty distribution). *Is this large or small?* That is a relevant question and it is not easy to answer because the Waste Index (WI) is only used in the case studies presented here. We can compare it to the UT 705 Platform Supply Vessel (PSV) discussed in Chapter

6, for example, which generated 3,350,000 pWU in 1995. Hence, the Interface plant in Troup County is generating 7.3 times the amount of waste as one single UT 705 PSV. We can also compare it to the Westnofa Industrier waste generation, which is done in Chapter 8.

Such a comparison is often not possible using the ISO 14040-43 LCA approach because of the usually incomparable results. Of course, some may argue that there is no point of making such comparisons between widely different industries. To refute that we can argue from many perspectives. Take, for example, a policy maker perspective: if a government were to put taxation (or subsidy) on environmental performance, we are convinced that this government would be very interested in finding out where taxation/subsidy would be the most effective. Also, how can we avoid comparisons if we think in terms of the value-chain or life-cycle? Ultimately, a PSV is a part of the life-cycle of oil and gas mined by the platforms that the particular PSV is servicing, and this oil and gas are input materials for making nylon 66, one of Interface's most important yarn materials.

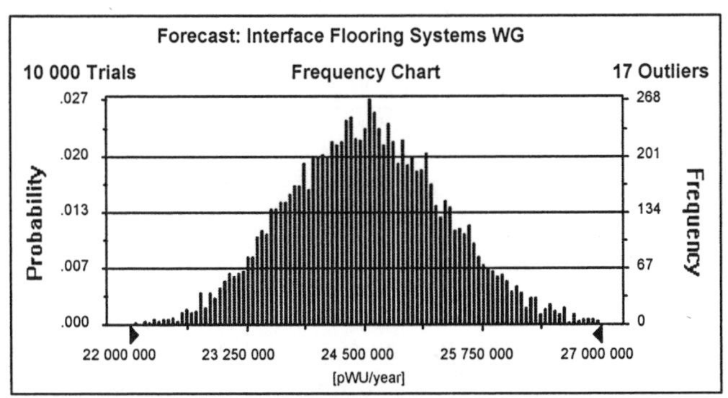

Figure 18 - Interface Flooring Systems Waste Generation.

The critical success factors appear in the sensitivity chart presented in Figure 19. Again, the waste content (similar to energy content) and the materials usage are the main factors. The yarn waste content is particularly important along with the largest consumers of that yarn; the of 1/10 Slat Tiles product line. The 1/10 Slat Tiles are so important simply because they are produced in such a high volume (4,902,591 yd^2 in 1997, which is 48.8% of total production volume). What is particularly interesting is to note that the I-Bond Tiles show up on the chart at all since they are produced in such a low volume. This indicates that the fusion bonded product lines are generating, comparatively, a lot of waste. This is investigated more in Section 4.2.7. First, however, we discuss the waste generated by the various facilities. Because of the importance of the material related factors, nothing that has to do with the facilities showed up in Figure 19.

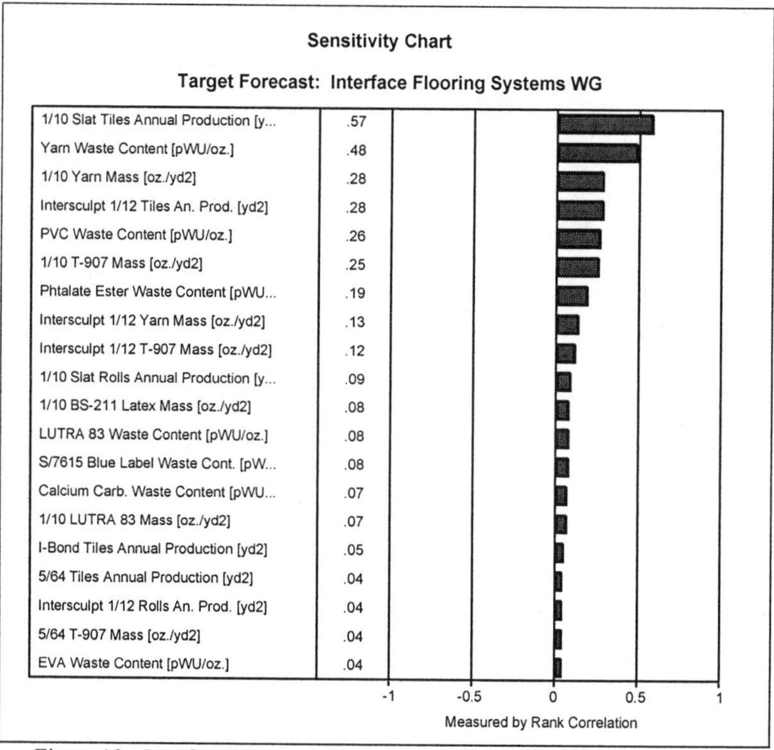

Figure 19 - Interface Flooring Systems Waste Generation Sensitivity Chart.

4.2.2 Aggregated Production Facilities Waste Generation

We find that the deterministic waste generation from all the facilities is 702,694 pWU in 1997. The corresponding uncertainty distribution is shown in Figure 20. Compared to the PSV, this 4.8 times lower! In other words, based on the Waste Index results, a single PSV generates about 5 times more waste [pWU] than the waste that is *directly* attributable to an entire Interface business unit when material waste contents are excluded (!).

The critical success factors can as usual be read from the sensitivity chart in Figure 21. Again we see that the electricity plays a major role both by its consumption level and by the waste content of electricity. Of releases that Interface, Troup County, is directly responsible for, we see that the CO_2 releases from the R.C.A. Plant are significant with a correlation factor of 0.14. In fact, all of Interface, Troup County, releases roughly 5,288,850 kg CO_2, according to our model. The two main sources of these releases are the consumption of natural gas and propane, as can be seen from Figure 21. One should always be careful when interpreting the lower part of the sensitivity chart due to the possibility of random effects, especially when the correlation coefficients are 0.05 or lower. Hence, if we were to explore more on this issue, we should remove the top 8 or 9 critical success factors and the rerun the model so that the next critical success factors can be detected reliably.

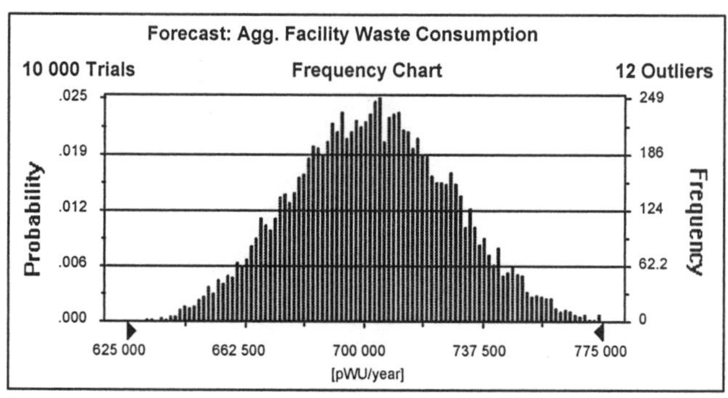

Figure 20 - Aggregated Facility Waste Generation.

Sensitivity Chart

Target Forecast: Agg. Facility Waste Consumption

Variable	Rank Correlation
Electricity Waste Content [pWU/MJ]	.85
Kyle Electricity Consumption [MJ/year]	.39
R.C.A. Electricity Consumption [MJ/y...	.20
Kyle 2 Electricity Consumption [MJ/y...	.15
Natural Gas Unit CO2 Release [kg/kg]	.14
R.C.A. Gas Consumption [kg/year]	.08
G.S.T.C. Electricity Consumption [MJ...	.07
Kyle Gas Consumption [kg/year]	.06
Kyle Propane Consumption [kg/year]	.05
Kyle 2 Gas Consumption [kg/year]	.04
Natural Gas Unit NOx Release [kg/kg]	.04
Propane Unit CO2 Release [kg/kg]	.04

-1 -0.5 0 0.5 1

Measured by Rank Correlation

Figure 21 - Aggregated Facility Waste Generation Sensitivity Chart.

4.2.3 Quality from a Waste Perspective

The quality performance measures are the same as before. Recall from Section 4.1.3 that the quality loss is 3.40% of total energy consumption. From a waste perspective, the quality loss is 4.21% of the total waste generation, or about 1,030,000 pWU, see Figure 22. Interestingly enough, the waste content of the quality loss is higher than all the waste generated from the facilities. This is due to the high waste content of the purchased materials.

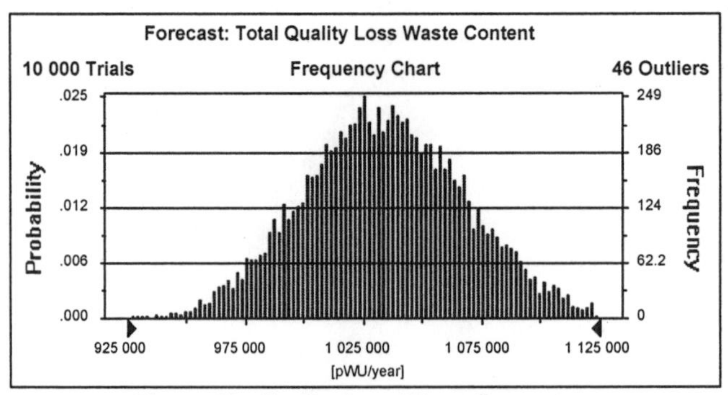

Figure 22 - Quality Loss Waste Generation.

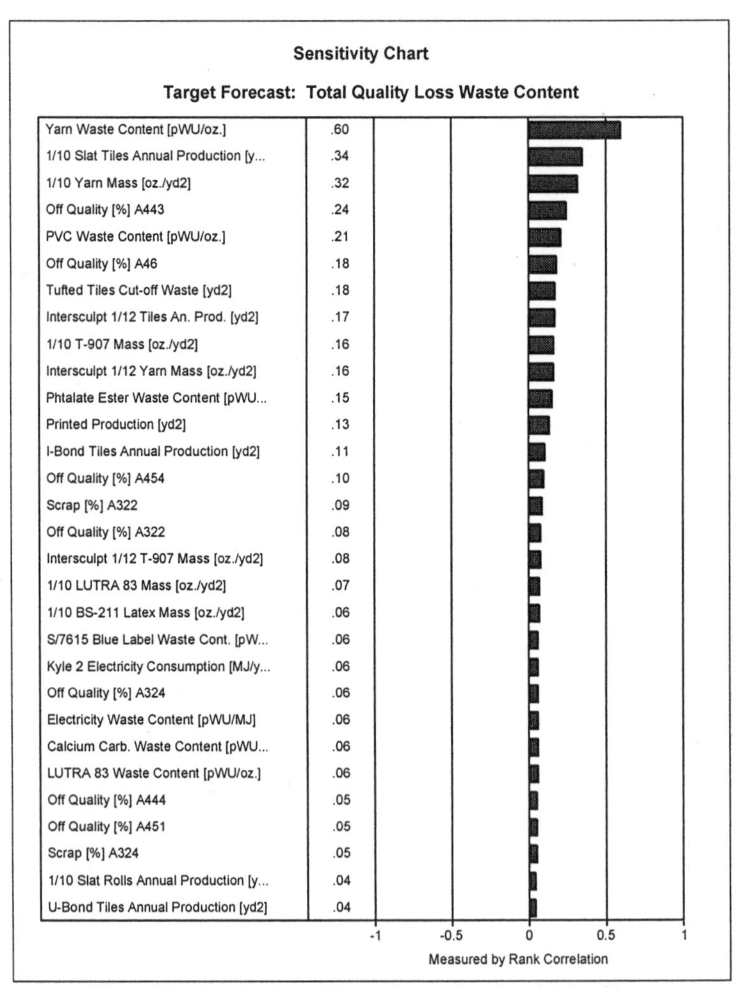

Figure 23 - Quality Loss Waste Generation Sensitivity Chart.

To find ways to reduce the quality loss we look at Figure 23. The similarities with the same chart for the energy consumption are numerous.

However, there is one thing we find particularly interesting. From the energy dimension, A46 (the 'Print Pattern and Package Products' activity) off-quality problems yielded the highest quality loss. Here, however, we see that it is the off-quality at activity A443 (the 'Produce on Backing Line 4' activity) that produces the highest quality loss. This occurs even though the off quality percentage at A443 is only 1.16% while it is 48.63% for A46. The reason is that the materials stream through A443 has a much higher waste content than the material stream through A46, while for the energy perspective it is the energy consumption at the activities that matters. In this way we see that our activity-based approach is capable of going beyond looking at scrap percentages and the like and in fact realistically captures the loss of quality regardless if it is energy consumption or waste generation. Clearly, in this case Interface may never have thought of A443 off-quality as the largest source of waste generation, from a quality perspective, because there are four activities with larger off-quality scrap percentages than the A443 activity. But if we add in the waste content of the materials, A443 is the most critical to focus on.

4.2.4. Product Lines Waste Generation - Trend Chart

In Figure 24, we present a chart of the unit waste generation of the product lines.

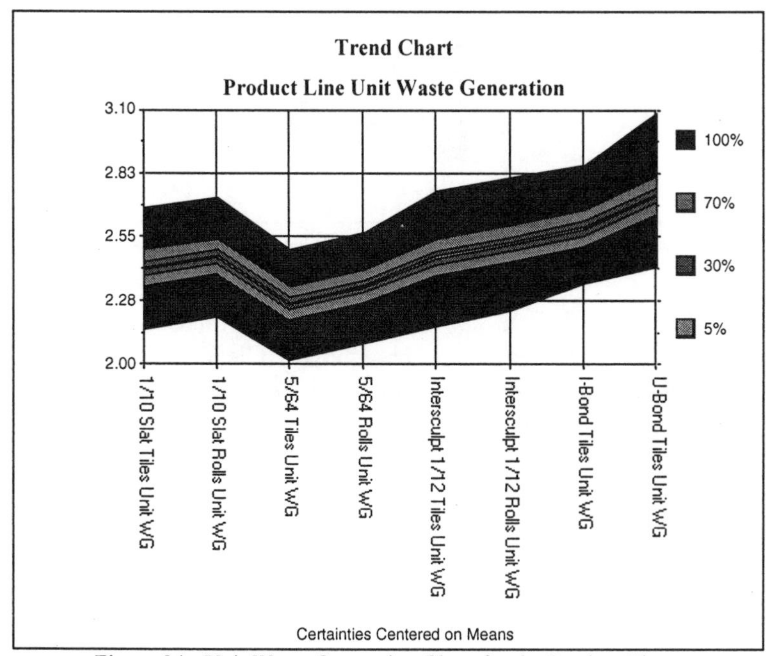

Figure 24 - Unit Waste Generation Chart for the Product Lines.

As expected, the bonded product lines perform worse than the rest. In many ways the waste generation trend chart is very similar to the energy consumption trend chart. The only clear trend difference is between the 5/64 and Intersculpt 1/12 product lines and that the difference between rolls and tiles is less than what is the case with respect to energy consumption. That should come as no surprise since many activities that consume a lot of energy often generate a lot of waste.

4.2.5 1/10 Slat Tile Waste Generation

The 1/10 Slat Tiles generate 2.43 pWU/yd^2 with a range from 2.20 pWU/yd^2 to 2.65 pWU/yd^2. This makes the 1/10 Slat Tiles among the product lines that generate the least waste *per unit* and it is therefore also among the most environmentally friendly product lines. If we look at how much the 1/10 Slat Tiles contribute to the total amount of waste, we get 48.3%. Hence, as for energy consumption, the 1/10 Slat Tiles is the dominating product line due to the high production volume even though the 1/10 Slat Tiles on a unit level performs among the best.

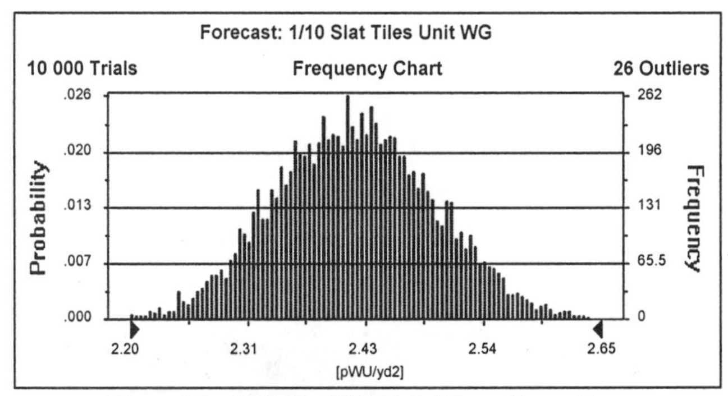

Figure 25 - 1/10 Slat Tile Unit Waste Generation.

To identify the critical success factors we, as always, turn to the sensitivity charts, see Figure 26. We see that only materials related factors such as mass and waste content, are shown. Clearly, focusing on more material efficient designs and using materials with lower waste content are the two most important changes that should be made at Interface, Troup County, for the 1/10 Slat Tiles. It is interesting to notice that the annual production for the 1/10 Slat Tiles did not show up at all. That means that there are no significant opportunities for simple economies-of-scale gains.

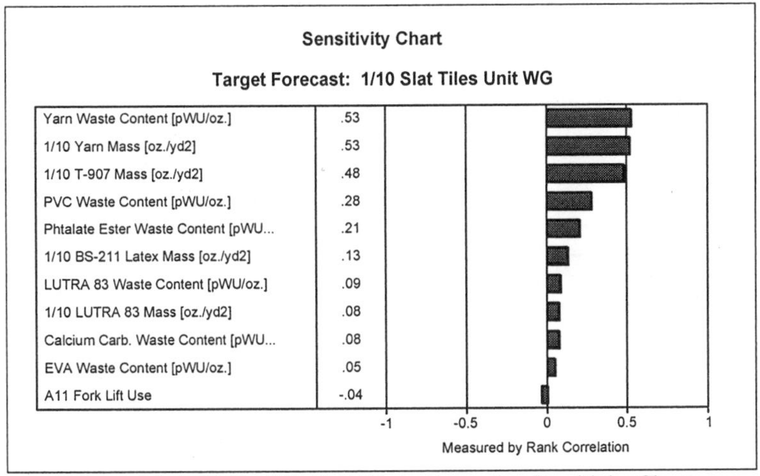

Figure 26 - 1/10 Slat Tile Unit Waste Generation Sensitivity Chart.

4.2.6 5/64 Tile Waste Generation

The 5/64 Tile product line is the most environmentally friendly product line, in terms of waste generation, with the lowest unit waste generation: 2.25 pWU/yd^2, or 6.6% lower than for the 1/10 Slat Tiles (see Figure 27). This is not very much, but it is interesting that it occurs because the 5/64 Tile product line is produced in so much lower volume and is therefore susceptible to diseconomies-of-scale, i.e. the opposite of economies-of-scale. Regardless of that disadvantage, it does not show up in the sensitivity chart, see Figure 28. This can only be interpreted in one way. The factors related to materials are so dominating that even economies-of-scale effects, which are often important, do not make it to the sensitivity chart for the first run. We would probably have to run the model several times in order to detect any economies-of-scale issues. Note however, that because we have only included a fraction of the overhead resources economies-of-scale effects are less important.

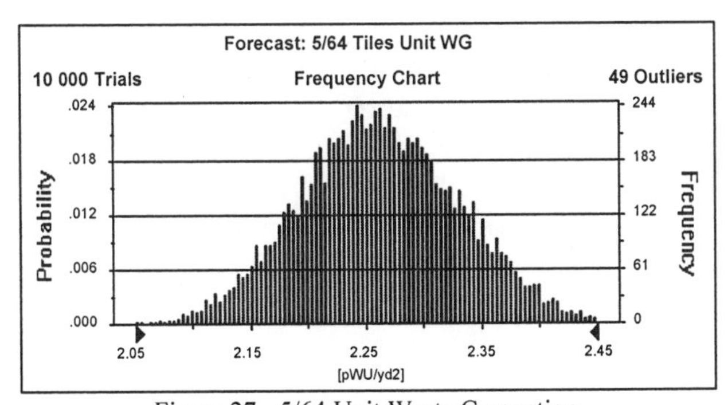

Figure 27 - 5/64 Unit Waste Generation.

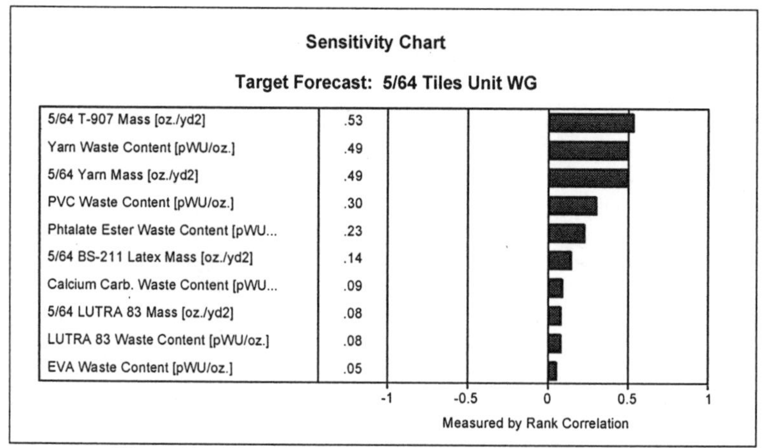

Figure 28 - 5/64 Unit Waste Generation Sensitivity Chart.

4.2.7 I-Bond Waste Generation

According to the trend chart in Figure 24, the I-Bond product line is the second most waste generating product line (after the U-Bond Tiles). The mean waste generation is 2.59 pWU/yd^2, or 14.6% higher than for the 5/64 Tiles. In Figure 29, the uncertainty distribution of the I-Bond Tile product line is shown, and we see that the band around the mean is roughly ±0.25 pWU/yd^2 given the ±10% variation in the assumption cells. This bandwidth is about the same for all the product lines, hence, all the product lines are quite similar when it comes to sensitivity to changes in the assumption cells. That information is valuable to ensure that none of the product lines can perform greatly negatively if unforeseeable and possible unwanted changes occur.

And here, finally, we see that an economies-of-scale effect is present, see Figure 30. The reason is that the I-Bond Tiles are produced in an exceptionally low number compared to the tufted product lines. Other than that, there is little new in the sensitivity chart in Figure 30 as well. The only thing that may be of interest, although the correlation factors are very low - indicating a possible random effect, is that on the bottom of the sensitivity chart we see that if we increase the PHR (Per Hundred Ratio) of PVC and ATH in the T-926 mix the energy consumption will *decrease*. The explanation for this is that the PVC and ATH waste contents must be comparatively lower than the other chemicals in the T-926 mix. Hence, we are essentially saying that one should also investigate the mixture of the various chemicals.

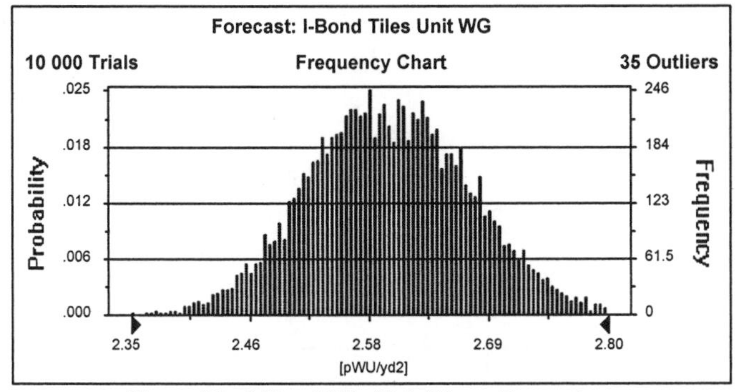

Figure 29 - I-Bond Tile Unit Waste Generation.

Sensitivity Chart

Target Forecast: I-Bond Tiles Unit WG

Variable	Rank Correlation
Yarn Waste Content [pWU/oz.]	.58
I-Bond Yarn Mass [oz./yd2]	.56
PVC Waste Content [pWU/oz.]	.28
I-Bond T-926 Mass [oz./yd2]	.24
Phtalate Ester Waste Content [pWU...	.21
I-Bond T-902 Mass [oz./yd2]	.17
I-Bond T-900 Mass [oz./yd2]	.17
ATH Waste Content [pWU/oz.]	.10
Fumed Silica Waste Content [pWU/...	.08
I-Bond Tiles Annual Production [yd2]	-.08
I-Bond T-925 Mass [oz./yd2]	.08
Fumed Silica (Glass) PHR T-926	.07
Electricity Waste Content [pWU/MJ]	.07
Kyle Electricity Consumption [MJ/year]	.05
Poly Vinyl Chloride PHR T-926	-.05

-1 -0.5 0 0.5 1

Measured by Rank Correlation

Figure 30 - I-Bond Unit Waste Generation Sensitivity Chart.

5. RESULTS IN RELATION TO THE PLETSUS PROGRAM

In (Anderson 1998) the program called Practices Leading Toward SUStainability (PLETSUS) is presented as Interface's approach towards sustainability. PLETSUS incorporates a large variety of various practices, of which many cannot be related to Activity-Based Cost and Environmental Management, e.g. 'use experimental learning techniques to explain complex concepts'. The ones that are related to manufacturing are presented in Table 4 where we have tried to rate their importance as indicated by our findings. A similar exercise can also be conducted for other functions. A rating of 5 for a suggestion indicates that we have great belief in it, while a rating of 1 indicates little belief. The rating scale is discrete in the range of 1 through 5. Please note that we may have misunderstood some of the PLETSUS suggestions and therefore rated on wrong premises.

Table 4 - PLETSUS and Activity-Based Environmental Management Findings.

PLETSUS Suggestions	Activity-Based Environmental Management Findings: Rating	Reason
Reclaim waste heat from processes, furnaces, air compressors and boilers.	3	The resource expenditures needed for reclamation may outweigh the benefits.
Systematically review all electric motor systems to minimize installed horsepower and maximize motor efficiency.	2	There are better, and more pressing, ways of utilizing the manpower needed to conduct such reviews unless costs are of no importance.
Design pumping systems with big pipes and small motors.	2	The benefits of easing the flow are not very substantial in relation to total resource usage.
Design pumping systems by laying out pipes first (to minimize distance and elbows), then motors and other equipment.	3	The benefits of easing the flow and reducing the heat transfer is not very substantial in relation to total resource usage.
Lay out plants to minimize distance materials travel.	2	The cycle time is determined by the processing machinery and transport resources are insignificant, except the 13 mile transport between West Point and La Grange that should be avoided for new plants.
Research product formulations to reduce process temperature requirements.	5	This is important since water has a very high specific heat, and thus requires a lot of energy during heating.
Minimize the number of times materials are heated an cooled.	5	This is important since water has a very high specific heat, and thus requires a lot of energy during heating.
Install multiple small motors to handle varying volumes rather than one big motor.	2	Not much to gain from this due to the relatively low resource consumption from motors combined with the potential higher energy content and waste content of many motors.
Design systems for expected operating conditions rather than maximum expected capacity.	4	Over-dimensioned systems are resource inefficient, as are under-dimensioned.
Stage plant flows and energy peaks to maximize efficiency.	3	If possible, it will contribute substantially. The problem is to actually make it possible.
Use computer modeling techniques to minimize energy usage.	3	Depends a lot on what is being minimized. In general, one should seek to eliminate a problem; not minimize it.
Research Energy Miser technology on motors	2	Running a plant on peaks is probably extremely difficult and may even be counterproductive due to reduced overall robustness.
Install power sub-meters on all processes to continuously monitor efficiencies.	3	Sub-metering is important, but not on all processes. Also, continuous monitoring is not necessary.
Install automatic switches to turn off equipment at a determined time of inactivity.	3	Automatic switching system has probably a higher energy- and waste content than the process benefits.
Research and adopt alternative energy sources consistent with local surroundings, such as hydroelectric, biofuel, solar, wind power etc.	3	Depends a lot on the energy- and waste content of the chosen local fuel.
Negotiate Green Energy contracts with utilities.	1	First determine what is 'green', because green today has almost lost its meaning.
Research soft starting and control motor technologies.	3	Soft starts are important for large machinery. The question is to what extent this is a problem for Interface.
Research energy storage technologies such as flywheels.	2	Storing energy kinetically is indeed a possibility, but it seems difficult to get it working in the plants.
Adopt a zero-waste mentality; design processes to create no waste.	5	Waste, widely defined, is extremely important to reduce since a zero-waste mentality affects the entire company in terms of design of products and processes, operation and management.
Adopt a zero defect mentality; most defects become waste.	2/4	If done reactively we have little faith because the defects account for only 0.70% of total resource utilization. If done throughout as a new philosophy for everything, we have high hopes.
Eliminate all smokestacks, effluent pipes and hazardous waste.	4	Eliminating hazardous waste is important although in this case study we cannot tell due to lack of data.

Adopt high efficiency planning and scheduling practices to minimize waste.	3	The waste, conventionally defined, accounts for only 3.6% of total resource utilization.
Network with other companies to find waste streams that can become inputs for other processes.	4	Such networking can be effective, but it depends on the resources used in the networking.
Buy raw materials in bulk to minimize packaging.	4	Can be important; depends on the options.
Carefully segregate waste materials for reuse and recycling.	3	First should the effects of reuse and recycling be identified.
Develop processes to utilize internal scrap material.	2	The defects account for only 0.70% of total resource utilization.
Develop quick stop technology to minimize waste created by off-quality processes.	2	Such development and the technology itself will probably outweigh the potential savings of 3.6%.
Take corrective action on quality problems as far upstream as possible to minimize waste.	3	Corrective actions are rarely useful for quality purposes. Proactive measures are what counts.
Closely measure all materials streams to monitor materials efficiency.	3	Material efficiency is important, but it is more important to find out what materials count.

Undoubtedly, there are many very good PLETSUS suggestions in Table 4. However, we feel that there are too many reactive and too few proactive measures. Also, the number of measures is very large. With budget constraints and lack of ranking, it is virtually impossible to find out what to do when. We therefore, believe that the strong attention directing capabilities of Activity-Based Cost, Energy, and Waste models are crucial along with the more pure assessment capabilities. We can pinpoint the measures that are likely to succeed currently, or we can find out where the problems are. It is important to notice that some of the measures we rated low can be effective later when the more pressing matters are solved. Quoting King Salomon (Ecclesiastes 3:1):

"To every thing there is a season, and a time to every purpose under the heaven"

6. WHAT CAN WE LEARN FROM THE RESULTS?

If you believe in what you are doing, then let nothing hold you up in your work. Much of the best work of the world has been done against seeming impossibilities. The thing is to get the work done.

Dale Carnegie

"So what?" you may ask. How did all this help Interface? More importantly, what could you learn from this?

The first thing to learn from this is that an energy and waste assessment of a manufacturing facility is not at all outside the scope of one's abilities. It took us less than two months from start to finish to complete the model and present the results to Interface.

Secondly, it is quite clear that the inclusion of uncertainty makes tracing very effective and efficient especially for larger models such as the Interface model. The sensitivity charts shown in Section 4 pinpoint all the critical success factors. And it is not something that is beyond your reach; commercial software for this is available.

Thirdly, notice the extreme importance of materials as a source for both energy and waste content. This may seem strange at a first glance. On one hand, we think that their importance may be somewhat overrated, primarily because overhead resources are not included and we may have, therefore, too much emphasis on direct resource consumption. On the other hand, though, we find this result quite logical because the value chain is quite long for most materials. Preceding the Interface part of the value chain, the following links in the value chain are involved in various degrees:

1. Raw material extraction of all chemical compounds used directly or indirectly to manufacture nylon.
2. Transportation of raw materials.
3. Refining of oil and gas products.
4. Transportation of the actual chemical compounds used to manufacture nylon.
5. Manufacturing of nylon.
6. Transport of nylon to Interface.

To reduce the contribution of the energy and waste content of the purchased materials, Interface could do the following:

1. Redesign both products and processes to reduce the need for material, electricity and water, in the first place, and to make the manufacturing process more resource efficient, in general. This is indeed an active strategy of Interface.
2. Get their suppliers involved and 'integrate backward' in the value chain. Preferably the suppliers should conduct similar analyses to ours which will produce data that is valuable input in Interface's Activity-Based Energy and Waste assessment implementation.

From an in-house process point of view, we recommend to sub-meter core process activities to better quantify the material and energy flows. The activities that would be particularly important are A441 ('Produce on Backing Line 1'), A443 ('Produce on Backing Line 4') and A444 ('Cut Off-line'). Virtually the entire production goes through these activities and better energy and waste data will result in better assessments, better insight, and better management.

And now the big question: *Did Interface like our activity-based approach?* Yes. Founder and Chairman of the Board, Ray Anderson, served on the dissertation committee of Jan Emblemsvåg and approved the approach. According to him "It is a major step forward in the field of environmental management". In addition, we worked closely with Dave Gustashaw, then Director, now Vice-President, of Engineering. The models provide him a long awaited capability of a unified assessment system that can efficiently and effectively deal with the various material, waste and energy streams. He clearly endorsed the work by asking and funding a follow-up project (together with Interface R&D) in which a detailed Activity-Based Cost and Environmental Management model will be developed using current data. This model will include all dimensions (cost, energy consumption, and waste generation), as well as feedback based on factory sensor data.

7. IN CLOSING

Wilderness is the raw material out of which man has hammered the artifact called civilization.

Aldo Leopold

What are the main points that we wanted to show you with this case study? First of all, although the cost dimension is missing (but currently being added in a joint project), the case study undoubtedly shows that energy consumption and waste generation assessments can be performed in an integrated fashion; the same model set up (resources, activities, drivers, assessment objects, etc.) is used. The results are also presented in the same format (uncertainty distributions, sensitivity charts, same level of absorption and the like) and this makes the whole assessment consistent and understandable.

Furthermore, the models directed attention towards environmental management and design efforts on a level of detail not possible before (at least not for Interface). From the results it is clear that the primary focus of Interface should be to involve the suppliers and start undertaking major product and process design to both reduce the need for materials in the first place, and to make the manufacturing process more resource efficient, in general. Particularly, water reduction is crucial due to its high specific heat and its effect on carpet drying - something that a monetary analysis would not have indicated because of the low cost of water. Our assessment did. And not only did our activity-based approach direct attention, but it also provided a prioritized list of critical factors. Existing environmental management systems such as PLETSUS typically cannot answer questions such as:

- Which efforts give the highest reduction in environmental impact?
- What should we do next?

The Activity-Based Energy and Waste model allowed Interface to answer these questions, and more.

8. ACKNOWLEDGMENTS

We would like to make some acknowledgements here. First of all, we would like to thank the Chairman and Chief Executive Officer of Interface, Ray C. Anderson, for allowing us to use his company as a case study. Furthermore, we thank Institute Professor C. S. Kiang at the School of Earth and Atmospheric Sciences at the Georgia Institute of Technology for establishing the contact in the first place. We would also like to thank the President of Interface Research Corporation, Dr. Michael D. Bertolucci, for setting us up with the manufacturing plant in Troup County. The inspiring help from then Director and now Vice-President of Engineering David H. Gustashaw has been invaluable. Without his help and continuing enthusiasm this case study would have been very difficult to complete. Also, the patience of Cost Accounting Manager John Daniel Purgason in explaining the intricate costing system used in textile industry, providing the Bill-Of-Material and specific process information has been most helpful. Other people who contributed substantially are Industrial Engineering Manager Phyllis Woodson for her in depth explanation of the various production activities and Director of Process Development Stuart A. Jones who provided a clear and meaningful translation of all the trade names for the various chemical compounds used in production into a layman's language. Thank you all.

Chapter 8

THE WESTNOFA INDUSTRIER CASE STUDY

When a superior man understands the Way, he follows it with zeal.
When an average man understands the Way, he adopts parts of it, but leaves the rest.
When an inferior man understands the Way, he bursts out laughing.
If he did not laugh the Way would not be the Way.

Lao Tzu
From "Tao Te Ching"

In this chapter we show how Activity-Based Cost and Environmental Management (ABCEM) has been implemented in a medium sized manufacturing company in Norway called Westnofa Industrier AS - denoted Westnofa. Again a real company with real problems. The most important purpose of this case study is to illustrate the fact that an actual implementation was made that has been used by management since 1996. This case study also illustrates the comprehensiveness of Activity-Based Cost and Environmental Management. This case outgrew the Crystal Ball® software and it is our largest activity-based model. In this book the highlights of this case study are presented. The most complete presentation is in (Emblemsvåg 1999).

1. DESCRIPTION OF WESTNOFA INDUSTRIER

Natural selection cannot possibly produce any modification in a species for the good of another species, though throughout nature one species incessantly takes advantage of and profits by the structures of others.

Charles Darwin
In "The Origin of Species"

1.1 Company Overview

Westnofa Industrier AS, to which we refer as 'Westnofa', is a medium sized company in the heartland of Norway. Located in the end of a long fjord with steep mountains around, one would believe that doing business there is rather hard. Nevertheless, this company is doing well, and since it was founded in 1968 it has grown steadily to become one of the leading mattress and bed manufacturers in Norway. In 1995 it had 163 full-time employees and the annual sales were NOK[1] 163 million with a profit of NOK 8 million.

[1] Norske Kroner, NOK 7 ≈ USD 1

Westnofa is owned 100% by Stokke Gruppen AS in Ålesund which is one of Norway's leading furniture manufacturers with 540 employees and annual sales of NOK 540 million. In addition to Westnofa, Stokke Gruppen consists of two other major business units: Stokke Fabrikker AS and Fora Form AS. ABCEM was also implemented in Stokke Fabrikker, but there is nothing new to learn from that except that implementations are repetitive, systematic and always work. This is, however, important in itself because it clearly adds credibility to our statements and approach.

1.2 Organization

Westnofa is small in international context, but it has all the challenges of larger companies, such as multiple production departments and large product variety, which makes Westnofa an excellent case study for us. There are more than ten departments in the company. Six departments are production departments, which are organized according to their products, see Table 1.

1.3 Facilities

There are three main facilities, all located in the town of Åndalsnes in the county of Rauma. The facilities cover 17,000 m^2 of production, storage and office areas. Transportation between the three facilities 'Main Building Øran East', 'BNC' (main storage) and 'Kammen' (the Wonderland production facility) is needed. The separation between the facilities is a result of the expansion of the company.

From Department 01, which is located in the main building, Wonderland gets the foam plates for their beds and mattresses. Wonderland is also receiving parts at their facility, but from a logistics point of view, the transportation between the main building and the Wonderland facility is the most critical. The finished goods from Wonderland are then mostly stored in the main storage. It is located roughly 500 meters from Wonderland, so a truck is needed. When the Wonderland products are shipped, they must then be transported to the main building - another 500 meters away - prior to the shipment. Obviously, this set up requires 'unnecessary' transportation, which in 1995 amounted to NOK 300,000. This transportation activity is caused partly by the high volume of products combined with a Make-To-Storage (MTS) production philosophy. Consolidation into one building would cost much more. Another option is to eliminate Department 14 (Receiving and Shipping) and let the main building and the Wonderland facility be responsible for their own logistics. Also, a change in production philosophy from MTS to MTO (Make-To-Order) to eliminate storage is an option too.

1.4 Product Overview

Westnofa has two main product lines. The first is the production of industrial foam (polyurethane). These products, referred to as 'Industrial Products', are mainly sold domestically to furniture companies. The second product line is the production of spring mattresses and beds. These products are marketed and sold throughout Scandinavia under the Wonderland brand.

More specifically, the different production departments and the corresponding product lines are as shown in Table 1. Out of the hundreds of different products that Westnofa produces, only a few are presented here. For example, Department 03 has several thousand different products such as various armrests for chairs, chair cushions and so on. In our model most of these products are grouped into the various product lines they belong to, and the entire product line is assessed instead of each product. In this chapter, we will only focus on Department 08 (Wonderland) products and in the following paragraphs some of the typical Wonderland products are presented.

Table 1 - Production Departments and Their Respective Product Lines.

Production Departments	Product Lines
Department 01	• Foam
	• Cut foam for the industry
	• Mattresses without springs
	• Foam plates for Wonderland
Department 03	• Thousands of different injection molded foams for the industry
Department 04	• Integral molds
Department 05	• Special foam for ski manufacturers
Department 06	• Foam molds for Department 03
Department 08 (Wonderland)	• Wonderland products

In Figure 1 one of the simplest Wonderland products, called the '0010 Mattress', is pictured. The 1995 sales price for the 0010 was NOK 771, and with an annual production of over 21,000 units, this was one of the most profitable products.

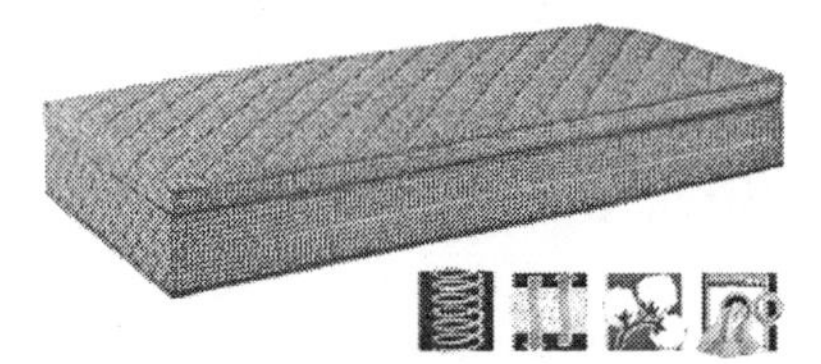

Figure 1 - The 0010 Mattress with 0010 Top-Mattress.

The four small symbols represent some important features about the mattress - from left to right:

1. The spring is a so-called LFK spring, which is not encased in a cylinder, as the pocket springs in Figure 2. This gives lower comfort but for people who like a harder mattress, the LFK is preferred.

2. The second symbol signifies that the mattress is breathing and ventilating, due to a special type of foam, called Celltex foam, produced solely by Westnofa. Celltex also improves comfort and is very durable.
3. The flowers indicate that the textiles are made of 100% cotton.
4. The washing machine means that the mattress cover can be removed, washed and dried.

Another product is the '4000 Bed' with pocketed springs which is used in the 4000 Continental Bed, see Figure 2. The 4000 Bench was sold for NOK 1,407 and 4,800 units were produced in 1995. This is a new product mostly positioned for the Swedish and Danish markets where simple mattresses as the 0010 are less competitive than the more complex beds. The second symbol means that the cover of the thin top mattress is made of perforated natural latex. The manufacturer claims it enhances comfort and mattress life.

Figure 2 - The 4000 Continental Bed with Pocketed Springs.

The last product presented here is the frame bed, such as the '5000 Frame Bed' shown in Figure 3. Like the 4000, the 5000 beds are mainly sold in Sweden and Denmark. In 1995 only 500 units of the 5000 Frame Bed were produced which were sold for NOK 1,782 per unit to retailer.

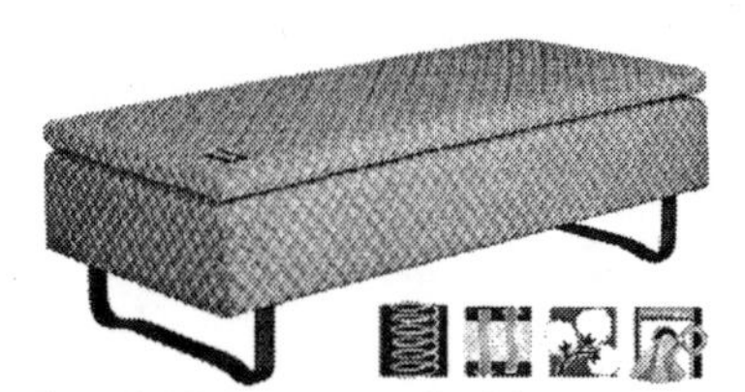

Figure 3 - The 5000 Frame Bed with 5000 Top-Mattress.

1.5 EMAS Certification

An important aspect of this case study is that Westnofa is EMAS (Eco-Management and Auditing Scheme) certified. This aided our implementation as Westnofa had many environmental data readily available for us.

EMAS is a voluntary certification procedure administered by the European Organisation for Testing and Certification (EOTC) and under contract to the European Commission (Council Regulation (EEC) No 1836/93 1993). As the name indicates, EMAS is really only an auditing approach that is useful to establish an inventory of inputs and outputs. It promotes:

a) The establishment and implementation of environmental policies, programs and management systems by companies in relation to their sites.
b) The systematic, objective and periodic evaluation of the performance of such elements.
c) The provision of information of environmental performance to the public.

To become EMAS certified a company must comply with all articles and annexes and a certification bureau must approve the company. EMAS is not usable as an assessment approach, but it is a useful way to perform an inventory analysis in most cases. The policy and strategic part of EMAS are similar to the ISO 14001 standard.

1.6 What Did Management Want?

Management was particularly interested in learning more about the following problems:

- What is the profitability of the different products?
- What are the critical success factors?
- How much does the scraping of foam costs annually?
- What is the utilization of the block production and how can block production be improved?
- What are the critical environmental aspects of Westnofa?
- How can the environmental performance of products and processes be improved?
- How can the 5000 Frame Bed be improved?
- Will Make-T0-Order production be better than Make-To-Storage?

To answer these questions we implemented our approach. Moreover, it was implemented as a *full absorption* model. That is, we trace <u>all</u> costs to the products, except things like gifts, newspapers, etc. However, these costs ('Facility Overhead Costs') are only a fraction of the total costs, see Table 4.

2. DEVELOPING AN ACTIVITY-BASED ENERGY AND WASTE MODEL FOR WESTNOFA

My advice is to look out for engineers. They begin with sewing machines and end up with nuclear bombs.

Marcel Pagnol

As usual, we follow the steps outlined in Chapter 4, Section 2.

2.1 Step 1 - Identify the Assessment Objects

In this case, the management at Westnofa chose the system boundaries to be the same as the boundaries of the current Westnofa management systems, that is, all product, process and organizational issues within the company. The marketplace is considered an external variable beyond their control. Based on this, a total of 50 assessment objects were identified:

- Westnofa Industrier as a whole
- Aggregated Industry Deparments
- 6 separate departments
- 42 products and product lines

2.2 Step 2 - Create an Activity Hierarchy and Network

When we created the activity hierarchy we simply broke down all the operations, or actions, performed at Westnofa and gathered them into activities, see Table 2 and Table 3.

Table 2 - Activity Hierarchy of Westnofa Industrier Levels 1 through 3.

Level 1		Level 2		Level 3	
Industrial Production	A1	Dep. 01 Production	A11	Purchase Raw Material	A111
				Produce Products	A112
				Compact Residual Foam	A113
				Ship to Wonderland	A114
				Packaging of Products	A115
		Dep. 03 Production	A12	Purchase Raw Material	A121
				Change Molds	A122
				Produce by Machines	A123
				Produce with Frame	A124
				Cool down	A125
				Crushing	A126
				Finishing Process Products	A127
				Package Products on Machine	A128
				Service the Department	A129
		Dep. 04 Production	A13		
		Dep. 05 Production	A14		
		Dep. 06 Production	A15	Support Department 01	A151
				Maintenance for Dep. 03	A152
				Form/model Integral Mold	A153
				Form/model Soft Foam	A154
		Administrate Industry	A16	Aid Production	A161
				Carry out Projects	A162
				Purchase	A163
				Manage Production	A164
				Design Products	A165
				Sell Products	A166
		Welfare Program Industry	A17		
Wonderland Production	A2	Receive Raw Materials	A21	Handle Raw Material	A211
				Store Raw Material	A212
		Production	A22	Produce Cover	A221
				Produce Frame	A222

		Pull Cover on Frame A223
		Assemble Frame/Bench A224
		Package Product A225
	Forwarding via BNC A23	
	Administrate Wonderland A24	Plan Production A241
		Purchase A242
		Manage Production A243
		Design Products A244
		Sell Products A245
	Welfare Program Wonderland A25	
Run Mechanical Shop A3	Service Dep. 01 Equipment A31	
	Service Dep. 03 Equipment A32	
	Service Dep. 04 Equipment A33	
	Service Dep. 05 Equipment A34	
	Service Dep. 06 Equipment A35	
	General Industry Service A36	
	Service Wonderland A37	
Receive and Ship off A4	Store Products A41	
	Ship Products to Customer A42	
	Administrate Forwarding A43	
Administrate Westnofa A5	Manage A51	
	Run Ordering A52	
	Run Accounting/Wages/Debt A53	
	Run Switchboard A54	

Table 3 - Activity Hierarchy of Westnofa Industrier Level 4 and 5.

Level 4	Level 5
Purchase Liquid Raw Material A1111	
Purchase Fabric, etc. A1112	
Produce Foam Blocks A1121	
Produce on Cutting Line for Ind. A1122	
Produce on Cutting Line for Wond. A1123	
Crushing of Top-Mattress to Wond. A1124	
Cut for the Furniture Industry A1125	
Produce Foam Plastic Mattresses A1126	Sort Residual Plates A11261
	Adhere Residual Plates A11262
	Make Grooves in Plates A11263
	Lime/Roll Package/Label on Line A11264
	Sew Cover A11265
Produce by Machine 4 A1231	
Produce by Machine 2/3 A1232	
Produce by Machine 1 A1233	
Produce by PU-Machine A1234	
Lubricate/Change Frame A1241	
Put Frame in Machine 4 A1242	
Put Frame in Machine 2/3 A1243	
Put Frame in Machine 1 A1244	
Cut and Fit Edges A2211	
Sew Edges A2212	
Sew in Zipper A2213	
Cut and Fit Plates A2214	
Sew in Velcro A2215	
Sew Cover A2216	
Mold Frame A2221	
Cut Frame A2222	
Pull on Cover A2231	
Fill up Top-Mattresses A2232	
Semi-manufacture Product A2241	
Nail Product A2242	
Package on Line A2251	
Package Top-Mattress A2252	
Transport Internally A2253	

The gray shaded cells are the activities that we can find in the activity networks. Here, we only present the Wonderland activity network see Figure 4. The 'Support Activities' are activities that are not in the flow of materials, Work in Progress (WIP) or products. Due to the information systems at Westnofa, we are unable to estimate the WIP, but the error is negligible. The other Westnofa activity networks can be found in (Emblemsvåg 1999).

The insight of Chief Operations Officer Tore Lillebostad, Chief Technical Officer Ove Søvik and Health, Safety and Environment Coordinator Roger Grande were invaluable in this process. We should note that Westnofa has used the model for over four years now without much change. This shows that we managed to define the activities in such a way that redefinition has not been necessary.

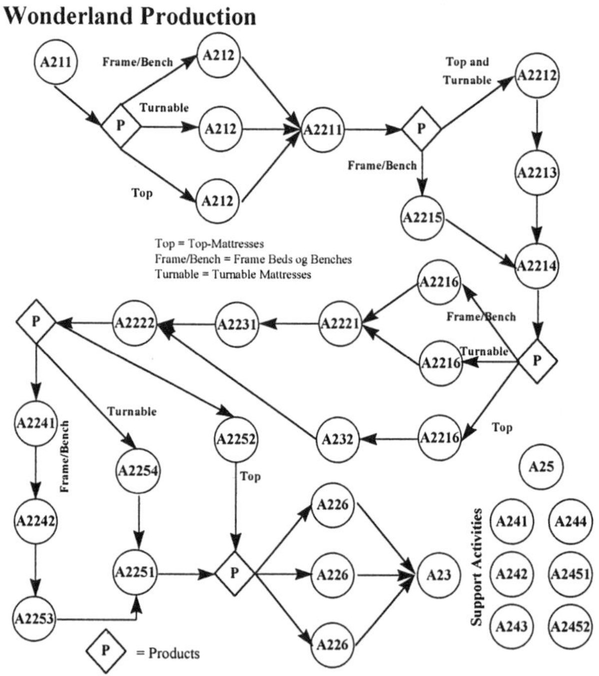

Figure 4 - Wonderland Activity Network.

2.3 Step 3 - Identify the Resources

2.3.1 Resource Overview

Our model's scope was equal to the company's financial management system. If we compute the total costs from the General Ledger we get 144,619,266 NOK/year, which is 428,581 NOK/year less than the total in the model, see Table 4. This is because Westnofa does not have an information system to gather the data in an activity-based manner. Thus, we had to convert some information and in many cases actually go to the factory floor and measure. Nevertheless, our total cost estimate is only 0.30% too high, which in our opinion - and in the opinion of the management at Westnofa - is more than 'good enough'. A 100% match was not justified because of the cost of information versus the slightly increased accuracy.

Clearly, we cannot describe all Westnofa's resources here. Hence, we only present the main resources elements in Table 4 through Table 6. In Table 4, the resource elements associated with environmental impact analyses are also presented. These numbers are associated with high degrees of uncertainty and are probably underestimated. In some cases, information is totally lacking, and consequently 'not available' (N/A). Hence, we have omitted these resource elements. An example of this is the waste content of purchased parts for materials used by support.

Table 4 - Resources at Westnofa Industrier.

Main Resource Elements	Cost [NOK/year]	Energy content [MJ/year]	Waste content [pWU/year]
Activity Wages (Corrected for Car Usage)	37,042,232	N/A	N/A
Work Environment	335,818	N/A	N/A
Deprecations and Maintenance	8,617,771	3,418,489	16,972
Buildings	6,400,338	21.308.687	11,489
Facility Overhead	131,143	N/A	N/A
Insurance	263,511	0	0
Material Usage in Production (Corrected for Internal Sale)	71,863,604	228,308,517	0
Material Usage Support	111,106	N/A	N/A
Training	58,870	N/A	N/A
Sales Resources	20,223,455	N/A	N/A
Total Costs	**145,047,847**	**253,094,693**	**28,454**

The waste generation is measured using the Waste Index (WI) whose unit is Waste Units (WU), see Chapter 2. We see that the generated waste has two sources: 'Depreciation and Operation' and 'Buildings'. These are aggregated resources, similar as for the Interface case study's buildings. In Table 5 and Table 6 we show how the WI is calculated. The same calculation procedure is followed as in the previous case studies. Please note the units in Table 5. The cars' gas consumption is assumed to be 1 liter/10 km.

Table 5 - Waste Index Calculations.

Resource	Waste Element Annual Release		Waste Content (Appendix A)	Waste Release from Use	Annual Waste Generation (WI)
Means of Transportation					
Van (Dep. 07)	Car Fuel	3,000 km	0.0670 pWU/kg	9.88E-03 pWU/km	44 pWU
Truck (Dep. 14)	Car Fuel	7,000 km	0.0670 pWU/kg	9.88E-03 pWU/km	103 pWU
Machinery					
Block Production	Heptane	8,000 kg	N/A pWU/kg	1.76E+00 pWU/kg	14,080 pWU
	n-Pentane	350 kg	N/A pWU/kg	1.75E+00 pWU/kg	612 pWU
	CO_2	100,000 kg		2.15E-02 pWU/kg	2,150 pWU
Buildings					
Main Building Øran East	Diesel	50,400 kg	0.0644 pWU/kg	1.31E-01 pWU/kg	9,848 pWU
Kammen Ind. Building	Diesel	8,400 kg	0.0644 pWU/kg	1.31E-01 pWU/kg	1,641 pWU

Table 6 - Unit Waste Index Calculations.

Resource Waste Element	Release	Degrades into	Unit Release	$V.T_N/2A_N$ Ratio (Appendix A)	Unit Waste Release from Use [pWu/kg]
Car Fuel					**9.88E-03**
	CO_2	Appears naturally	2.09E-01	2.15E-02	4.49E-03
	CO	Appears naturally	2.30E-03	2.21E-01	5.08E-04
	SO_2	Appears naturally	2.00E-04	6.28E-01	1.26E-04
	CH_4	Appears naturally	4.20E-05	1.72E+00	7.22E-05
	NO_x	Appears naturally	2.60E-04	1.80E+01	4.68E-03
Heptane	C_7H_{16}	Smaller polymers - CH_4	1.00E+00	1.76E+00	**1.76E+00**
n-Pentane	C_5H_{12}	Smaller polymers - CH_4	1.00E+00	1.75E+00	**1.75E+00**
CO_2	CO_2	Appears naturally	1.00E+00	2.15E-02	**2.15E-02**
Diesel					**1.31E-01**
	CO_2	Appears naturally	3.20E+00	2.15E-02	6.88E-02
	CO	Appears naturally	1.35E-02	2.21E-01	2.98E-03
	SO_2	Appears naturally	6.00E-03	6.28E-01	3.77E-03
	CH_4	Appears naturally	6.00E-04	1.72E+00	1.03E-03
	NO_x	Appears naturally	3.00E-03	1.80E+01	5.40E-02

2.3.2 Information Assumptions

It was assumed that heptane and pentane will decompose into methane (CH_4) in air. Using (Mackay, Shiu *et al.* 1993), we can calculate that after only 238 hours, 99.9939% of the initial pentane (and similar for heptane) is decomposed into methane. Thus, we can assume that the balance time for pentane and heptane is roughly the same as for methane since 238 hours is orders of magnitude smaller than 10 years. Other data was obtained from the following sources:

- Energy Content and Waste Content for the various resource elements are found in the IDEMAT software.
- The data regarding the 'Waste Elements' are found in (Westnofa Industrier 1997).
- Diesel data is from the Opplysningskontoret for Energi og Miljø AS in Norway, David Hart, Centre for Environmental Technology, Imperial College, London, and Günter Hörmandinger, Energy Policy Research, London. Car fuel data was obtained from the latter two sources.

2.4 Step 4 - Identify and Quantify the Resource and Activity Drivers and their Intensities.

In a model of this size, there are far too many resource and activity drivers to present them all. We therefore only present the most frequently used ones, see Table 7. The same drivers may be associated with costs, energy consumption, waste generation or a combination of these three performance measurement dimensions. As mentioned before, we believe that this is a major strength of our approach because it enables an integrated performance measurement system for an organization, thus, reducing the learning curve and costs of implementation and maintenance and it enhances efficiency.

Table 7 - Main Resource and Activity Drivers.

Resource Drivers	**Activity Drivers**
Area	Involved Products
Involved Activities	Labor Hours
Labor Hours	Machine Hours
Time Percentages	Number of Units
	Product Volume

We worked with Westnofa management to select the drivers. Decisions were not always obvious. For example, we used 'number of units' as an activity driver because we assumed that the consumption of an area is strongly related to the number of units produced. But this is a simplification because a product often needs a minimum amount of space to be produced at all. The alternative is to say that given an area, all resources related to that particular area are traced equally among the products that consume this area. That means in theory, that a product produced once a year is attributable to the same degree as a product produced in huge quantities, which definitively is not a very accurate approximation. Clearly, both approaches have their weaknesses, but we think that the first is the least weak.

This discussion illustrates three fundamental problems with performance measurement in general:

1. There is no single correct way of doing performance measurement.
2. The crux of performance measurement is to choose the approach that is the least wrong.

These two lead to a third problem in performance measurement, which we think is hard for many engineers, in particular, to accept:

3. *There is <u>no single correct answer</u> or set of equations that can describe a solution.*

The intensities are found the usual ways. If Westnofa had reliable capacity estimates for each activity, we could also have used the hourly wages as consumption intensities for production labor. This would have facilitated computation of the cost of unused capacity and identification of excess capacity. Unused capacity could also have been studied by investigating the throughput of each activity in relation to the time spent of each activity. But there is an inherent problem of all productivity and capacity studies due to the adaptation of workers. That is, if there is little to do, workers will work slower than if there was a lot to do, and vice versa. We therefore run into the problem of estimating standard times: how fast should a 'standard worker' work in the various activities? A related issue concerning productivity and capacity studies is the inevitable usage of time studies, which most employees strongly dislike. A bar code scanning system is often a better option, but that was not implemented at Westnofa in 1995.

2.5 Step 5 - Identify the Relationships between Activity Drivers and Design Changes

2.5.1 Possible Relationships

The relationships between activity drivers and design changes can be treated in two ways as explained in Chapter 4. The first way is to simply ignore the relationships in the model, and let the designers themselves figure out how to reduce for example the sewing time. In such a case, the model will only provide guidance in finding *what* to improve (e.g., sewing time), but not *how* (e.g. a mechanism to hold the fabric during sewing).

The second way is to create detailed *sub-models* where, e.g., the parameters of interest are material usage and time. This will result in accurate modeling and either mathematical functions or action charts can do this. The mathematical functions are the most accurate, but may be very hard to describe an activity in terms of equations. An activity that could have been described in terms of equations is 'Produce on Cutting Line for Wonderland' (activity A1123), however, due to the general lack of automation at Westnofa the parameters in the functions would have been somewhat unreliable. Action charts (see Glossary) are therefore the preferred approach at Westnofa, and possibly in most cases.

2.5.2 Sub-modeling Using Action Charts

To illustrate the use of action charts, we illustrate how to create a sub-model for a specific assessment object. The assessment object can be a product, a product line, a process or any combination, it all depends on what the purpose of the sub-model is. Here, we investigate how the Wonderland 5000 Frame Bed can be improved using the *existing* production equipment.

The action-based sub-modeling method is shown in Figure 5. The method consists of six steps that are fairly intuitive. It starts by identifying the assessment object(s). In our case we chose the 5000 bed because of its intricate seam, which is introvert, i.e. it cannot be seen from the outside like on most mattresses. Therefore from a design point of view, it would be interesting to find out how to reduce the sewing time in order to reduce the unit costs and unit energy consumption. However, it is not as easy as just doing a time study because activities consume resources in unequal amounts and consumption intensities. Thus, one second in one activity can cost more than one second in another activity. The point is to find ways to reduce the expensive seconds.

Step 2 is therefore to simply identify which activities are affected by changes in activity time consumption. That is, we identify those activities whose variable cost, variable energy consumption (EC) and variable waste generation (WG) will be affected by changes in the product. This is done in

Table 8, which forms the basis on which an action chart can be set up. The consumption intensity is found by dividing the cost of, e.g., activity A2211 on the total time per unit to perform activity A2211, which is 64.5 seconds. To compute the total time per unit, all the batch sizes must be accounted for. Note that there is no variable waste generation for any of the activities. That is caused by lack of data because when there is energy consumption, there will be in most, if not all, cases waste generation. Also note that the numbers in the 'Total' row of the CI and GI columns are the weighted averages.

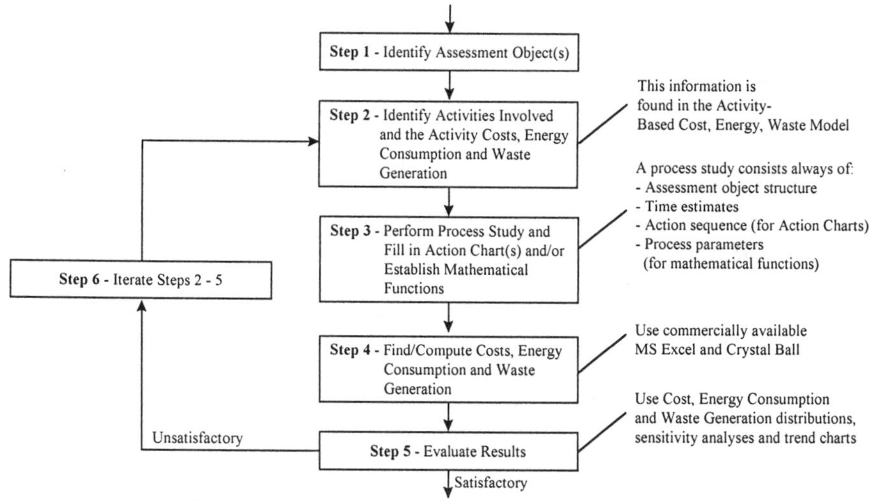

Figure 5 - The Action-Based Sub-Modeling Method.

Table 8 - 5000 Variable Unit Cost, Energy Consumption (EC) and Waste Generation (WG) and Corresponding Consumption and Generation Intensities.

Activity	Cost [NOK]	CI [NOK/sec.]	EC [MJ]	CI [MJ/sec.]	WG	GI [pWU/sec.]
A2211	6.11	0.10	0.15	0.00	0.00	0.00
A2213	5.38	0.07	0.00	0.00	0.00	0.00
A2214	2.32	0.15	0.03	0.00	0.00	0.00
A2216	119.52	0.09	0.00	0.00	0.00	0.00
A2221	32.30	0.15	108.79	0.51	0.00	0.00
A2222	2.47	0.00	0.00	0.00	0.00	0.00
A2231	28.20	0.31	3.84	0.04	0.00	0.00
A2251	6.68	0.08	6.56	0.08	0.00	0.00
Total	**202.96**	**0.07**	**119.37**	**0.04**	**0.00**	**0.00**

Excerpts from the 5000 Bed action chart are shown in Table 9. The shaded cells are assumption cells while the forecast cells are not put directly in the action chart. As one can see, some of the action numbers, i.e. 'Action No.', are in italics. These are actions that add no value to the customers, i.e. non-value added. These are actions that represent process inefficiencies and waste, which must be eliminated.

Table 9 - '5000 Bed' Action Chart Sample.

	Action			Batch	Time	Performance Dimensions			Material	Mass	Quality			
Act.	No.	Name	Freq	Size	[sec.]	[NOK]	[MJ]	[pWU]	Type	[kg]	%	[NOK]	[MJ]	[pWU]
A2211		**Cut and Fit Edges**												
	1	Stretch Out Fabric for Corner	1	10	100.0	0.95	0.02	0.00						
	2	Measure for Corner	1	10	45.0	0.43	0.01	0.00						
	3	Cut Corner	1	10	90.0	0.86	0.02	0.00						
	4	Transport Corner	1	10	10.0	0.10	0.00	0.00						
	5	Stretch Out Fabric for Side	1	5	60.0	1.14	0.02	0.00						
	6	Measure for Side	1	5	40.0	0.76	0.02	0.00						
	7	Cut Side	1	5	90.0	1.71	0.04	0.00						
	8	Transport Side	1	5	10.0	0.19	0.00	0.00						
			8		64.5	6.13	0.13	0.00				0.00	0.00	0.00
A2213		**Sew in Zipper**												
	9	Get Fabric Rolls	1	10	20.0	0.14	0.00	0.00						
	10	Set Up Machine	1	10	45.0	0.32	0.00	0.00						
	11	Sew in Zipper	1	1	45.0	3.15	0.00	0.00			2.0	0.06	0.00	0.00
	12	Put on Glider	1	1	5.0	0.35	0.00	0.00						
	13	Transport	1	1	20.0	1.40	0.00	0.00						
			5		76.5	5.36	0.00	0.00				0.06	0.00	0.00
A2214		**Cut and Fit Plates**												
	14	Get Fabric Rolls and Adjust	1	84	130.0	0.24	0.00	0.00						
	15	Place on Stand	1	84	10.0	0.02	0.00	0.00						
	16	Stretch Out Fabric	1	84	600.0	1.09	0.01	0.00						
	17	Measure	1	84	160.0	0.29	0.00	0.00						
	18	Cut Fabric	1	84	160.0	0.29	0.00	0.00						
	19	Transport	1	84	210.0	0.38	0.01	0.00						
			6		15.1	2.31	0.03	0.00				0.00	0.00	0.00

Using action charts is a very powerful, yet simple approach. It applies to *any* process that can be split up into discrete elements (i.e. actions) and to *any* product. Mathematical functions, on the other hand, apply in continuous processes such as in chemical plants, but are far less flexible since they require the establishment of reliable functions. In this book, studies using mathematical functions are not shown. Interested readers are referred to a vehicle case study in (Emblemsvåg 1995) where functions were used. In the demanufacturing model presented in (Emblemsvåg 1995), the usage of action charts is shown in the context of demanufacturing. In this book, we show that the very same approach is valid for manufacturing. We also run Monte Carlo simulations for performing a sensitivity analysis when action charts are used. The 5000 mattress results are presented later.

Compared to the action charts used in, e.g., (Emblemsvåg 1995; Bras and Emblemsvåg 1996), the redesigned action chart in Table 9 is simpler and purely time and quality based. We found that much of the information in the older type action charts was either inadequate or redundant for our purpose.

2.6 Step 6 - Model the Uncertainty

Similar to preceding case studies, we simply choose a bounded and symmetric uncertainty distribution with ±10% bandwidth, regardless of the 'real' uncertainty, to trace the critical success factors and to assess potential ranges for the assessments due to distortion in the models. In (Emblemsvåg 1999), however, an analysis of the real uncertainty was performed for Wonderland using the best knowledge from Westnofa about the actual uncertainty. The principles are the same for both cases. The only difference between an actual uncertainty analysis versus tracing is the uncertainty distribution in the assumption cells.

2.7 Steps 7, 8 and 9 - Find/Compute the Cost, Energy Consumption and Waste Generation of Activities and Objects, and Perform Sensitivity and Other Analyses

The model was implemented as three separate spreadsheet models that were run using the Crystal Ball® software. The reason for splitting the model up in three models was that the amount of information was too large for Crystal Ball to handle. The version of Crystal Ball we used is restricted to maximum 1,000 assumption cells and 10,000 trials (Decisioneering Inc. 1996). The cost model has 1,100 assumption cells and 92 forecast cells, so that model is beyond what is recommended. The energy model is smaller with 646 assumption cells and 92 forecast cells, while the waste model is even smaller with 540 assumption cells and 95 forecast cells. This yields 2,286 assumption cells and 279 forecast cells, and 2,565 variables in total.

It is hard to explain how a model of this size works beyond the fact that it is process oriented/activity-based, and all relevant process, product and organization information is included. However, on a conceptual level it works as shown in Figure 6.

In Figure 7 the cost model spreadsheet infrastructure is shown and this is probably the best we can explain the model in a book. As can be seen, there are a substantial number of feedback loops, see also Figure 6. The feedback loops are essentially making the model react dynamically to any changes in inputs. The feedback loops exist on all levels in the model and allow not only dynamic behavior but it also eliminate the need for fairly large cost pools, which are so common in commercial ABC software. We can, however, aggregate the information and present it for various cost pools. This is done in the 'Special Focus' spreadsheet where the foam production among others is under close scrutiny from a capacity utilization perspective.

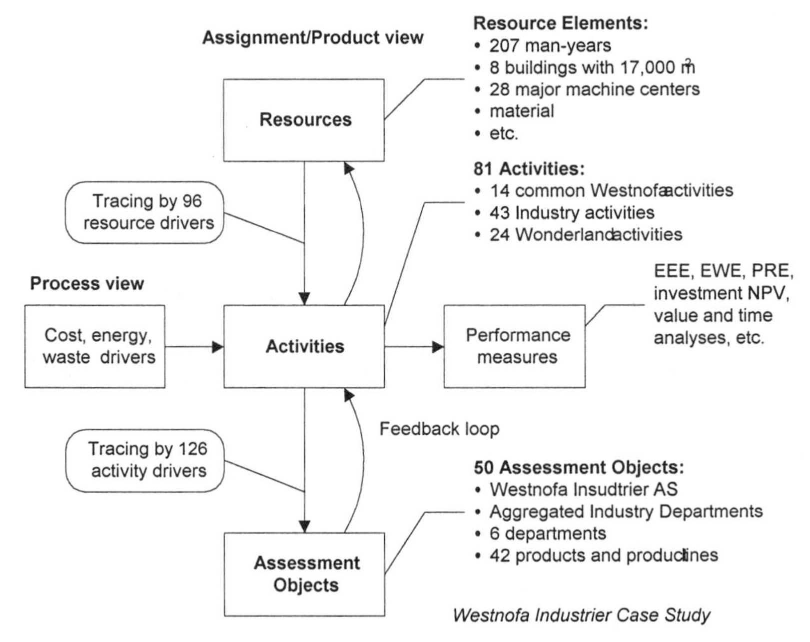

Figure 6 - Working Principles for the Westnofa Activity-Based Cost, Energy and Waste Model.

This complex infrastructure makes it possible to have assessment objects on product level, product line level, activity level, department level, aggregated department level and facility level. Furthermore, with charts we can go all the way down to action level and component level depending on what is beneficial and the budget.

The infrastructure is similar, but simpler, for the energy consumption and waste generation models. Some of the spreadsheets are also very large. The 'Wonderland Product' spreadsheet, for example, is about 1,000 rows by 15 columns. Others are small, such as the 'A4' spreadsheet that is just 10 rows by 15 columns. For the actual results we refer to Section 3.

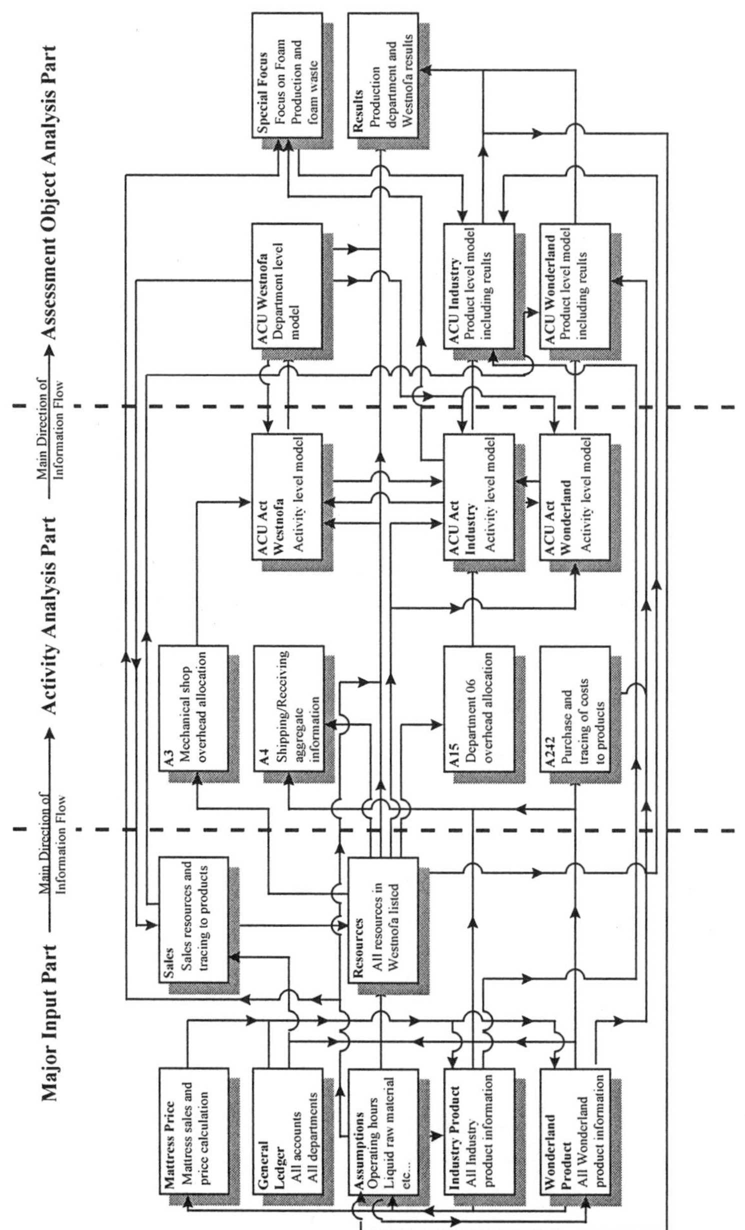

Figure 7 - Westnofa Activity-Based Cost, Energy and Waste Model Infrastructure.

2.8 Step 10 - Interpret the Results and Iterate if Necessary

It took a couple of runs before the model worked properly. Initially, logical errors were found by critically investigating the sensitivity charts and the models themselves. In this process it is vital to understand the production and business processes. We found, for example, an error in how inventory was handled and affected profitability.

The first version of the model was made in 1996, but there were several issues that could be improved. This was mostly due to the fact that we used an older approach that can be found in (Emblemsvåg 1995). We therefore redid the entire implementation yielding the model presented here. More iteration was not necessary.

3. RESULTS FOR WESTNOFA INDUSTRIER

The most exciting phrase to hear in science, the one that heralds new discoveries, is not "Eureka!", but "That's funny ...".

Isaac Asimov

Since the model is divided into three models due to the software limitations, each handling a performance measurement dimension, the results are organized accordingly. We start with the economic results. Furthermore, in order to focus this case study we discuss a limited number of assessment objects. For a complete presentation, see (Emblemsvåg 1999).

3.1 Economic Results and Discussion

There are over 270 forecast cells and we cannot present all of them here. We have therefore chosen to focus on the following two assessment objects: 1) Westnofa Industrier and 2) the 5000 Frame Bed. In addition, other assessment objects will be discussed briefly and various comparisons made. There are two things that are important to understand when looking at the results from such a model.

- First, everything affects everything; that is, if you eliminate a product the costs associated with that product will be transferred to other products unless overhead costs and direct costs are reduced correspondingly.
- Secondly, the assessments include uncertainties. The uncertainty is caused by model uncertainty due to driver choices and by numerical uncertainty that arises whenever numbers in the model are not based on recent measurements or validated numbers from accounting systems.

3.1.1 Westnofa Industrier Economic Performance

The economic result is presented in Figure 8 as an uncertainty distribution. This distribution is a result of the ±10% variation in all the assumption cells. It is interesting to see that a ±10% variation can cause fluctuations in the profitability from -880,998 NOK to 18,927,143 NOK. This means that with fairly small changes in market and cost management, huge improvements in profitability are possible. From this we understand that the butterfly effect from chaos theory, see (Gleick 1987; Mosekilde and Feldberg 1994), is also present in business. Consequently, it is vital for managers and designers to understand the processes of their company and the critical success factors.

The mean is NOK 8,115,012. The Profitability Resource Efficiency (PRE), defined as profitability divided by costs, is 7.24%. That is, for every NOK 1 spent, in 1995 NOK 0.0724 was made. Compared to the results for 1997, which was close to NOK 20 million, this is significantly lower. According to CFO Jørn Nes, the very good 1997 result was mostly attributable to an improved market situation, but also because they now consider for every aspect how their decisions will turn out in the ABCEM model. The model has therefore changed behavior, and that is important as it ensures awareness of 'doing the right things right'. That performance measurement drive behavior is also discussed in, e.g., (Brown 1995).

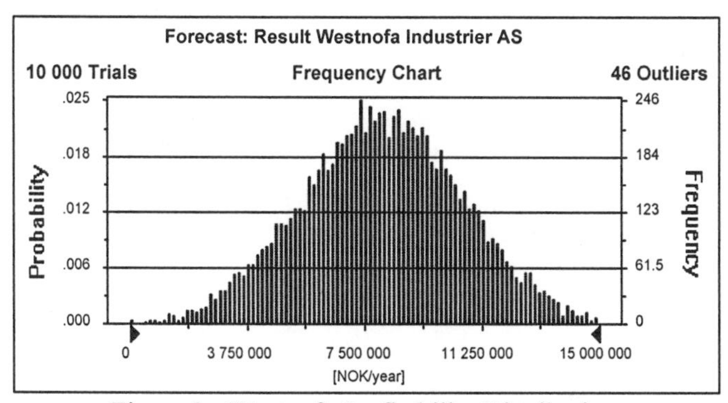

Figure 8 - Westnofa Profitability Distribution.

The capability of tracing the critical success factors is one of the most important features of our method, and here the sensitivity charts are indispensable. The sensitivity chart for Westnofa as a whole is presented in Figure 9. It should be noted that when the absolute value of a correlation coefficient is lower than 0.05, it is increasingly unreliable. This can be verified by looking at the sensitivity chart to see if it makes sense or not. This is also substantiated by experience which shows that 0.05 to 0.02, depending on the size of the model, is the region where random effects come into play. Since the Westnofa model is a very large model 0.05 seems a likely cut-off. If the management at Westnofa was interested in establishing a comprehensive ranking of the critical success factors, this can be achieved by simply running the model many times and eliminating the assumption cells that have a correlation value of over roughly 0.07.

Understanding the sensitivity chart requires training, but most of all, understanding of the processes within the company and how the model works.

We see that Department 03's products top the list. This is probably mostly due to the fact that we had to model them as *one* single product line.

However, it should be mentioned that Department 03 is large and that all the products there are indeed very important.

Sensitivity Chart

Target Forecast: Result Westnofa Industrier AS

Variable	Rank Correlation
An. Prod. Dep. 03	.42
Mean Sales Price Dep. 03 Products ...	.41
Purchase Liquid Raw Material [NOK...	-.32
Sales Price Cut for Furniture Industry [	.29
Sales Rebate Wonderland	-.21
Price 0010 [NOK/unit]	.21
Liquid Raw Material [m3/year]	.21
Various Sale Wonderland	.20
An. Prod. Cut for Furniture Industry	.19
Price 2000 Top-Mattress[NOK/unit]	.12
Price 2000 [NOK/unit]	.12
Fabric, Top Surface Midas 160 [NO...	-.11
Price 0010 Top-Mattress[NOK/unit]	.11
Unit Volume Cut for Furniture Industry [	-.11
Price 007 [NOK/unit]	.11
Price 4000 Bench [NOK/unit]	.10
An. Prod. 007	.09
Price 6000 Top-Mattress[NOK/unit]	.09
Annual Production 2000 Top-Mattress	.08
Sales Price Foam Plates for Wonderl...	.08
Account 6210	-.07
Annual Production 0010	.06
Price 6000 [NOK/unit]	.06
Pocket Spring Price	-.06
A245 Labor Cost [NOK/year]	-.06
Unit Volume Foam Plates for Wonder...	-.06
A2216 Labor Cost [NOK/year]	-.06
Annual Production 2000	.05
Account 8150 Department 13	-.05
Mean Sales Price Integral Mold [NOK]	.05
Annual Production 0010 Top-Mattress	.05
Wonderland Freight [NOK/year]	-.05
Foam Plates 3,0 cm Price	-.05
Annual Production 6000 Top-Mattress	.05
An. Prod. Integral Mold	.05
Fabric, Top Surface Castor 380 [NO...	-.05
Price Tilbud [NOK/unit]	.04
A1125 Labor Cost [NOK/year]	-.04
5000 Top-Mattress Number of Comp...	.04
A1233 Labor Cost [NOK/year]	-.04
Annual Production 4000 Bench	.04
Department 01 Freight [NOK/year]	-.04

-1 -0.5 0 0.5 1

Measured by Rank Correlation

Figure 9 - Westnofa Profitability Sensitivity Chart.

The purchase price of liquid raw material for the foam production is the single most important cost factor for the entire company. Thus, it is vital for Westnofa to use the raw material efficiently; that is, eliminate waste and also purchase the material at lowest possible price. This makes perfect sense because all the mattresses and beds have foam in them and the foam cost is a direct activity cost that affects the bottom line directly, that is, NOK 1 saved in material costs is NOK 1 added to profitability. It is also a process and product design related issue; products should be designed to reduce the need for foam and processes should be designed to reduce foam consumption.

The sales price for the 'Cut for Furniture Industry' product line is the second most important marketing variable. Increasing this sales price will (obviously) increase profitability given that the sales volume remains the same. It is important that the marketing department knows the market well, because we cannot estimate how the market will react on increased prices from the model.

That the sales rebate is such an important cost factor may seem surprising, and indeed, this point is missed by many volume-based costing systems that focus on material and labor costs. The sales rebate is a double-edged sword that is hard to avoid. Particularly in the US, the steady appearance of sales blitzes has made customers virtually immune to ordinary pricing and sales without some kind of rebates involved, see for example (Drucker 1995). On one hand, the sales rebates are used to increase market shares, but on the other hand, the same sales rebates erode profitability. For Westnofa it is important not to fall for the same trap. This means that the sales rebates should not be increased, but employed conservatively.

The 0010 mattress sales price is the single most important sales price for a single product. That means that the 0010 in the most important single product that Westnofa got in their product portfolio. But again, increasing the price may not be smart. It all depends on what the competitors do. Hence benchmarking is needed. The 0010 must constantly be revisited for quality purposes and to ensure that the production of the mattress is efficient. These are typical product and process design activities.

The volume of the purchased liquid raw material is an important revenue factor as increased volume gives higher profitability *if* the price remains the same. Basically, this tells us the same as before: liquid raw material is a critical factor.

'Various Sale Wonderland' is sale of accessories for their products such as blankets. Such sales are not attributable to any product and are therefore treated separately. The importance of this sale is according to the model significant and this shows that keeping a complete product specter is important, but at the same time other findings suggest that the product specter must be managed carefully due to the possibility of high overhead costs. For Harley-Davidson, this type of sale kept them alive during the troubled years around 1980 (Barrington 1994).

The annual production of 'Cut for Furniture Industry' relates strongly to the previously mentioned sales price for the 'Cut for Furniture Industry' product line. However, since the price is ranked over the production in importance (higher correlation coefficient), we understand that the profitability increase is potentially bigger by increasing the price compared to increasing the annual production. This is probably due to diminishing effectiveness of economies-of-scale caused by an already low amount of overhead costs per unit.

The second most important single product variable is the 2000 Top-mattress sales price closely followed by the 2000 sales price. Hence it is vital for Westnofa to keep or increase the 2000 Top-mattress and 2000 competitive. Further down the list we also spot the 0010 Top-mattress sales price as being very important.

The purchase price of the Midas 160 fabric is the second most important direct activity cost. In 1995 the unit purchase price was 61.38 NOK/m. It is important for Westnofa to negotiate better prices or avoid increase in prices, but it is also important for Westnofa to look at the process and product designs to reduce the need for this fabric.

The next interesting success factor is the unit volume of the 'Cut for Furniture Industry'. This is due to logistics costs, and we see that Westnofa should try to reduce the volume of the packages to facilitate more cost-effective distribution.

In the entire list, the 007 is the only Westnofa mattress that is considered important for the overall profitability. That means that design improvements of the Westnofa mattresses should primarily be focused on the 007 mattress.

This list can in principle be extended to include all 1,100 cost related assumption cells, but for now this will do since we have already pointed out several areas of potential process and product design improvements along with some good information to marketing.

3.1.2 Foam Production and Residual Foam Press Economic Performance

As already shown, the cost of foam production is one of the most important areas to manage and improve. From the model we see that the annual effective machine hours are only 87.9 hours, which gave a machine hour consumption intensity of 21,386 NOK/h. From this we immediately see that there is a lot of unused capacity. For Westnofa, we can use 1739 hours as a man-year. That means that the foam production machinery is only utilized with 5.05% of theoretical capacity. It should be noted that since the foam production requires a lot of storage space, the real capacity utilization is much higher, but definitively not high enough. We therefore suggested increasing foam production and selling to other companies. This was done and in 1997 the foam production machinery ran 100.9 hours, lowering the machine hour consumption intensity to 18,567 NOK/hour. Because this will not affect the overhead costs much, it will increase the overall profitability as long as Westnofa charges a price equal to the cost of raw material and transportation. This is evident from the fact that the alternative overhead cost is the same.

Another issue that Westnofa wanted us to shed light upon was the cost of the wasted foam that was sent to the residual foam press and then to recycling. In 1995, 258,000 kg foam waste was created and sent to a recycling plant, which paid Westnofa 3.83 NOK/kg. Including this revenue, Westnofa lost NOK 2,870,078 caused by this waste, or 11.12 NOK/kg waste. Basically, this means that if Westnofa can eliminate this activity by making a NOK 2.8 million investment, it would be paid back in one year! The question is *how* to do that. Due to the foam production technology, it is unlikely to eliminate all this waste at the source, that is, the foam production. One viable solution, however, is to create a new product that uses this waste as raw material. That can be done by cutting up the waste in smaller pieces and create a cheap mattress, and some of the mattresses indeed serve as such products. The danger by doing this is that it can cannibalize sales of some of their cheaper Westnofa mattresses. However, selling, or even giving, these mattresses to people in unexplored market segments *may* be a solution. One option is to give such mattresses to charity organizations. This will not produce any revenues for Westnofa, but it will eliminate this activity and expand some of the other activities. The result is that the bottom line for Westnofa remains unchanged. Before something like this could be undertaken, however, the logistics need to be studied more carefully.

3.1.3 5000 Frame Bed Economic Performance

The 0010 mattress is the most successful product of Westnofa even though it is a low-end product for Wonderland, while at the other end we have the 5000 Frame Bed that is basically a luxury version of the 0010 with more expensive materials and intricate sewing process. But the 5000 performance is bad, as shown in Figure 10; it can under no circumstances be profitable. This is also illustrated by the very low PRE of -52.51%. This clearly illustrates that the traditional rule of thumb that a high margin yields high profitability can be wrong, see also (Cooper 1990a).

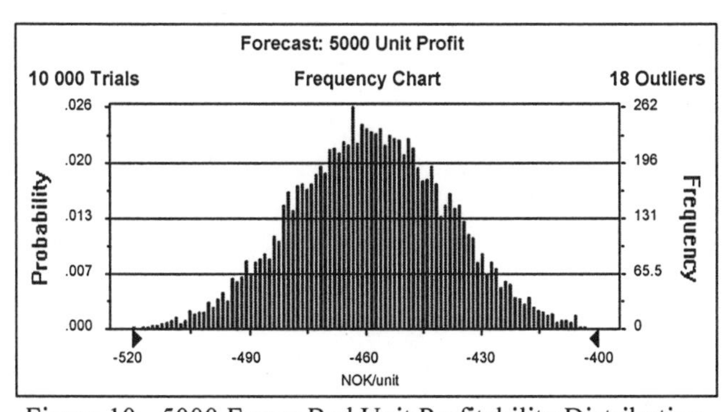

Figure 10 - 5000 Frame Bed Unit Profitability Distribution.

From the sensitivity chart in Figure 11 we see that to improve the 5000 profit, we need to increase sales price - rather obvious. We also see, for example, that activity A2216 - Sew Cover - causes costs, mainly by being too time consuming. The single most important design change of the product would be to change fabric to lower the purchase cost, but this is probably not a feasible option due to the luxury image of the product. A better option is to improve the overall efficiency of activity A2216 by reducing the sewing time.

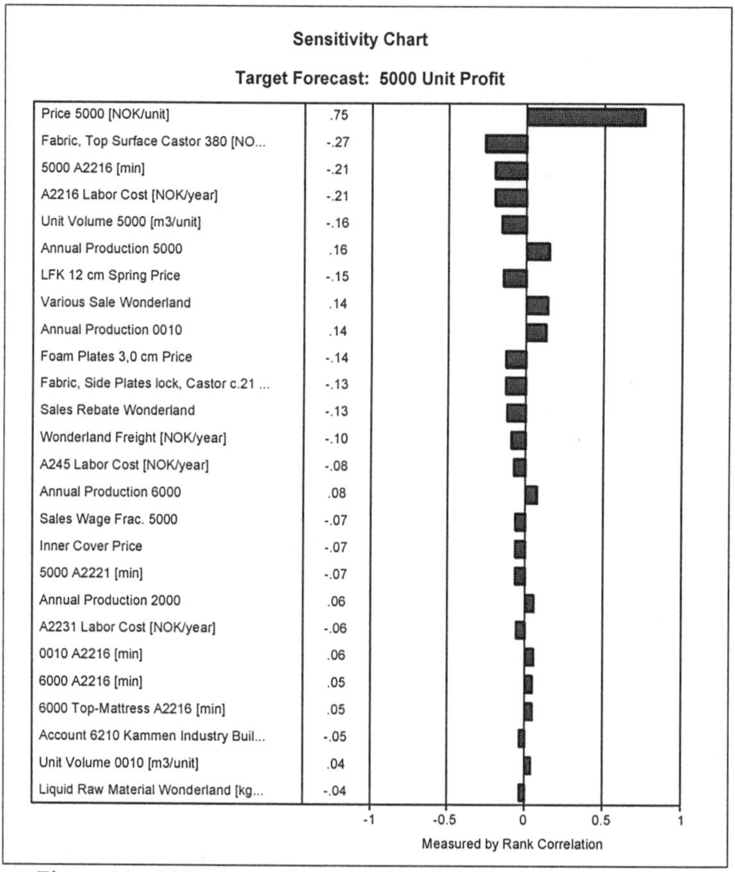

Sensitivity Chart

Target Forecast: 5000 Unit Profit

Assumption	Rank Correlation
Price 5000 [NOK/unit]	.75
Fabric, Top Surface Castor 380 [NO...	-.27
5000 A2216 [min]	-.21
A2216 Labor Cost [NOK/year]	-.21
Unit Volume 5000 [m3/unit]	-.16
Annual Production 5000	.16
LFK 12 cm Spring Price	-.15
Various Sale Wonderland	.14
Annual Production 0010	.14
Foam Plates 3,0 cm Price	-.14
Fabric, Side Plates lock, Castor c.21 ...	-.13
Sales Rebate Wonderland	-.13
Wonderland Freight [NOK/year]	-.10
A245 Labor Cost [NOK/year]	-.08
Annual Production 6000	.08
Sales Wage Frac. 5000	-.07
Inner Cover Price	-.07
5000 A2221 [min]	-.07
Annual Production 2000	.06
A2231 Labor Cost [NOK/year]	-.06
0010 A2216 [min]	.06
6000 A2216 [min]	.05
6000 Top-Mattress A2216 [min]	.05
Account 6210 Kammen Industry Buil...	-.05
Unit Volume 0010 [m3/unit]	.04
Liquid Raw Material Wonderland [kg...	-.04

-1 -0.5 0 0.5 1

Measured by Rank Correlation

Figure 11 - 5000 Frame Bed Unit Profitability Sensitivity Chart.

3.2 Energy Consumption Results and Discussion

As noted before, the numbers we have for energy consumption are probably too low due to the lack of energy accounting, however, by using IDEMAT we have tried to remedy this to the extent possible.

3.2.1 Westnofa Industrier Energy Consumption Results

According to our model, Westnofa Industrier consumed 246,450,000 MJ, in 1995, see Figure 12. We find an EEE of 1.733 MJ/NOK or 12.131 MJ/USD. WagonHo! had an EEE of 4.540 MJ/USD, which is 2.7 times better. Does that mean that WagonHo! is 2.7 times more energy efficient? No, because there are sources of distortion such as the lack of energy accounting, which lead to various degrees of accuracy when it comes to the energy consumption numbers. However, we do have reason to believe that WagonHo! is indeed more energy efficient since the organizational structure is much simpler. When it comes to Interface Flooring Systems, see Chapter 7, we cannot make an EEE comparison because the cost dimension was omitted.

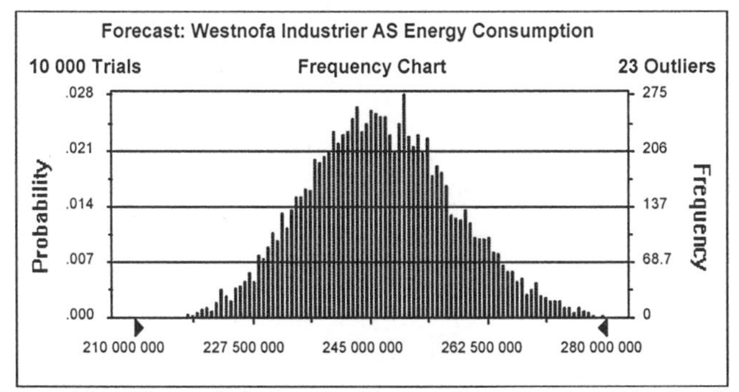

Figure 12 - Westnofa Industrier Energy Consumption Distribution.

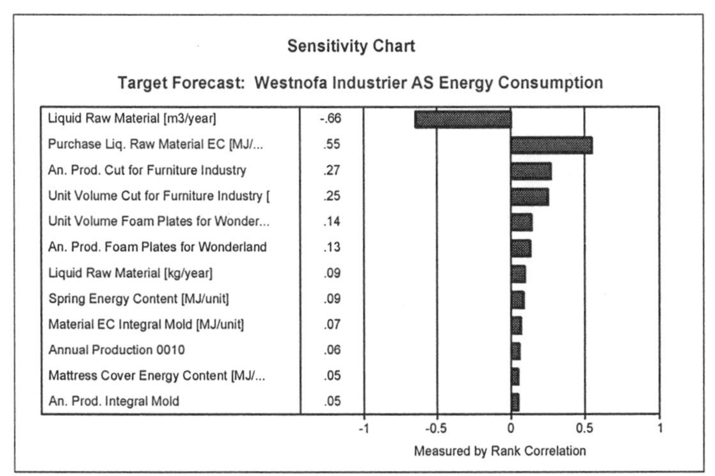

Figure 13 - Westnofa Industrier Energy Consumption Sensitivity Chart.

In Figure 13 we can find the critical success factors for the energy consumption. The ten most significant energy consumption factors are all related to the foam production in some fashion:

1. We see that by increasing the amount of liquid raw material (volume wise), the overall energy consumption will decline sharply. This is evident since this basically means that we purchase liquid raw material with less unit energy content.
2. Similarly for factor number 2; if we buy liquid raw material with higher energy content, the overall energy consumption will rise.
3. Numbers 3, 4, 5 and 6 are all related to the consumption of foam which drives the purchase of liquid raw material. Basically, the bigger the 'Cut for Furniture Industry' products and 'Wonderland Plates' are, the more foam is produced and the more energy is needed.
4. Number 7, which is the mass of liquid raw material consumed annually, also drives up energy consumption. That is, the more liquid raw material is used, the more energy is consumed. It is interesting to note that the energy content [MJ/kg] of the liquid raw material is ranked much more important than the actual consumption [kg/year] of the liquid raw material. Hence, it is more important to reduce the energy content than reducing the consumption of the material. Clearly, this must be the case when the liquid raw material consists mainly of polyols, isocyanates and amines, which are highly refined oil products.
5. Number 8 - the energy content of the springs used in all the Wonderland beds - is also an important energy driver since the spring steel is highly processed to give high strength and low mass yielding a relatively high specific energy content. This is the first energy driver not related to the foam production in any fashion.
6. Number 9 is particularly interesting as the integral molds are produced in a relatively low volume. However, each mold not only consists of foam with a high energy content, but also has metal (steel) integrated; hence the name. The production of integral molds is therefore very energy intensive compared to the rest of the Westnofa's production activities.
7. Number 10 is the annual production volume of 0010. Since 0010 is the top-selling Wonderland product in terms of units sold, it is of no surprise that the 0010 production level is an important energy driver for Westnofa.

Further down the list, assumption cells that should have no impact of the overall energy consumption appear, such as the 'Sales Wage Frac. 5000 Top-Mat'. This is a result of random effects, and it means that 0.04 is probably the lowest correlation coefficient that can be trusted in this sensitivity chart.

As with the economic analysis, possible design changes can be read straight out of the sensitivity chart. For example, the energy consumption can be reduced if we design mattresses without springs. Also, a similar benefit can be achieved by using other liquid raw materials with less energy content or by simply designing the mattresses and beds to need less foam.

3.2.2 5000 Frame Bed Energy Consumption Performance

From a unit energy consumption point of view, the 5000 Frame Bed is the best product with a unit energy consumption of 1663 MJ, see Figure 14. The EEE is 1.907, which is 19.3% better than the 0010. In contrast, we find that the 5000 was a very bad product from a profitability point of view. Thus, profitability and energy efficiency may not correlate well. This illustrates the need for doing energy consumption assessments.

We allow ourselves to make this particular comparison because the 0010 and the 5000 are similar products, and the energy accounting problems are therefore similar. The existence of energy accounting problems can be illustrated by the fact that the unit energy consumption of the 0010 and the 5000 is about the same, while the unit cost for the 5000 is 23.5% higher. It is very likely that the energy numbers in the model are much more distorted than the monetary numbers due to the absence of energy accounting in the world. By doing energy accounting along the lines of monetary accounting the bias caused by the distortions will be very similar and thus the EEE will be meaningful also as a mean of comparison *between* products, see (Emblemsvåg and Bras 1998a). Unfortunately, due to software limitations, tracing of the PRE, EEE and EWE critical success factors was not possible. However, by looking at the various sensitivity charts presented here, one can get a good idea of what will lead to success.

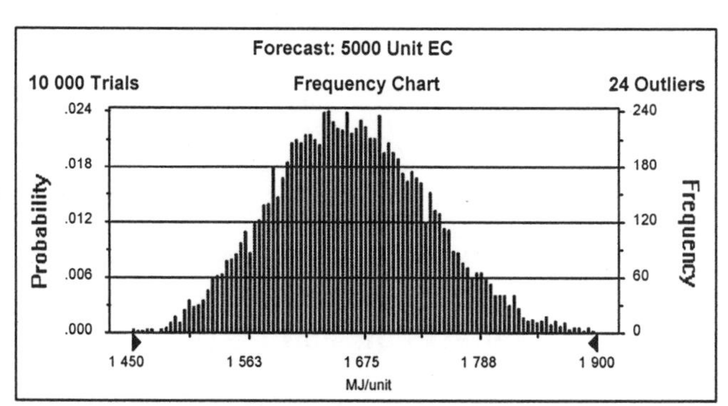

Figure 14 - 5000 Frame Bed Unit Energy Consumption Distribution.

The sensitivity chart for the 5000 is presented in Figure 15, and all the foam production related assumption cells show up as very important. It is interesting to note that the annual production of 0010 (and some other products) is deemed more important than the annual production of the 5000 itself. This is clearly a result of 'economies-of-scale' in the sense that if the 0010 production is increased, then the 0010 production will carry a higher fraction of the overhead energy consumption and hence lowering the unit energy consumption for the 5000. Evidently, everything affects everything and the interplay between products is important to capture in a model, because ultimately the whole company is *one*. Hence, comprehensiveness in an Activity-Based Cost, Energy and Waste model is indispensable.

If we investigate how to improve the 5000 EEE we realize that the annual production seems to be in a compromising situation; if production is increased, the unit energy consumption will increase while the costs will decrease. But which effect is dominant, if any? That can be resolved by comparing the correlation coefficients: From Figure 11 we see that the absolute value of the correlation coefficient is 0.16 (assuming that the correlation coefficient is similar for profitability and costs) while in Figure 15 it is 0.05. Hence, an increase in production will give a higher EEE *unless* process and product improvements are made, e.g., by reducing the A2221 time and reducing the amount of foam in the product.

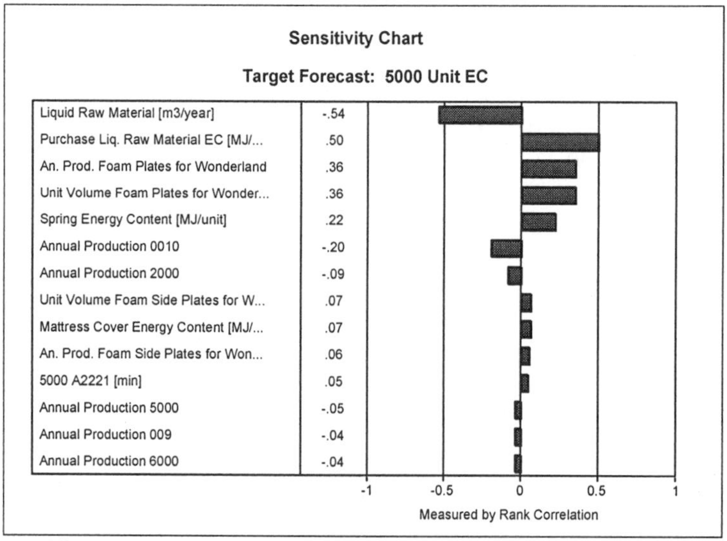

Figure 15 - 5000 Unit Energy Consumption Sensitivity Chart.

3.3 Waste Generation Results and Discussion

Waste generation is measured using the Waste Index (WI) as in all of our case studies. The units, Waste Units (WU), are unfamiliar for many of us, but since we have performed three substantial WI analyses we can start to get a feel for the WU. The results are presented in the same order as before.

3.3.1 Westnofa Industrier AS Waste Generation Performance

In Figure 16 the forecasted distribution for Westnofa's waste generation is shown. From the statistics, which are not shown here, we can find a more accurate mean, 28,209 pWU/year.

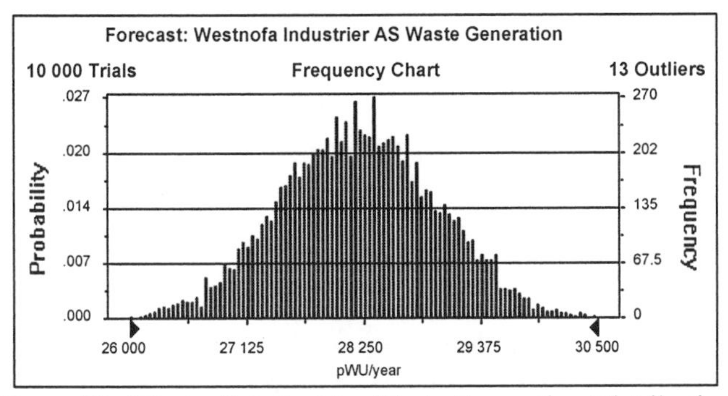

Figure 16 - Westnofa Industrier Waste Generation Distribution.

Compared to WagonHo!, which has a WI of 225 pWU, Westnofa generates significantly more waste. That is no surprise since Westnofa is a much bigger company even though WagonHo! uses electric power from a modern light coal power plant, while Westnofa uses either clean hydroelectric power or diesel; both are probably 'cleaner' than coal in all respects. Westnofa, on the other hand, has foam production that generates a substantial amount of waste. The Westnofa facilities are also much larger. In this way we see that the WI can provide useful comparisons between widely different companies due to the generic WI. The EWE is 1.983E-04 pWU/NOK for Westnofa. Compared to WagonHo!, which has an EWE of 0.0113 pWU/USD, or circa 1.614E-03 pWU/NOK, Westnofa has only a roughly 8 times worse EWE even though the WI was over 100 times worse. Hence, Westnofa generate less waste with similar resource use.

To find out what are the major waste factors for Westnofa, we again turn to the sensitivity charts, see Figure 17. We see that the release of heptane is the single most important source of waste generation. Also, use of diesel for heating, when spot electricity is more expensive, is very significant, mainly due to the CO_2 and NO_x releases and the diesel's waste content. Then, the releases of CO_2 from foam/block production appear, and so on.

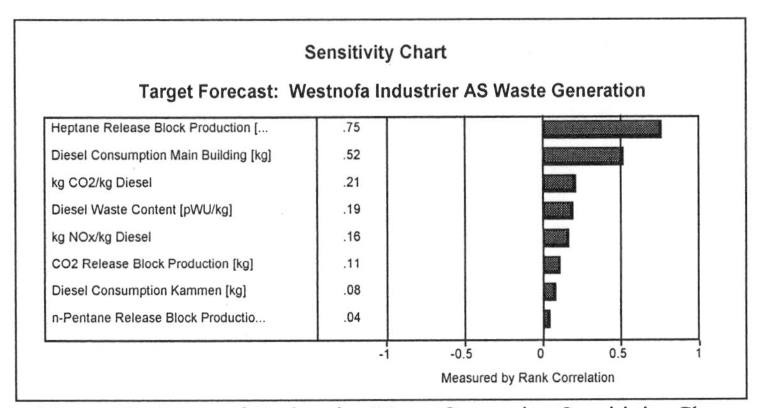

Figure 17 - Westnofa Industrier Waste Generation Sensitivity Chart.

3.3.2 Product Waste Generation Comparison

From the trend chart in Figure 18, we can see that the 0010 is among the worst (high unit waste generation) performing Wonderland products and this is because the 0010 is the best selling product for Wonderland with 21,700 units in 1995. That means that the 0010 must carry a lot of overhead waste generation, because the building overhead is using 'area' as resource driver and 'number of units' as activity driver. The reasoning for the latter is that products produced in high volume tend to also use more space. However, this ignores the fact that the 5000 is a more complex product than the 0010. In fact, although they seem equal in Figure 18, the 5000 probably generates more waste per unit than the 0010 due to this higher complexity.

By closely investigating Figure 18, we can establish the unit waste generation of all the Wonderland products, including the uncertainty range. The various bands are uncertainty intervals, e.g., 100% means that there is 100% probability for the forecast to be within that interval, *given* the assumptions in the model. We see that all the top-mattresses have very low waste content relative to the other products. This is due to the fact that the top-mattresses only consist of fabric and, for example, a Latex plate in the middle. Hence, the unit waste generation attributable to the top-mattresses is due to overhead waste generation.

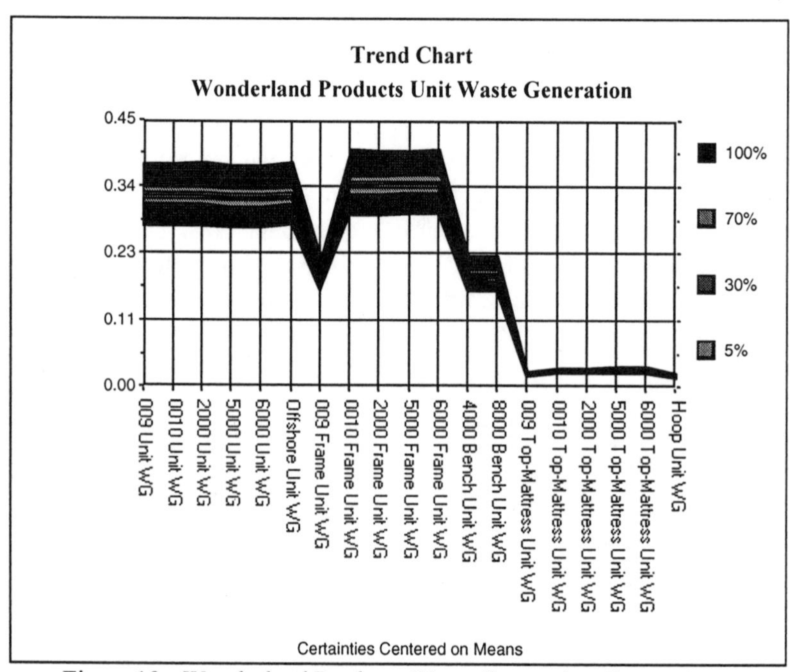

Figure 18 - Wonderland Products Waste Generation Comparison.

3.3.3 5000 Frame Bed Waste Generation Performance

The 5000 Frame Bed with an unit waste generation of 0.14 pWU (see Figure 19) generates more waste per unit than the 4000 Bench Bed, probably because 5000 consumes a lot more foam than the 4000 Bench Bed.

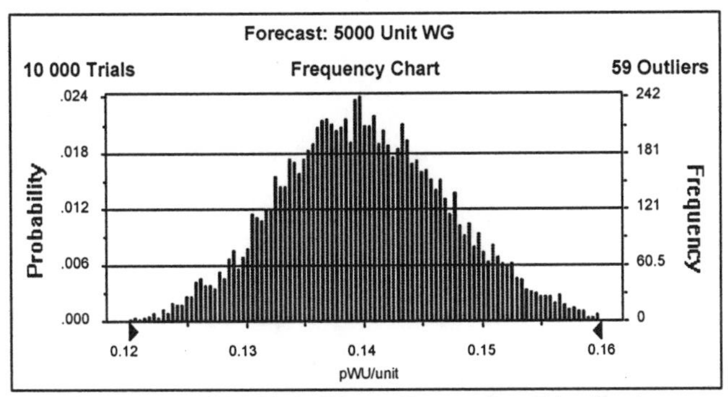

Figure 19 - 5000 Unit Waste Generation Distribution.

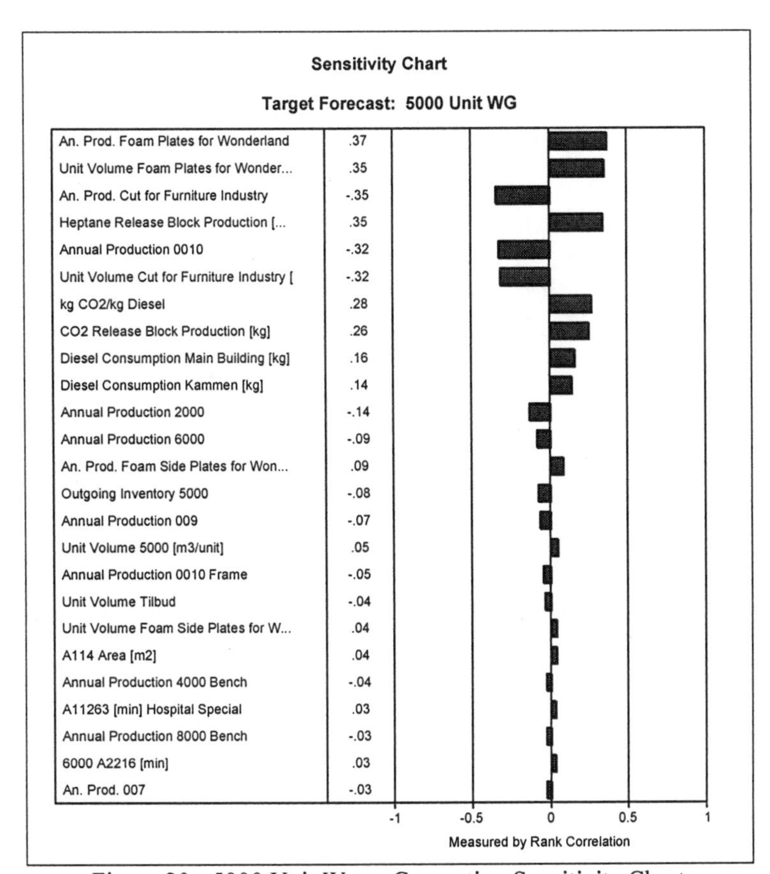

Figure 20 - 5000 Unit Waste Generation Sensitivity Chart.

The sensitivity chart is given in Figure 20. The importance of the overhead waste generation is less for the 5000 than for the 0010 because of the lower production volume. The 4000 Bench Bed actually has a higher consumption of overhead waste than the 5000, even though the 5000 is produced in higher volume. This is because 4000 Bench Bed causes extra activities to be performed, that neither the 0010 nor the 5000 cause, and the 4000 Bench Bed has therefore a relatively high overhead waste generation.

In this fashion we see that the Activity-Based Cost and Environmental Management model captures all the subtleties of Westnofa and gives very good guidance in finding the critical success factors and in making the assessments. The results from the 5000 mattress sub-modeling, which is discussed next, further justify this point of view.

3.4 5000 Frame Bed Sub-Modeling Results and Discussion

The purpose of the action chart approach (see Section 2.5) is primarily to identify the critical success factors for an assessment object. A sensitivity chart is therefore needed. In Figure 21 the cost sensitivity chart for the 5000 Frame Bed action chart is shown. From that we see that activity A2216 (Sew Cover) is the single most expensive activity. Moreover, Action 30 of A2216 takes too much time. Action 30 is identified as 'Stitch Top and Bottom' (not shown in Table 9).

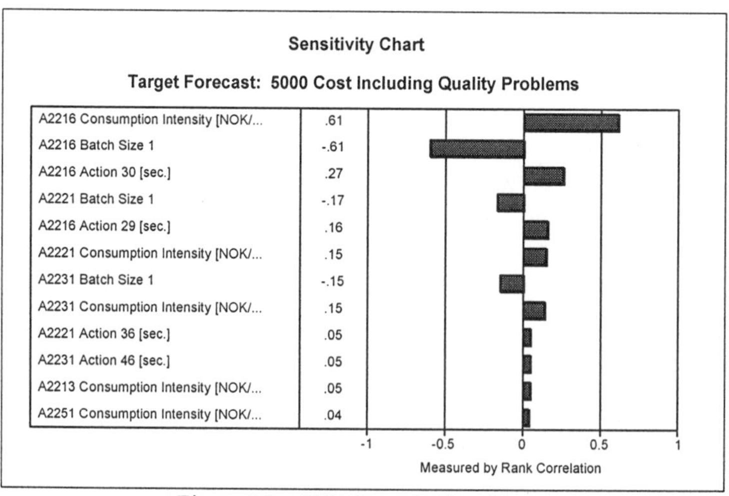

Figure 21 - 5000 Cost Action Chart.

When this was presented to Westnofa, management said that eliminating this action entirely was indeed an option. Furthermore, the operator that was timed said that if there were a mechanism to separate the fabrics better, she could perform Action 30 much faster. In this fashion we see that this type of analysis can be used to identify the need of both potential process changes and potential product changes. Finding *how* these changes can be brought about is up to the various decision-makers and employees. Thus, action charts can be used in the design of both products of processes, but only to find out *what* needs to be changed; not *how*.

From Figure 21 we also learn that an increase in the batch size, which currently is one, would yield lower costs. However, sewing more than one cover concurrently is impossible with the current technology. By investigating the rest of the action chart, similar analyses can be made regarding other activities and actions.

From a cost perspective activity A2216 is by far the most critical one, however, from an energy consumption perspective we see from Figure 22 that Action 36 in activity A2221 (Mold Frame) is the most important one. Again, increasing the batch sizes can reduce the unit energy consumption.

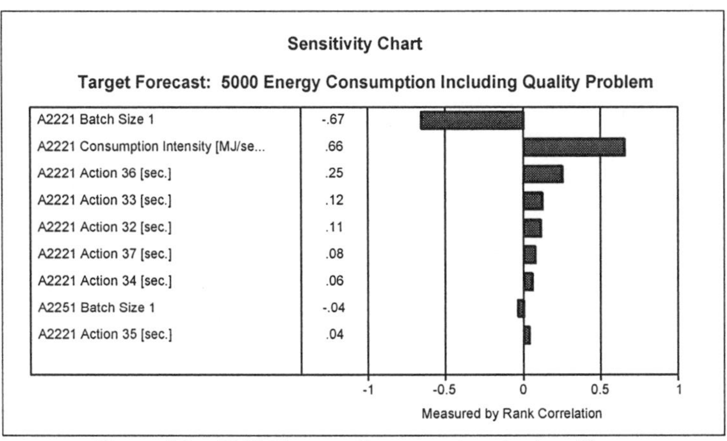

Figure 22 - 5000 Energy Consumption Action Chart.

If the issue of investigation is a pure process analysis, it can be useful to also look more explicitly at the actions that do not add value, i.e. non-value-added actions. For the 5000 Frame Bed, out of the total costs, 35%, or NOK 70 adds no value. From an energy consumption point of view, it is even worse; 76% (91 MJ) adds no value. To find the causes for, e.g. non-value added cost, we employ the sensitivity charts again, see Figure 23, from which we can basically read where the process problems are. All these non-value added activities should preferably be eliminated or at least reduced. To reduce the activities we can either reduce the consumption of the activity, which is a matter of improving the process or product designs, or share/distribute the activities better so that the overall consumption is reduced. Similar analyses can be done for energy consumption and waste generation. For more discussion on action charts see (Emblemsvåg 1995).

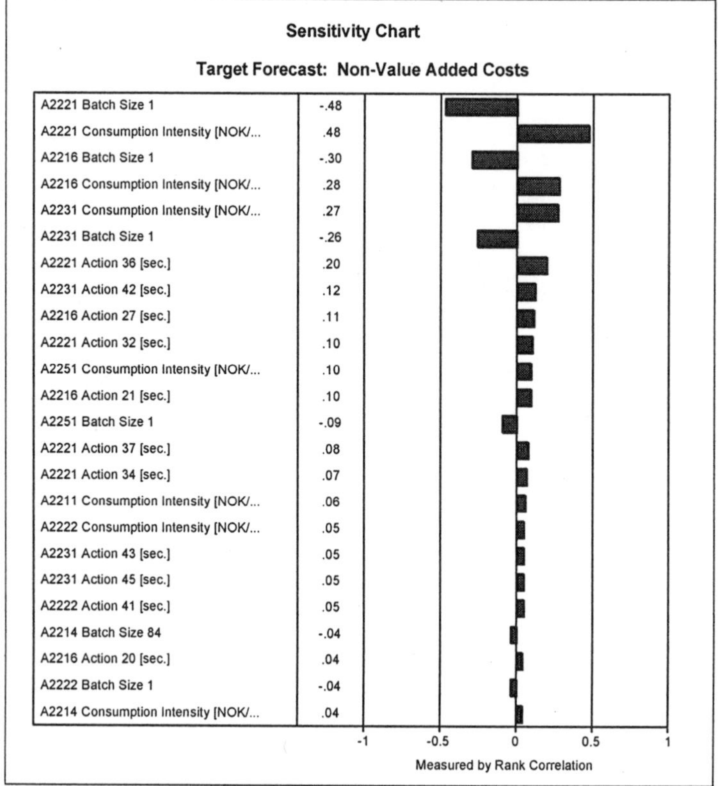

Figure 23 - Non-Value-Added 5000 Costs Sensitivity Chart.

4. WHAT CAN WE LEARN FROM THIS CASE?

Education costs money, but then so does ignorance.

Sir Claus Moser

This case has a number of unique features that were not present in the preceding case studies. These are as follows:

Full business implementation: First of all, this case was a full implementation with all three dimensions (cost, energy, and waste) for a real company. More importantly, Westnofa has used the model since 1995 in managing their business. They can do this because model and Westnofa management systems are mapping virtually one to one on the monetary side: our model captured 100.3% of the total costs as reported by the accounting system. The energy and waste dimensions have clearly lesser accuracy, but do provide additional support in environmental improvements compared to Westnofa's EMAS certification.

Detail and size: This is the largest model that we have constructed to date. The entire model is built up around what (Turney 1992) would refer to as 'micro-activities', activities that represent individual process steps in great detail. Because of that, the Westnofa model has many more activities and drivers than many large (or so-called large) commercial ABC models (Sharman 1991). We also completely avoided the use of large cost pools that can induce serious distortions. The use of cost pools relies upon the fact that the activities within the cost pool are homogeneous - an assumption that becomes harder and harder to satisfy as the size of the cost pool increases (Roth and Borthick 1991). Hence, our models have even better accuracy than most commercial ABC implementations as the problem of distortion is reduced by better activity and driver definitions.

Link to product and process design: How our approach handles design issues is particularly noteworthy in this case study. From the action chart analysis of the 5000 Frame Bed (section 3.4) there should be no doubt that our approach can indeed facilitate integrated product and process design, if desired by accurate assessments and most of all; *detailed tracing*. With respect to process design, an action chart can also be used to identify the non-value-added actions using sensitivity charts. The overall worst actions can also be identified and this may give an impetus and/or decision support for product changes. Furthermore, because an action chart analysis is time based, it follows directly that cycle time can be focused upon as well, if desired. This can be done at any level: product level, process level, department level, main product line level, and overall for Westnofa. What facilitates the tracing of every little parameter, if desired, is the combination of a) detailed process and product description due to the activity-based approach, b) the use of detailed BOMs, action charts, and c) the Monte Carlo method.

Economic and environmental win-win situations: The integration of economic and environmental aspects produced some very interesting results regarding the correlation between costs, energy consumption and waste generation. For Westnofa, there is 92% correlation between costs and energy consumption, and 86% correlation between costs and waste generation. Moreover, there is 98% correlation between energy consumption and waste generation! The situation is similar in the Farstad Shipping case study presented earlier. From this, we can start to speculate whether this is indeed the case on a more general basis. If so, it means that integrated Activity-Based Cost and Environmental Management has a tremendous potential for promoting environmentally sound cost management.

5. THE EFFECT ON WESTNOFA

Disruptive people are an asset.

Robert Lutz

What we found most interesting in this case study, however, is not related to the case itself, but to the effect it has had on Westnofa. Because Westnofa started using the model in 1995, we have here a case study that we have been able to follow over an extended period of time and this gave us some interesting insight on its effect on the company.

Since 1996 Westnofa has used the model as a supplement to their other performance measurement systems, and the results from this model are taken just as seriously (if not more) than the results from the other systems. This is possibly because the model captures very subtle aspects of the company. Something that always startles people who have never seen such models before is the fact that a change in one single variable often affects the profitability for *all* products. Then, once they realize that everything in a company is ultimately related to everything else, the doubt is turned into appreciation. There is no 'his problem' and 'her problem' anymore because everybody realized that everything indeed affects everything. By demanding input for the model from every department and group in the organization, ownership and cooperation is promoted and cross-functional and multi-disciplinary teamwork fosters.

For CFO Jørn Nes the usefulness of the results are beyond doubt:

> *Even though we are not yet able to use the model by ourselves; now, we always ask ourselves when making a decision 'how will this [i.e. decision] affect the ABC[2] analysis'.*

He basically said that the Activity-Based Cost, Energy and Waste model had influenced their way of thinking. This is a well-known phenomenon in the literature, see e.g. (Brown 1995) where it is concluded that 'measurements drive behavior'. CEO Steinar Loe expressed the following:

> *We either have to accept this [ABCEM] way of thinking and take the consequences, or stop doing these analyses that just brings bad news about many of our products.*

Westnofa is therefore facing a dilemma; they got a model that gives them the decision support they need, but it is challenging the ruling paradigm of thinking. For us, the single biggest achievement of the model is not the results themselves, but that the model has managed to impact the behavior at Westnofa. Only when a model can affect the behavior, true and lasting change for the better can be brought about. A more profound effect by an assessment system is hard to find.

---

[2] They like to call our models ABC models. This is due to the similarities, even though our models can do many things not possible for an ordinary ABC model, such as uncertainty analyses, sensitivity analyses and design parameter tracing.

6. IN CLOSING

The universe is not required to be in perfect harmony with human ambition.

Carl Sagan

What are the main points that we wanted to show you with this case study? The Westnofa Industrier AS case study's purpose was to illustrate a complete implementation of our method in an industrial setting. The accuracy of assessments and tracing is beyond doubt and supports not only cost management, but also product and process design plus environmental management. The degree of integration is clearly well beyond any conventional environmental management systems as they would (still) require a separate and different costing system to handle cost issues. And the importance of costs as the primary performance dimension in business is still indisputable. Our integration of Activity-Based Costing and Environmental Management makes it possible to present all assessment results in a similar and consistent format for every assessment dimension, and easily identify economic and environmental win-win situations. In our opinion, this is needed if environmental management is ever going to become more than a back-room activity.

7. ACKNOWLEDGMENTS

We would like to thank the management and board at Stokke Gruppen and at Westnofa Industrier AS for inviting us to perform such an extensive case study and for having faith in our work. The total commitment from Westnofa Industrier AS in providing us with information was extremely valuable. CFO Jørn Nes provided an extensive amount of accounting information. COO Tore Lillebostad and CTO Ove Søvik provided an extensive amount of process and product related information. Health, Safety and Environment Coordinator Roger Grande provided most of the environmental information. In the end we would like to thank CEO Steinar Loe and his Leader Group for critically evaluating the results and the model and for giving us valuable feed-back. We also greatly acknowledge the feedback from CFO Geir Løseth at Stokke Fabrikker.

Chapter 9

CLOSURE

Before you study Zen, mountains are mountains and rivers are rivers; while studying Zen, mountains are no longer mountains and rivers are no longer rivers; but once you have had enlightenment, mountains are once again mountains and rivers again rivers.

Zen saying

The work presented in this book consists of a series of thoughts that so far has culminated in what we call Activity-Based Cost and Environmental Management (ABCEM). Understanding these thoughts and ideas is crucial, but often difficult for the first-time reader. In this chapter we will therefore look back upon the cornerstones of our work, its benefits, its future, and the things that ABCEM cannot accomplish.

1. ACTIVITY-BASED COST AND ENVIRONMENTAL MANAGEMENT - CORNERSTONES REVISITED

Simplicity of character is the natural result of profound thought.

Chinese fortune cookie

Activity-Based Cost and Environmental Management is essentially built upon some very simple constructs. Indeed, so simple are the constructs that we are amazed that nobody else thought of them before us, even though the first construct (ABC) is not our idea. In this section, we will crystallize our basic constructs as clearly as we can. And by doing so (given that you have read the rest of this book) we are convinced that you will agree with us when we say that 'Activity-Based Cost and Environmental Management is a different approach to the ISO 14000 compliance; only much better'.

It all started in fall 1993 when we read how Activity-Based Costing (ABC) could be used to manage environmental costs in (Brooks, Davidson *et al.* 1993) from which we identified the first cornerstone.

1.1 Activity-Based Costing and Management

Practitioners as well as academics embraced Activity-Based Costing initially due to its highly increased accuracy compared to the traditional, volume-based costing systems. Today, the accuracy advantages of ABC and ABM are well documented, see e.g. (Drucker 1995), but a more important advantage, namely *tracing*, is now eclipsing the accuracy advantage. The traditional volume-based costing systems simply cannot establish the causal relationships between what is being done and the associated costs. ABC can. Companies that use the traditional, volume-based costing systems are therefore often running their company in a 'management-darkness'. Depending on the industry and company specific variables, such as what constitutes the competitive advantage, the degree of automation, product mix, degree of overhead costs, periodicity and repetitiveness, etc., different companies will value ABC differently. But in the general case, ABC is a need, not an option (Cooper 1990a) and the question 'what makes ABC better' has been answered extensively in Chapter 3.

ABC is intuitive and logical, and we found this logic and clarity very appealing to our work. If we are to become sustainable, we must be able to assess and manage economic resources effectively. In fact, we believe that good cost management is a prerequisite of sound environmental management. But as we pointed out in Chapter 1, many resources cannot be measured well economically, hence we realized that we must expand ABC to include environmental dimensions as well.

1.2 Expansion of ABC with Environmental Measures

The expansion of ABC into the environmental domain is remarkable simple as we have shown, as long as we can find a couple of reliable measures. We have argued for using energy and waste in this book, but other measures could be employed as well. Given these measures, expanding ABC into environmental dimensions can be done by merely adding a couple of columns in our spreadsheets. Energy drivers and waste drivers work exactly the same way as cost drivers when it comes to describing and tracing activity/resource consumption and energy/waste assignment.

With this expansion, energy and waste are treated the same as costs. That means that resources not only have a monetary price, but also a price in terms of energy content and waste content in the well-known value chain. Correct accounting of energy and waste, thus, becomes as important as correct accounting of costs in the value chain. We cannot emphasize enough that the beauty of building upon cost accounting is that the principles for energy and waste accounting are the same as for cost accounting. Therefore, the interesting question regarding the second cornerstone is not whether ABC can be expanded or not, but rather what measures should be used in order to make such as expansion useful.

1.3 Measuring Impact by Benchmarking Nature

After realizing the huge benefits of utilizing ABC and expanding it for environmental management purposes, we were really only lacking one thing to make our method useful; we needed a good measure of environmental impact in addition to energy.

In the literature we can find many environmental indices, but they all suffer from some sort of incomparability. Some are too simplistic, such as mass indicators, while others are based on impact categories (i.e., real environmental problems such as global warming, smog and so on) that are inherently incomparable. Therefore, any environmental impact indicator that is to produce comparable results must avoid the simplicity of counting something *and* avoid the confusion of categorization, evaluating, and weighing perceived environmental problems. In other words, we can neither be too 'simplistic' or too 'scientific'.

In our opinion, the answer to this seemingly impossible situation is to listen to Nature for advice. Environmental problems are after all simply a misallocation of resources that can be waste in one system but 'food' in another system. This misallocation results in imbalance in Nature and these imbalances manifest themselves as environmental problems and the like. Hence, an environmental impact indicator should measure the imbalance of Nature caused by our misallocation (releases) of resources that we (as of today) perceive as waste. This imbalance can only be detected by benchmarking Nature, leading to our Nature Knows Best axiom (Chapter 2):

Environmental impact can only be measured relatively by benchmarking Nature

In other words, we want to measure a release against the amount the release exists naturally in Nature and not according to the latest political debate. And using the Waste Index (WI), which essentially is the time integral of a release in relation to itself (Nature), where the function and numbers are solely derived from the Laws of Nature, we can do that. This way of measuring environmental impact is a paradigm shift; away from measuring actual environmental impact to measuring relative environmental impact - away from scientific theories and debates to Nature and its laws. This paradigm shift can potentially open up the way for what we call energy and waste accounting since the Waste Index is consistent and comparable on the practitioner level.

1.4 Explicit Inclusion of Uncertainty

We purposely include uncertainty in ABCEM, allowing us to do effective and efficient tracing and uncertainty analysis. It is important to emphasize the simplicity of this approach. The generality of Monte Carlo methods and the associated sensitivity analysis allow us to use the same approach repetitively and accurately regardless of what we want to investigate. We strongly believe that the use of uncertainty distributions (instead of single numbers) adds extra value to the results. As you look at Figure 1, you will realize that your tendency to trust a single number is suddenly a lot less. An uncertainty distribution can take into account that things are uncertain, whereas a number is just a single representative of a possible outcome. Then, when decisions are made, the uncertainty will be a part of the decision. As expressed before in (Cooper 1990d): 'The premise for designing a new ABC system should be: "It is better to be approximately right than exactly wrong"'. Our use of uncertainty distributions undoubtedly supports this premise strongly.

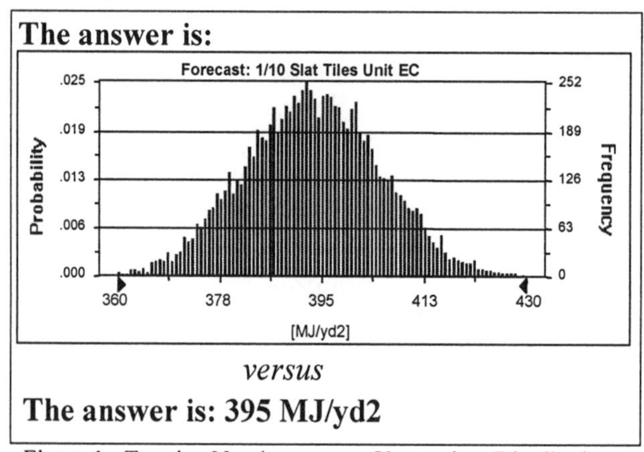

Figure 1 - Trusting Numbers versus Uncertainty Distributions.

2. ACTIVITY-BASED COST AND ENVIRONMENTAL MANAGEMENT IN PERSPECTIVE

A cultivated person does not promote people on account of what they say, nor ignore what is said because who is saying it.

Confucius
Analects 15:23

We have revisited the cornerstone on which our work was built, but how good is it? In this section, we try to put Activity-Based Cost and Environmental Management (ABCEM) in perspective in order to answer that question.

2.1 ABCEM - Critical Evaluation

There is a fine balance between claiming too much on one hand and claiming too little on the other hand. Both are equally bad because they show lack of understanding, to the extent understanding exists. With that in mind, we want to look at some aspects that should be critically evaluated before accepting an approach or method as sound and good, namely:

- Internal logic (does it work properly?).
- Effectiveness and efficiency (how well does it work and how easy is it to use?).
- External usefulness (can it be used in real-life and produce results that people find useful?).
- Repetitiveness (is it applicable to a wide range of implementations?).

These issues, and many more, are discussed extensively in (Emblemsvåg 1999) and will not be repeated here, but some critical evaluation is warranted.

The internal logic is derived from a proven and tested cost management approach - Activity-Based Costing and Management. Hence, we have no reason to believe that our ABCEM approach would not be logical, although illogical mistakes can be made during model development and implementations. Furthermore, ABCEM has been developed and improved over several years, starting in 1993.

Clearly, the ABCEM approach has been used repetitively in a wide range of applications, ranging from supply vessels to carpets and furniture. Clearly, ABCEM is effective. Results pinpoint areas of improvement. They also give valuable information on how uncertainty has an impact in regard to distortion problems and forecasting/simulation.

From an efficiency point of view, we can look at the cost of implementation. According to (Vigon 1997), Procter & Gamble has made some assessments on how much only ISO 14001 certification will cost them: roughly $100,000 per site. Procter & Gamble concluded to <u>not</u> seek third party registration for ISO 14001 as they felt that their existing environmental management systems met or exceeded the requirements of the ISO standard. Such a high price may scare away any company interested in being environmentally proactive, especially Small and Medium-sized Enterprises (SMEs). ABCEM, in contrast, can do both environmental and cost management in one single framework by just adding two extra columns in the cost management sheets, so to speak. Clearly, this must be a more cost efficient approach.

2.2 ABCEM and conventional ABC/ABM

We have elaborated quite sufficiently on ABC (see Chapter 3) and many may already know ABC. But we like to point out that our way of implementing ABC is different that what is done commercially by, for example, large management consulting companies such as Arthur Andersen, KPMG and Deloitte & Touche. Particularly, because our method employs

Monte Carlo methods in calculations and sensitivity analyses, more variables can be handled. We can consequently create models of much higher accuracy and much better tracing capabilities than what is common. In (Sharman 1991), for example, the author describes briefly a model that had more than 100 cost drivers as 'very detailed'. In comparison, the Westnofa Industrier model presented in Chapter 8 has 222 drivers. Furthermore, (Turney and Stratton 1992) have added a new type of activities they call 'micro activities'. However, based on their examples, this seems to be what we refer to as low-level activities. Hence, our activity definition is also more detailed than what is done in commercial implementations presented in the literature. Also, none of the ABC models in the literature handle uncertainty beyond simple risk assessments and 'what-if-analyses'.

2.3 ABCEM and ISO 14000

The interesting question, however, is whether ABCEM is the preferable option for ISO 14000, given that the companies have several choices. This relates directly to the subtitle of this book; is Activity-Based CEM a preferable approach to ISO 14000 compliance?

We have one real-life implementation of ISO 14040-43 LCA to compare our approach to, namely, the Farstad Shipping case study (Chapter 5). With respect to that case study, our approach outperformed the conventional approach by far on equal grounds. The conventional LCA produced neither useful results nor indications for improvements. This case study strongly supported the preference for our method.

It is important to note that Activity-Based CEM does not exclude ISO 14000. On the contrary, Activity-Based Cost and Environmental Management is a better way, we believe, of doing many (but not all) aspects of environmental management that *can* be a part of an ISO 14000 certification.

We are therefore confident when we say that our method is an improvement over the environmental management system and practices outlined in the ISO 14000 standards. Interestingly, the ISO 14000 standards are so loosely defined that almost any environmental management system can become ISO 14000 certified as long as it can be documented and fitted into the ISO 14000 overall framework. Our Activity-Based Cost and Environmental Management approach with its Activity-Based Cost, Energy, and Waste models fits in the framework, but is at the same time a different approach to the ISO 14000 compliance - only better.

2.4 ABCEM and Continuous Improvement

Even though results may be satisfactory, the constantly changing market conditions make it important to continuously improve whatever there is to improve. It is also vital to keep in mind that success stems from the famous Total Quality Management dictum 'doing it right first time', see (Dillton-Hill and Glad 1992), hence the improvement efforts must be directed towards where they are most needed. That is the primary goal of our approach and to Deming's concept of 'profound knowledge'. Profound knowledge signifies that gathering the *right data* from the *right sources* at the *right time*, then using the *right resources* to make the *right interpretations*, leads to the *right actions* to improve quality (Hronec and Hunt 1997). Deming understood early (at least by the 1950s) the importance of continuous improvement, so he devised what later became known as the Deming Cycle, see Figure 2, which is essential in the continuous improvement thinking, see (Brassard 1989). The Deming Cycle (and the process it represents) is as follows:

1. The company should **plan** what to accomplish over a period of time and what it is going to do to get there. Our ABCEM models can be helpful in this process to identify the critical success factors upfront and the impact of uncertainty can be assessed.
2. The company should then **do** what furthers the goals and strategies developed previously. Using ABCEM models to identify the critical success factors is crucial in this stage because it allows proactiveness.
3. Then the company should **check** the results of its actions to ensure a satisfactory fit with the goals. This requires only a simple after-calculation with ABCEM models, and all of our case studies include this.
4. The company should then **act** to eliminate possible differences between actual performance and the goals stated up front.

Please notice that we have added the four sectors 'Simulation', 'Proactive', 'Assessment' and Reactive' in Figure 2 to relate the Deming Cycle towards what we are doing. Activity-Based Cost and Environmental Management is employing 'simulation' and 'assessment' to direct attention for the other two sectors (proactive and reactive).

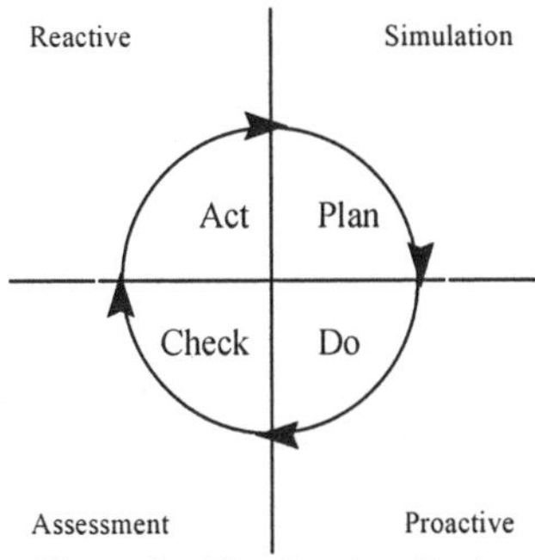

Figure 2 - The Deming Cycle.

3. ACTIVITY-BASED COST AND ENVIRONMENTAL MANAGEMENT AND INDUSTRY

Management is a practice rather than a science... and the ultimate test of management is performance.

Peter F. Drucker

Performance is the ultimate test according to the above quote. How well did our approach do in industry, our target group? That is the focus of this section.

3.1 ABCEM Applications

In this book, as in (Emblemsvåg 1999), we have used case studies to show that our method has noteworthy performance. We did not show all case studies that we performed. The complete list of case studies and their respective foci (summarized) is as follows:

- *Farstad Shipping*: Use, maintenance and repair, all dimensions (cost, energy and waste), assessment and general recommendations.
- *WagonHo!*: Manufacturing (incl. raw material acquisition), all dimensions (cost, energy, waste), assessment and general recommendations, specific recommendations regarding product, process, and organizational design.
- *Westnofa Industrier*: Manufacturing (incl. raw material acquisition), all dimensions, assessment and general recommendations, and specific recommendations regarding CAP, budgeting, product and process design, organizational changes, etc.
- *Interface Flooring Systems*: Manufacturing (incl. raw material acquisition), energy and waste only, assessment and general recommendations, specific recommendations regarding product and process design.
- *Stokke Fabriker*: Manufacturing (incl. raw material acquisition), cost only, assessment and general recommendations, and specific recommendations regarding product and process design.
- *Recycling facility*: Demanufacturing, cost only, assessment and recommendations for product and process design (Emblemsvåg 1995).

Given that we also were able to model a recycling facility in (Emblemsvåg 1995), we can conclude that our case studies are covering fairly well the entire life-cycle. This is important because ABCEM can, thus, also be used as an LCA approach, if so desired.

Clearly, we do not have the same depth in all case studies. This is due to practicalities such as time, budgets and, most importantly, the focus and information systems of the various companies. This is, however, not a significant problem since we were able to perform various things in various case studies. Moreover, since the set up of ABCEM and the results are similar, we can transfer findings, regarding the method, from one case study to another.

It is important, however, that in all of the case studies the results are useful, that is, the results are accurate, capable of tracing critical success factors and comparable. This can be easily checked since the feedback from the companies where our method is implemented is unanimously positive.

3.2 Industry Feedback

The bottom-line is; did our industry partners like the ABCEM approach, models and results? Although we have not performed a formal survey, we did poll key company people who have been involved in the implementations at their respective companies.

Technical managers at Farstad Shipping found the cost model satisfactory. The environmental model was never presented to them because the project was long finished by the time the model presented in this case study was completed. These managers have the skills and knowledge to really probe the model, and their approval is indeed an indication of quality. For the costing part, the technical managers supplied data, checked the model logically and approved the model in 1996 when the comprehensive version in (Fet, Emblemsvåg *et al.* 1996) was designed. We were told that the model influenced decision-making.

One clear indication that Westnofa and Stokke Fabrikker liked the approach is that they asked for additional implementations and model developments since the initial model, resulting in subsequent updates. We solicited written feedback from Stokke Fabrikker's CFO who was also involved to some extent in the Westnofa Industrier implementation and has seen two successive implementations. On the question, would you recommend this type of modeling to other companies, his response was: "Yes, the methodology, the model and the results of this work is paramount to any company."

The initial Interface Flooring Systems model was for environmental evaluation only (i.e., cost information was not included) and was based on 1997 data. Founder, CEO and Chairman of the Board, Ray Anderson, served on the dissertation committee of Jan Emblemsvåg and approved the approach. According to him "It is a major step forward in the field of environmental management". In addition, we worked closely with Interface's Vice-President of Engineering, Dave Gustashaw. He clearly endorsed the work by asking and funding a follow-up project (together with Interface Research) in which a detailed ABCEM model is being developed using current data. This model will include all dimensions (cost, energy consumption, and waste generation). In addition, we are investigating how we can achieve not only a model for quarterly assessments, but also a *dynamic* model that can give real-time cost, energy, and waste by including feedback and data gathering capabilities using factory sensor data and web-based information technologies.

3.3 The Reason 'Why'

One could argue that the feedback is anecdotal, but in our mind the clearest 'evidence' that these companies approve of our approach is the fact that two asked us for additional implementations and follow-up projects.

We have noticed that a primary appeal of our approach is that it is simple to understand. That does not mean that it is easy to implement, in fact, 'simple' Activity-Based Costing can be very hard. However, the idea and the way of linking environmental impact assessment in terms of energy consumption and waste generation with an accurate economic assessment is very easy to understand by any manager in industry. Even if environmental issues are secondary to economic issues, the ABCEM approach provides them the framework for at least having an accurate cost assessment to which they can add energy and waste assessment using the same model structure.

4. THE BENEFITS OF ACTIVITY-BASED COST AND ENVIRONMENTAL MANAGEMENT

The more extensive a man's knowledge of what has been done, the greater will be his power of knowing what to do.

Benjamin Disraeli

There are, in our opinion, many benefits to our approach. Many have been given before and also in (Emblemsvåg 1999). Here we revisit the most significant ones.

4.1 Integration of Economics and Environmental Impact

Conventional environmental management does not facilitate cost management of any noteworthy sort. Hence, the most important dimension for companies is simply ignored. We are afraid of this approach because we believe that this will lead to environmental management becoming left-hand work. Important success factors for implementations will suffer, such as 1) organizational buy-in, 2) cross-functionality, 3) slope of the learning curve, 4) capability to utilize available assets, 5) cost of implementation, 6) tradeoff capabilities and 7) consistency. If these factors suffer, the implementation is in most cases bound to become unfruitful and the company will probably not be able to harvest economic benefits from their environmental management.

We believe that environmental management will only become a success if it enables companies to do well (economically) by doing good (environmentally). Conventional approaches simply cannot facilitate this because there is no link between economics and environment.

Clearly, having a generic and integrated cost and environmental management framework that gives significant decision-support, that is, decision support beyond the obvious, is extremely advantageous. The fact that ABCEM is generic is clear from the variety of case study implementations presented. The ABCEM method also handles them all in the same way. The cost, energy consumption and waste generation results are the same in terms of form, presentation, and interpretation, with the same procedures (uncertainty distributions, sensitivity charts, trend charts, etc.). The content is of course different because the case studies are different

This is a major benefit because it will enable organizations to manage environmental issues as they manage costs and to perform *integrated* environmental and cost management. Then, benefits can come not only in terms of improved environmental track record, 'greener' products and so forth, but companies will actually *know* if they are doing well by doing good. In fact, a survey of the 300 largest U.S. companies suggests that improved environmental management systems can increase stock prices (Feldman, Soyka *et al.* 1996). A condition for this is that improved environmental management and improved environmental performance must be clearly articulated to the investment community. ABCEM can have a pure financial payback in addition to cost savings and improved environmental performance.

4.2 Comparability

As argued by us in this book and elsewhere, (Ayres 1995) has also shown that many conventional environmental performance evaluations and LCAs suffer from deficiencies, among them, incomparability. '*So what?*' you may ask. Well, without producing comparable results, the environmental performance evaluation cannot give any rank prioritization of products or product improvements, and ultimately cannot be used for decision-making.

We, however, can compare, for example, a Platform Supply Vessel to a carpet manufacturer. This should be evident from Table 1 where we compare the key numbers for our case studies. EC stands for 'Energy Consumption' while WG is short for 'Waste Generation'. The N/A for Profit Resource Efficiency (PRE), Economic Energy Efficiency (EEE), and Economic Waste Efficiency (EWE) in the Interface Flooring Systems case study is due to the fact that the cost dimension was deliberately excluded. The N/A for Farstad

Shipping case study's PRE number is due to the fact that revenues were not included here. The numbers for WagonHo! are from *after* the strategy change. Similar tables to Table 1 can be made regarding any assessment object.

Table 1 - The Comparability of the Overall Company Results in the Case Studies.

Case Study	Costs [USD]	PRE [%]	EC [MJ]	EEE [MJ/USD]	WG [pWU]	EWE [pWU/USD]
WagonHo!	2,489,071	9.11	10,152,764	4.54	25,366	0.012
Farstad	5,269,760	N/A	147,564,029	28.00	3,359,506	0.638
Interface	N/A	N/A	3,985,434,271	N/A	24,533,736	N/A
Westnofa	20,721,000	7.24	253,094,693	12.13	28,545	0.001

Being able to compare results allows us to direct attention to issues that may even interest policy makers. For example, with respect to Table 1, note that running a single supply vessel costs a relatively high amount of energy compared to the operation of a whole furniture factory. Some reasons for the seemingly highly different results are:

- Size of company - Interface Flooring Systems is largest in all respects.
- Power supply - Westnofa Industrier has 100% hydroelectric power while Farstad's PSV has large machinery running on marine gas oil.
- Type of industry - Interface manufactures carpets and a lot of chemicals are involved while WagonHo! produces wooden toys.

In all cases, the factors that drive energy consumption were similar to those that drive waste generation. This indicates that running an energy efficient business also reduces waste generation. Clearly, ABCEM is capable of actually dealing with widely different cases and all their products, while giving comparable results and directing attention for both designers and managers in an integrative fashion.

4.3 Inclusion of Indirect/Overhead Impact

The environmental impact of overhead resources, such as buildings and production equipment, is not mentioned in any noteworthy degree by, say, conventional LCA researchers. This issue seems to be totally ignored. Hence, tracing of overhead resources is not even attempted. Such practices are not sound when it is known that overhead costs are often over 30% of total costs. There may be a similar relationship for the environmental impact of overhead resources. ABCEM, on the other hand, handles such issues as a matter of fact.

4.4 Improved Tracing

Tracing is important because it allows the users to find the answers to the (in)famous '*why?*' questions. Without aid in answering such questions, the method cannot really be used in design or any other improvement process. Tracing in conventional environmental performance evaluation and LCA *is* possible, but practitioners are left to their own devices to do the tracing. In fact, 'most companies have far too many performance measures - and far too few that are relevant' (Keegan, Jones *et al.* 1991). In ABCEM, in contrast, the tracing is virtually automatic by using Monte Carlo simulations and supporting software. The tracing capabilities of our models (e.g., by using the sensitivity charts) make it possible to identify the most significant success factors.

5. CHANGE MANAGEMENT AND HUMAN BEHAVIOR

One whose troops repeatedly congregate in small groups here and there, whispering together, has lost the masses. One who frequently grants rewards is in deep distress. One who frequently imposes punishments is in great difficulty. One who is at first excessively brutal and then fears the masses is the pinnacle of stupidity.

Sun-tzu
In "The Art of War"

Change management is one of the new phrases that everybody like to use, but according to two major surveys conducted in Europe by A.T. Kearney and worldwide by Atticus, only one in five change management projects succeeds (The Economist 2000a). Evidently, change is something we all know that we must be able handle, yet very few actually can do it. According to Craig Baker who leads A.T. Kearney's European 'enterprise transformation' program; "the largest gap between companies that were good and bad at change arose because some learnt from change and institutionalized their knowledge, building it into their culture and performance assessment". Thus, implementing new performance measurement systems well and handling behavioral issues wisely are extremely important in order to successfully bring about change. In fact (Shields and Young 1989) have designed a model called the 'Sevens Cs' for behavior variables that most prominently influence a cost management system implementation, see Figure 3. The seven Cs are:

1. Culture. Culture has two aspects to it. First, the local work culture determines what is considered important, what is desirable, etc. Second, the various national cultures that further enhance the local work culture are important. Culture affects the whole implementation process but in an unclear way, see Figure 3, which makes cultural issues difficult to handle.
2. Champion. A champion is someone in a high level position that has wide support in the organization and is good at motivating people. The champion should be closely involved in the entire implementation process to secure/improve anchoring in the organization.

3. Change process. The change process involves managing employee time orientations, decision-making rights and job profiles.
4. Controls. Controls concern how the power structures work and issues like teaming versus individualism, formality versus informality.
5. Compensation. Compensation simply revolves around 'what is in it for me?' Nobody will implement a system that erodes their own existence.
6. Commitment. Commitment is the degree to which the organization willing to do what is required and stick with it.
7. Continuous education. Continuous education is a necessity because a new system inevitably will have problems in the beginning either because the system needs adjustments (education for the implementers *and* the users) or the organization needs to learn how it works (education for the users).

Organizational buy-in is vital in order to get acceptance and commitment in the organization with respect to information flow, ownership of results, etc. Having a champion and top-management support along with a broad-based implementation can ensure this. We have experienced this first-hand. But we also know first hand that buy-in from lower management, and even laborers and crews, is essential for continuous success. Shields found ABC success to be strongly correlated with behavioral and organizational variables but not with technical variables, such as the software used (Shields 1995). Employees may feel that the new performance measurement system can prohibit wage raises, reduce possibilities for promotions, change the direction of management attention unfavorably and the like.

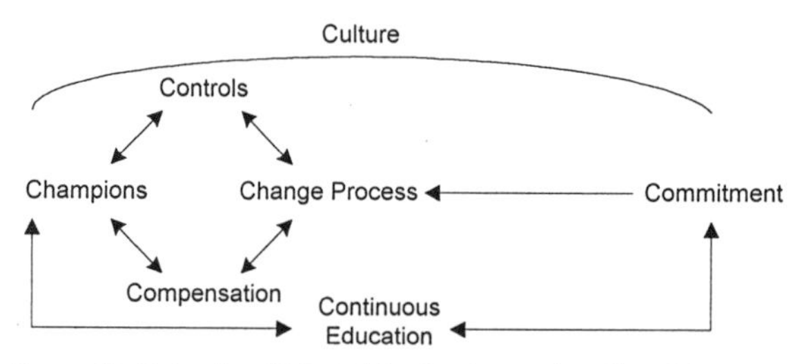

Figure 3 - The Seven Cs Behavioral Model for Implementing Cost Management Systems (Shields and Young 1989).

The Seven Cs also apply to our approach. A conventional ISO 14000 implementation, however, may not get the support it needs, because conventional environmental performance evaluation and management (like ISO 14000) are very different from cost management practices. The sole focus on environmental issues makes the whole approach alien to most of management. Activity-Based Cost and Environmental Management, in contrast, can provide management with accurate cost information, and *at the same time*, can also provide environmental information. This makes it more likely that the approach gets the necessary support. Plus, the approach is less different than what companies are doing today. In fact, the results can be interpreted as cost results.

6. WHAT COULD BE IMPROVED?

Who has no faults? Excellence is a matter of reforming them.

Chan Master Jiantang Ji

The future seems in many ways bleak, as there appears to be no end to the environmental problems. Thus, companies and society will inevitably have to face the fact that 'business as usual' is not sustainable and will have to change. How much change? Nobody can really tell. Hence, the single most important aspect of companies as well as the society as a whole in the future, is agility: the readiness for change. Since change is a process, we need process-oriented methods to enable agility. This includes everything from accounting systems, assessment systems, and design methods, to more qualitative methods used in, for example, organizational issues. Here, we list a few changes and improvements that are needed from an ABCEM perspective as well as accounting and marketing perspective. If you feel under-whelmed, realize that we could easily over-whelm you, and even write another book about what can be done, but we have to draw a line somewhere.

6.1 Some Specific ABCEM Improvements

Currently, much work is done in the area of Activity-Based Management (ABM) which includes an increasing variety of activity-based approaches. Our work has further added value to ABM by enabling companies to also do integrated activity-based cost and environmental management and by also bringing design more easily into the picture so that proactive initiatives can replace reactive efforts. But some specific improvements are:

- Better software is needed to handle ABCEM models better than the combination of MS Excel® and Crystal Ball®. In fact, at the Georgia Institute of Technology we are currently working on advanced software tools that facilitate a) model development, and b) performance monitoring (even in real time and over the Internet). This will also facilitate sophisticated statistical analyses within the Activity-Based Cost, Energy and Waste models, leading to what we call Statistical Business Control. Much more data will be handled much faster. In fact, we do not really know what *can* be possible because the ABCEM framework allows many different types of assessments; types that we may not even have realized ourselves yet.

- We need to develop techniques that allow identification and establishment of internal benchmarks that can be used in continuous improvement, that is, to identify what is the potential *internal* 'Best Practice' at any point in time. This can be very useful in Activity-Based Budgeting (ABB), possibly in product mix decisions and the like. Coupled with external benchmarks, it would also give insight on what is the maximum possible performance from an environmental perspective.
- Find more methods, such as the action charts, which can be incorporated into an ABCEM 'toolbox'. For example, we are thinking of Function Analysis, see e.g. (Yoshikawa, Innes *et al.* 1994), an approach that allows tradeoff analyses between functions and perceived customer value. Although there are several limitations with this approach by itself, e.g. the reliability of the market research needed, it may be an interesting approach in combination with the overall ABCEM method.
- The Waste Index needs to be tested more and expanded. For example, could the Waste Index handle biological entities by incorporating reproduction, and such issues, into the balance time and natural amount estimates? Nonetheless, more case studies are needed.
- The use of other sound indicators in ABCEM is also interesting to research. Conceptually, we see no problem (see Chapter 4), but (again) 'the devil is in the details'. We are currently investigating more closely how other environmental impact metrics can be used as well. On one hand, we want to avoid being subject to the 'metric du jour' syndrome. On the other hand, we want flexibility and maximum decision support.

There are many more improvements possible and the aforementioned points are only a representative few. ABCEM may become better than what even we think is possible. Only the future can reveal its full potential.

6.2 Energy and Waste Accounting

But sustainability is much more than methods and theories. It is just as much a matter of vision and will because we feel that there are several pressing matters to be solved at political level before environmental management can really become the driving force in industry that it should and must be in order for humanity to reach sustainability. As a start, schemes for energy and waste accounting must be investigated and tested.

With energy and waste accounting we envision basically the same type of accounting as today's monetary accounting performed using the very same principles as found for example in Generally Accepted Accounting Principles (GAAP). That is, organizations must file reports of what they used in terms of energy and waste and also send this information further to their customers as a 'price'. So in the future, companies will know exactly how much energy and waste they create and also the energy and waste content of their products and services. Customers can then makes choices based on a product's cost, energy consumption and waste generation. Then, the problem of externalities can be reduced, albeit not eliminated because there are externality costs not related to cost, energy and waste at all, e.g., the social dimension. However, without the political vision of enforcing, at least, energy accounting, companies will be left to their own devices to handle these complex issues alone. Needless to say, that will take time, particularly with heavy vested interest, see (Bradbrook 1994), lack of political will, see e.g. (Meserve 1998), and the massive subsidies of raw material extracting activities, see e.g (Brown, Flavin *et al.* 1999), that both consumers and industries have become so accustomed to.

We hope that large Original Equipment Manufacturers will use their influence for the better to at least promote energy and waste accounting along their supply chain, and not as a threat to suppliers, but as a genuine strategic initiative to facilitate economic and environmental win-win situations. Having energy accounting in place, ideally worldwide, would to some extent reconnect commerce to Nature. Only then do we have a chance to become sustainable.

6.3 Changing the Role of Marketing

Similarly, companies must change the role of marketing. Today, many goods are sold by heavy sales promotion and the goal is to have high sales volume due to economies-of-scale benefits. This is an unsustainable path because the sole purpose of such marketing is to *increase* consumption. In the future, marketing should rather aim towards selling high quality products that last long and are easy to repair (or upgrade). The slogan could be as St. Augustine said:

It is better to need less than to have more.

This is also something we, the customers, must understand and demand, because regardless of methods and tools and rationale knowledge; it is all about change - the rest is supplemental.

Thus, we cannot expect governments to lead only, and according to ABB CEO Percy Barnevik (Gupte 1998), industry is "the strongest force these days in improving the environment". With Activity-Based Cost and Environmental Management - an approach based upon expansions of innovations in cost management - we are convinced that progressive and proactive companies can be motivated and aided towards better environmental management so that they can *do well* by *doing good*. This is perhaps the greatest potential benefit of Activity-Based Cost and Environmental Management.

7. CLOSING WORDS

It is better to light one small candle than to curse the darkness.

Confucius

Sustainability is not about technology; it is not about methods; nor is it about population growth. Mahatma Gandhi expressed the challenge so wisely once:

> *There is enough in the world for everyone's need;*
> *there is not enough for everyone's greed.*

Greed and sustainable development are not combinable, regardless of technology level and population. As long as there is greed, problems are just a matter of time. Moreover, this statement by Mahatma Gandhi contains *three* crucial aspects of humanity: 1) Enough versus not enough, 2) world versus everyone and 3) need versus greed. The problem is that all these three aspects are related and complementary, thus one cannot address one without addressing the others. Sustainability is a way of thinking and acting when making decisions, not a way of *re*acting after wrong decisions have been made. Thus, society and companies alike must promote change in a positive direction and become proactive and agile as opposed to reactive and hostile towards change. If not, there is little hope for a better future. This is particularly true on the personal level because if humanity does not realize how wrong it is to say 'it is not my problem' or 'this is their problem', we are in danger of doing what Chief Seattle (Sealth) warned us against:

> *This we know: All things are connected. Whatever befalls the earth befalls the sons of the earth. Man did not weave the web of life; he is merely a strand in it. Whatever he does to the web, he does to himself...*

This age-old wisdom has followed humanity since it was born. Now, however, in its rebellious adolescence, we seem to have forgotten this simple truth as we worship the new religion of science and technology. Most of all, we must start to use whatever wisdom we have, because, as E.F. Schumacher once expressed so eloquently:

> *Modern civilization can survive only if it begins again to educate the heart, which is the source of wisdom, for modern man is now far too clever to survive without wisdom.*

Thus, of all the challenges that humanity has been through, the quest for sustainability will be the greatest, because in the past challenges were usually overcome by technology. But now, we must overcome ourselves.

REFERENCES

The man who does not read books has no advantage over the man that can not read them.

Mark Twain

Albright, T.L., S. Sharpe and T. Smith (1998). Software for Activity-Based Management. Handbook of Cost Management. Edwards, J.A. Boston, MA, Warren, Gorham & Lamont: pp. C6-1 - C6-31.

Allvine, F. (1996). Marketing: Principles and Practices. Boston, MA, Irwin/McGraw-Hill.

Alting, L. and J. Joergenson (1993). "The Life Cycle Concept as a Basis for Sustainable Industrial Production." Annals of CIRP **42**(1): pp. 163-167.

American Conference of Governmental Industrial Hygienists (1996). Threshold Limit Values and Biological Exposure Indices for 1995 - 1996. Cincinnati, OH, American Conference of Governmental Industrial Hygienists.

Anderson, C.D. (1993). Self-Polishing Antifoulings A Scientific Perspective. Proceedings of Ship Repair & Conversion 93.

Anderson, R.C. (1998). Mid-Course Correction. Atlanta, GA, The Peregrinzilla Press.

Ashley, S. (1993). "Designing for the Environment." Mechanical Engineering **115**(3): pp. 53-55.

Ashby, M.F. (1992). Materials Selection in Mechanical Design. Oxford, Pergamon Press.

Ayres, R.U. (1995). "Life cycle analysis: A critique." Resources, conservation and recycling(14): pp. 199 - 223.

Ayres, R.U., W.H. Schlesinger and R.H. Socolow (1994). Human Impact on the Carbon and Nitrogen Cycles. Industrial Ecology and Global Change. Socolow, R., C. Andrews, F. Berkhout and V. Thomas. Cambridge, UK, University Press.

Baca, A.S. (1993). "Environmentally Conscious Manufacturing: An Integrated Demonstration." International Journal of Environmentally Conscious Design and Manufacturing **2**(4): pp. 19-22.

Baltz, H.B. and R.B. Baltz (1970). Fundamentals of Business Analysis. Englewood Cliffs, NJ, Prentice-Hall, Inc.

Bandemer, H. and W. Näther (1992). Fuzzy Data and Analysis. Dordrecht, The Netherlands, Kluwer Academic Press.

Barlas, Y. (1996). "Formal Aspects of Model Validity and Validation in System Dynamics." System Dynamics Review **12**(No. 3 Fall): pp. 183 - 210.

Barlas, Y. and S. Carpenter (1990). "Philosophical roots of model validation: two paradigms." System Dynamics Review **6**(No.2 Summer): pp. 148 - 166.

Barrington, S. (1994). Harley-Davidson; An Illustrated History. London, Bison Books Ltd.

Beitz, W., M. Suhr and A. Rothe (1992). Recyclingorientierte Waschmachine. Institut für Machinenkonstruktion - Konstruktiontechnik, Technische Universität, Berlin.

Borden, J.P. (1994). "Activity-Based Management Software." Journal of Cost Management for the Manufacturing Industry(Winter): pp. 39 - 49.

Boustead, I. (1996). "LCA - How it Came About, The Beginning in the UK." International Journal of Life Cycle Assessment **1**(3).

Bradbrook, A.J. (1994). "Environmental Aspects of Energy Law - The Role of the Law." Renewable Energy **5, part II**: pp. 1278 - 1292.

Bras, B. (1997). "Incorporating environmental issues in product design and realization." Industry and Environment - UNEP IE **20**(1-2): pp. 7 - 13.

Bras, B. and J. Emblemsvåg (1996). Designing For The Life-Cycle: Activity-Based Costing and Uncertainty. Design for X. Huang, G.Q. London, Chapman & Hall: pp. 398 - 423.

Brassard, M. (1989). The Memory Jogger Plus+, Featuring the Seven Management and Planning Tools. Methuen, MA, GOAL/QPC.

Brinker, B.J., Ed. (1994). Activity-Based Management. Emerging Practices in Cost Management. Boston, MA, Warren, Gorham & Lamont.

Brinker, B.J. (1997). Handbook of Cost Management. Boston, MA, Warren, Gorham & Lamont.

Brooks, P.L., L.J. Davidson and J.H. Palamides (1993). "Environmental compliance: You better know your ABC's." Occupational Hazards(February): pp. 41-46.

Brown, K.K. (1995). "Strategic Performance Measurements." Florida CPA Today(June): pp. 28 - 30.

Brown, L., C. Flavin, H. French, J. Abramovitz, S. Dunn, G. Gardner, A. Mattoon, A.P. McGinn, M. O'Meara, M. Renner, D. Roodman, P. Sampat, L. Starke and J. Tuxill (1999). State of the World 1999. New York, Worldwatch Institute/W.W. Norton & Company.

Brown, L.R. (1992). State of the World 1992. Washington D.C., Worldwatch Institute.

Brown, M.T. and S. Ulgiati (1998). Emergy, Environmental Loading, and Carrying Capacity of Production Systems. Proceedings of the 42nd Annual Conference of The International Society for the Systems Sciences, Atlanta, Georgia, International Society for the Systems Sciences (ISSS).

Burkhardt, H. (1998). Ecological Sustainability Through Alternative Energy. Proceedings of the 42nd Annual Conference of The International Society for the Systems Sciences, Atlanta, Georgia, International Society for the Systems Sciences (ISSS).

CAM-I/CMS, T.C.I.G. (1996). Target Costing; The Next Strategic Frontier, The CAM-I/CMS Guide to Profit and Cost Planning. New York, NY, Irwin Publishers.

Carson, R. (1962). The Silent Spring. Boston, MA, Houghton Mifflin.

Cascio, J., Ed. (1996). The ISO 14000 Handbook. Baltimore, Maryland, CEEM Information Services with ASQ Quality Press.

CEFIC (1996). Use of Tributyltin Compounds in Anti-fouling Paints - Effectiveness of Legislation and Risk Evaluation of Current Levels of Tributyltin Compounds in Coastal Waters, International Maritime Organization Marine Environment Protection Committee (IMO-MEPC).

Christiansen, K., R. Heijungs, T. Rydberg, S.-O. Ryding, L. Sund, H. Wijnen, M. Vold and O.J. Hanssen (1995). Report from Expert Workshop at Hankø, Norway on LCA in Strategic Management, Product Development and Improvement, Marketing and Ecolabelling and Governmental Policies, Østfold Research Foundation.

COESA (1976). U.S. Standard Atmosphere. Washington, DC, U.S. Government Printing Office.

Cokins, G. (1993). "Business Process Reengineering: A Blueprint for Change." Insights(April): pp. 17 - 21.

Cokins, G. (1996). Activity-Based Cost Management - Making it Work. Boston, MA, McGraw-Hill.

Cokins, G., A. Stratton and J. Helbling (1992). An ABC Manager's Primer. New York, McGraw-Hill.

Congress, U.S. (1992). Green Products by Design: Choices for a Cleaner Environment, Office of Technology Assessment.

Consoli, F., D. Allen, I. Boustead and J. Fava (1993). Guidelines for Life-Cycle Assessment: A 'Code of Practice'. The SETAC Workshop, Sesimbra, Portugal, 31 March - 3 April, Society of Environmental Toxicology And Chemistry (SETAC).

Cooper, R. (1989). "The Rise of Activity-Based Costing - Part Three: How Many Cost Drivers Do You Need, and How Do You Select Them." Journal of Cost Management for the Manufacturing Industry(Winter): pp. 34 - 46.

Cooper, R. (1990a). "ABC: A Need, Not an Option." Accountancy(September): pp. 86 - 88.

Cooper, R. (1990b). "Explicating the Logic of ABC." Management Accounting (UK)(November): pp. 58 - 60.

Cooper, R. (1990c). "Five Steps to ABC System Design." Accountancy(November): pp. 78-81.

Cooper, R. (1990d). "Implementing an Activity-Based Cost System." Journal of Cost Management for the Manufacturing Industry(Spring): pp. 33 - 42.

Cooper, R. (1997). Activity-Based Costing: Theory and Practice. Handbook of Cost Management. Brinker, B.J. Boston, MA, Warren, Gorham & Lamont: pp. B1-1 - B1-33.

Cooper, R. and R.S. Kaplan (1987). How Cost Accounting Systematically Distorts Product Costs. Accounting and Management Field Study Perspectives, Harvard Business School Press: pp. 49 - 72.

Cornelissen, R.L. (1997). Thermodynamic and Sustainable Development. Ph.D. Dissertation. Mechanical Engineering. Twente, The Netherlands, University of Twente.

Corson, W.H. (1994). "Changing Course: An Outline of Strategies for a Sustainable Future." Futures **26**(March): pp. 206 - 223.

Coulter, S., B.A. Bras and C. Foley (1995). A Lexicon of Green Engineering Terms. 10th International Conference on Engineering Design (ICED 95), Praha, Heurista, Zürich. pp. 1033 - 1039.

Council Directive 89/677/EEC (1989). Official Journal of the European Communities, No. L 398/19-23. Brussels.

Council Regulation (EEC) No 1836/93 (1993). Allowing Voluntary Participation by Companies in the Industrial Sector in a Community Eco-Management and Audit Scheme. Brussels.

Cunningham, V.L. (1994). "Cost Analysis as a Pollution Prevention Tool." Waste Management **14**: pp. 309 - 315.

D'Amore, R. (1997). Value Cost Improvements. Handbook of Cost Management. Brinker, B.J. Boston, MA, Warren, Gorham & Lamont: pp. F2-1 - F2-19.

Davies, I.M., S.K. Bailey and M.J.C. Harding (1998). "Tributyltin Inputs to the North Sea from Shipping Activities, and Potential Risk of Biological Effects." ICES Journal of Marine Sciences(February).

Decisioneering Inc. (1996). Crystal Ball® Version 4.0 User Manual. Denver, Colorado.

Department of Pesticide Regulations (1996). Copper- and Tributyltin-Containing Pesticides, DPR Regulation No. 96-001.

Derwent, R.G. (1990). Trace gases and their relative contribution to the greenhouse effect. Oxon, Atomic Energy Research Establishment, Harwell.

Devon, R. (1993). Sustainable Technology, Green Design and Engineering Education. International Conference on Technology and Society, Washington, DC, IEEE Service Center, Piscataway, New Jersey. pp. 61-65.

Dillton-Hill, K. and E. Glad (1992). "Activity-Based Costing Empowers Quality Management." Accountancy(June): pp. 164 - 169.

Dodd, A.J. (1997). The Just-in-Time Environment. Handbook of Cost Management. Brinker, B.J. Boston, MA, Warren, Gorham & Lamont: pp. A3-1 - A3-35.

Drucker, P. (1963). "Managing for Business Effectiveness." Harvard Business Review(May-June): pp. 54 - 55.

Drucker, P.F. (1995). "The Information Executives Truly Need." Harvard Business Review(January-February): pp. 54 - 62.

Dubois, D. and H. Prade (1978). "Operations on Fuzzy Numbers." International Journal of Systems Sciences **9**(6): pp. 613 - 626.

Dutton, J.J. and C.A. Marx (1997). Target Costing. Handbook of Cost Management. Brinker, B.J. Boston, MA, Warren, Gorham & Lamont: pp. D2-1 - D2-26.

Edwards, J.A., Ed. (1998). Handbook of Cost Management. Boston, MA, Warren, Gorham & Lamont.

Eiler, R.G. and C. Ball (1997). Implementing Activity-Based Costing. Handbook of Cost Management. Brinker, B.J. Boston, MA, Warren, Gorham & Lamont: pp. B2-1 - B2-33.

Emblemsvåg, J. and B. A. Bras (1994). Activity-Based Costing in Design for Product Retirement. Proceedings 1994 ASME Advances in Design Automation Conference, DE-Vol. 69-2, Minneapolis, American Society of Mechanical Engineers.

Emblemsvåg, J. (1995). Activity-Based Costing in Designing for the Life-Cycle. Master's Thesis. Mechanical Engineering. Atlanta, GA, Georgia Institute of Technology.

Emblemsvåg, J. (1999). Activity-Based Life-Cycle Assessments in Design and Management. Ph.D. Dissertation The George W. Woodruff School of Mechanical Engineering. Atlanta, GA, The Georgia Institute of Technology.

Emblemsvåg, J. and B. Bras (1997a). An Activity-Based Life-Cycle Assessment Method. 1997 ASME Design Engineering Technical Conference, Sacramento, CA, American Society of Mechanical Engineers (ASME).

Emblemsvåg, J. and B. Bras (1997b). "Life-Cycle Design Cost Assessments Using Activity-Based Costing and Uncertainty." Engineering Design and Automation **3**(4): pp. 339-354.

Emblemsvåg, J. and B. Bras (1998a). Energy Accounting - A Step Towards Sustainability. Proceedings of the 42nd Annual Conference of The International Society for the Systems Sciences, Atlanta, GA, International Society for the Systems Sciences (ISSS).

Emblemsvåg, J. and B. Bras (1998b). Financial Analysis, Critical Assumption Planning and Uncertainty in Product Development. 1998 International Conference on Achieving Excellence in New Product Development and Management, Atlanta, GA, Product Development & Management Association (PDMA). pp. 1 - 11.

Emblemsvåg, J. and B. Bras (2000). "Process Thinking - A New Paradigm for Science and Engineering." FUTURES **32**(7): pp. 635 - 654.

Encyclopedia of Climate and Weather (1996). New York, Oxford University Press.

EPA (1992). Total Cost Assessment: Accelerating Industrial Pollution Prevention through Innovative Project Financial Analysis. Washington, DC, US Environmental Protection Agency, Office of Research and Development.

EPA (1993a). Life-Cycle Design Guidance Manual. Cincinnati, OH, US Environmental Protection Agency, Office of Research and Development.

EPA, U.S. (1993b). Life-Cycle Assessment: Inventory Guidelines and Principles. Washington DC, US Environmental Protection Agency, Office of Research and Development.

Evans, S.M. (1995). "Tributyltin Pollution: A Diminishing Problem Following Legislation Limiting the Use of TBT-Based Antifouling Paints." Marine Pollution Bulletin **30**.

Fallon, W.K., Ed. (1983). AMA Management Handbook. New York, AMACOM Special Projects Division, American Management Association, Inc.

Feldman, S.J., P.A. Soyka and P. Ameer (1996). Does Improving A Firm's Environmental Management System And Environmental Performance Result in A Higher Stock Price? Fairfax, VA, ICF Kaiser International, Inc.

Ferdows, K. and A. DeMeyer (1991). "Lasting Improvements in Manufacturing Performance: In Search of a New Theory." The Journal of Operations Management(Fall): pp. 168-184.

Fet, A.M., J. Emblemsvåg and J.T. Johannesen (1996). Environmental Impacts and Activity Based Costing during Operation of a Platform Supply Vessel. Ålesund, Norway, Møreforsking.

Fet, A.M. and G. Oltedal (1994). Renere Produksjon i Verftsindustrien i Møre og Romsdal (Cleaner Production in the Shipyard Industry in Møre og Romsdal). Ålesund, Norway, Møreforsking.

Fiksel, J. (1996). Towards Sustainable Development. Design for Environment – Creating Eco-Efficient Products and Processes. Fiksel, J. New York, McGraw-Hill: pp. 23-33.

Fisher, D.A., C.H. Hales, W.-C. Wang, M.K.W. Ko and N.D. Sze (1990). "Model calculations of the relative effects of CFCs and their replacements on global warming." Nature(344): pp. 513-516.

Fishman, G.S. (1996). Monte Carlo Concepts, Algorithms and Applications. New York, Springer-Verlag.

Fowler, R.J. (1990). International Policy Responses to the Greenhouse Effect and their Implications for Energy Policy in Australia. Greenhouse and Energy. Swaine, D.J. Melbourne, C.S.I.R.O.

Fox, R. (1986). "Cost Accounting: Asset or Liability." Journal of Accounting and EDP(Winter): pp. 31 - 37.

Freeman, H., T. Harten, J. Springer, P. Randall, M.A. Curran and K. Stone (1992). "Industrial Pollution Prevention: A Critical Review." Journal of the Air and Waste Management Association **42**(5): pp. 618-656.

Frosch, R. and N. Gallopoulos (1989). "Strategies for Manufacturing." Scientific American **261**(3): pp. 144-152.

Gates, J.R. (1998). The Ownership Solution. Reading, MA, Addison-Wesley.

GEMI (1994a). Environmental Reporting in a Total Quality Management Framework: A Primer. Washington, DC, Global Environmental Management Initiative.

GEMI (1994b). Finding Cost-Effective Pollution Prevention Initiatives: Incorporating Environmental Costs into Business Decision Making: A Primer. Washington, DC, Global Environmental Management Initiative.

Gleick, J. (1987). Chaos. New York, Penguin.

Gloria, T., T. Saad, M. Breville and M. O'Connell (1995). "Life-Cycle Assessment: A Survey of Current Implementation." Total Quality Environmental Management **4**(No. 3): pp.33-50.

Goldratt, E.M. (1983). Cost Accounting Number One Enemy of Productivity. Proceeding of APICS Conference. pp. 433 - 435.

Graedel, T.E. and B.R. Allenby (1995). Industrial Ecology. New Jersey, Prentice Hall.

Greenwood, T.G. and J.M. Reeve (1992). "Activity Based Cost Management for Continuous Improvement: A Process Design Framework." Journal of Cost Management for the Manufacturing Industry **5**(No. 4 Winter): pp. 22-40.

Gupte, P. (1998). Environment and the CEO. Newsweek. (August 3), pp. 64.

Hamner, B. and C.H. Stinson (1995). "Managerial Accounting and Environmental Compliance Costs." Journal of Cost management for the Manufacturing Industry(Summer): pp. 4 - 10.

Hanssen, O.J. (1998). "Sustainable Product Systems - Experiences Based upon Case Projects in Sustainable Product Development." Journal of Cleaner Production **7**(1): pp. 27 - 41.

Hanssen, O.J., A. Rønning and T. Rydberg (1995). Sustainable Product Development. Methods and Experiences from Case Projects. Final Results from the NEP Project, Østfold Research Foundation.

Hardy, J.W. and E.D. Hubbard (1992). "ABC: Revisiting the Basics." CMA Magazine(November): pp. 24 - 28.

Hawken, P., A. Lovins and H.L. Lovins (1999). Natural Capitalism - Creating the Next Industrial Revolution. Snowmass, Colorado, Rocky Mountain Institute.

Holberton, S. (1997). Energy demand may double by 2020. Financial Times. London. pp. 6.

Howard, P.H., R.S. Boethling, W.F. Jarvis, W.M. Meylan and E.M. Michalenko (1991). Handbook of Environmental Degradation Rates. Boca Raton, FL, CRC Press LLC.

Hronec, S.M. and S.K. Hunt (1997). Quality and Cost Management. Handbook of Cost Management. Edwards, J.B. Boston, MA, Warren, Gorham & Lamont: pp. A1-1 - A1-42.

Huang, G.Q., Ed. (1996). Design for X. London, Chapman & Hall.

Interface (1998). Sustainability Report. Atlanta, GA, Interface Research Corporation.

IPCC (1993). Climate Change; The Scientific Assessment. Cambridge, UK, Cambridge University Press.

Jacobson, C.A. and C.A. Hampel, Eds. (1946 - 1959). Encyclopedia of Chemical Reactions. New York, Reinhold.

Janzen, D.H. (1988). "Tropical Ecological and Biocultural Restoration." Science **239**: pp. 243-244.

Jensen, A.A., J. Elkington, K. Christiansen, L. Hoffmann, B.T. Møller, A. Schmidt and F. van Dijk (1997). Life Cycle Assessment (LCA) - A guide to approaches, experiences and information sources. Søborg, Denmark, dk-TEKNIK Energy & Environment.

Johnson, H.T. (1992). "It's Time to Stop Overselling Activity-Based Concepts." Management Accounting(September).

Johnson, H.T. and R.S. Kaplan (1987). Relevance Lost: The Rise and Fall of Management Accounting. Boston, MA, Harvard Business School Press.

Joshi, D., L. Lemay and C. Perkins (1996). Teach Yourself JAVA in Café in 21 Days. Indianapolis, IN, Sams.net Publishing.

Jump, R.A. (1995). "Implementing ISO 14000: Overcoming Barriers to Registration." Total Quality Environmental Management **5**(No. 1): pp. 9 - 14.

Kaplan, R.S. (1984). "Yesterday's Accounting Undermines Production." Harvard Business Review(July/August): pp. 95 - 101.

Kaplan, R.S. (1990). "The Four-Stage Model of Cost Systems Design." Management Accounting(Feb.): pp. 22 - 26.

Kaplan, R.S. (1992). "In Defense of Activity-Based Cost Management." Management Accounting(November): pp. 58 - 63.

Keegan, D.P., C.R. Jones and R.G. Eiler (1991). "To Implement Your Strategies, Change Your Measures." Price Waterhouse Review(Number 1): pp. 29 - 38.

Kerin, R.A. and R.A. Peterson (1998). Strategic Marketing Problems. Upper Saddle River, NJ, Prentice Hall.

Krajewski, J. and B. Ritzman (1993). Operations Management. Reading, Massachusetts, Addison-Wesley Publishing Company.

Krumwiede, K.R. (1998). "ABC: Why It's Tried and How It Succeeds." Management Accounting (IMA-USA)(April): pp. 32 - 38.

Kaufmann, A. (1983). Advances in Fuzzy Sets - An Overview. Advances in Fuzzy Sets, Possibility Theory, and Applications. Wang, P.P. New York, Plenum Press.

Lang, K.R. (1992). Astrophysical Data: Planets and Stars. New York, Springer-Verlag.

Lashof, D.A. and D.R. Ahuja (1990). "Relative contributions of greenhouse gas emissions to global warming." Nature(344): pp. 529 - 531.

Lee, J.Y. (1987). Managerial Changes for the 90's. New York, Addison-Wesley.

Liebtrau, A.M. and M.J. Scott (1991). "Strategies for Modeling the Uncertain Impacts of Climate Change." Journal of Policy Modeling **13**(2): pp. 185 - 204.

Lindeijer, E. (1996). Part VI: Normalization and Valuation. Towards a Methodology for Life-Cycle Impact Assessment. Udo de Haes, H.A. Brussels, Society of Environmental Toxicology and Chemistry (SETAC) - Europe.

Lovelock, J.E. (1988). The Ages of Gaia: A Biography of Our Living Earth. New York, Norton.

Lovins, A.B. and L.H. Lovins (1997). Climate: Making Sense and Making Money. Old Snowmass, CO, Rocky Mountain Institute.

Mackay, D. and S. Paterson (1991). "Evaluating the Multimedia Fate of Organic Chemicals: A Level III Fugacity Model." Environment, Science and Technology **25**: pp. 427 - 436.

Mackay, D., W.Y. Shiu and K.C. Ma (1992a). Monoaromatic Hydrocarbons, Chlorobenzenes, and PCBs. Boca Raton, FL, CRC Press LLC.

Mackay, D., W.Y. Shiu and K.C. Ma (1992b). Polynuclear Aromatic Hydrocarbons, Polychlorinated Dioxins, and Dibenzofurans. Boca Raton, FL, CRC Press LLC.

Mackay, D., W.Y. Shiu and K.C. Ma (1993). Volatile Organic Chemicals. Boca Raton, FL, CRC Press LLC.

Mackay, D., W.Y. Shiu and K.C. Ma (1995). Oxygen, Nitrogen, and Sulfur Containing Compounds. Boca Raton, FL, CRC Press LLC.

Mackay, D., W.Y. Shiu and K.C. Ma (1997). Pesticide Chemicals. Boca Raton, FL, CRC Press LLC.

Maisel, L.S. and E. Morrissey (1997). Using Activity-Based Costing to Improve Performance. Handbook of Cost Management. Brinker, B.J. Boston, MA, Warren, Gorham & Lamont: pp. B4-1 - B4-26.

Martin, J. (1998). Emergy Analyses of River Diversions within the Mississippi Delta. Proceedings of the 42nd Annual Conference of The International Society for the Systems Sciences, Atlanta, Georgia, International Society for the Systems Sciences (ISSS).

Mates, B. (1972). Elementary Logic. Oxford, Oxford University Press.

Matthiessen, P., R. Waldock, J.E. Thain, M.E. Waite and S. Scrope-Howe (1995). "Changes in periwinkle (Littorina littorea) populations following the ban on TBT-based antifoulings on small boats in the United Kingdom." Ecotoxicology and Environmental Safety **30**: pp. 180 - 194.

McIllhattan, R. (1987). "The Path to Total Cost Management." Journal of Cost Management of the Manufacturing Industry(Summer): pp. 5 - 10.

McNair, C.J. (1997). The Hidden Costs of Capacity. Handbook of Cost Management. Brinker, B.J. Boston, MA, Warren, Gorham & Lamont: pp. E5-1 - E5-27.

Meserve, J. (1998). Environmental Legislation Going Nowhere Fast. Washington. DC, AllPolitics.

Meyer, C. (1994). "How the Right Measures Help Teams Excel." Harvard Business Review(May-June): pp. 95 - 103.

Miles, M.P. and G.R. Russel (1997). "ISO 14000 Total Quality Management: The Integration of Environmental Marketing, Total Quality Management and Corporate Environmental Policy." Journal of Quality Management **2**(No. 1): pp. 155 - 168.

Miller, J.A. (1990). "The Best Way to Implement an Activity-Based Cost Management System." Corporate Controller(September/October): pp. 8 - 13, 32

Miller, J.G. and T.E. Vollmann (1985). "The Hidden Factory." Harvard Business Review(September-October): pp. 142-150.

Mistree, F., W.F. Smith, B. Bras, J.K. Allen and D. Muster (1990). Decision-Based Design: A Contemporary Paradigm for Ship Design. Transactions, Society of Naval Architects and Marine Engineers. Jersey City, New Jersey. **98**: pp. 565-597.

Morrow, M. and G. Ashworth (1994). "An Evolving Framework for Activity-Based Approaches." Management Accounting (UK)(February): pp. 32 - 36.

Mosekilde, E. and R. Feldberg (1994). Nonlinear Dynamics and Chaos. Lyngby, Denmark, Polyteknisk Forlag.

National Academy of Sciences (1996). Material Concerns: Pollution, Profit, and Quality of Life. London, Routledge.

Navin-Chandra, D. (1991). Design for Environmentability. Third International Conference on Design Theory and Methodology, Miami, Florida, American Society of Mechanical Engineers. pp. 119-125.

Naylor, T.J., J.L. Balintfy, D.S. Burdick and K. Chu (1966). Computer Simulation Techniques. New York, John Wiley & Sons.

Odum, H.T. (1996). Environmental Accounting, Emergy and Decision Making. New York, John Wiley.

Odum, H.T. (1998). Energy Hierarchy of the Earth. Proceedings of the 42nd Annual Conference of The International Society for the Systems Sciences, Atlanta, Georgia, International Society for the Systems Sciences (ISSS).

OECD (1993). Environmental Data Compendium 1993. Paris, Organization for Economic Cooperation and Development (OECD).

O'Guin, M. (1990). "Focus The Factory With Activity-Based Costing." Management Accounting(February): pp. 36-41.

O'Guin, M. and S.A. Rebiscke (1997). Customer-Driven Costs Using Activity-Based Costing. Handbook of Cost Management. Brinker, B.J. Boston, MA, Warren, Gorham & Lamont: pp. B5-1 - B5-29.

Olsson, L.E. (1994). "Energy-Meteorology: A new Discipline." Renewable Energy **5 Part II**: pp. 1243 - 1246.

Ostrenga, M.R. (1990). "Activities: The Focal Point of Total Cost Management." Management Accounting(February): pp. 42 - 49.

Owen, J. (1993). "Environmentally Conscious Manufacturing." Manufacturing Engineering(October): pp. 44-55.

Park, C.S. and G.P. Sharp-Bette (1990). Advanced Engineering Economics. New York, NY, John Wiley & Sons.

Player, R.S. (1993). "The Top Ten Things That Can Go Wrong With An ABM Project (And How To Avoid Them)." As Easy As ABC(Summer): pp. 1 - 2.

Population Crisis Committee (1990). Population Pressures: Threat to Democracy. Washington, DC.

Population Crisis Committee (1992a). The International Human Suffering Index. Washington, DC.

Population Crisis Committee (1992b). World Access to Birth Control. Washington, DC.

Population Reference Bureau (1993). 1993 World Population Data Sheet. Washington, DC.

Porter, M.E. (1985). Competitive Advantage: Creating and Sustaining Superior Performance. New York, The Free Press.

PRé Consultants (1995). The Eco-indicator 95. Amersfoort, The Netherlands, Product Ecology Consultants.

Raffish, N. and P.B.B. Turney (1991). The CAM-I Glossary of Activity-Based Management. Arlington, TX, CAM-I.

Reed, R.J., Ed. (1978). North American Combustion Handbook. Cleveland, OH, North American Manufacturing Company.

Ricklefs, R.E. (1990). Ecology. New York, W.H. Freeman and Company.

Robert, K.-H., J. Holmberg and K.-E. Eriksson (1994). Socio-ecological Principles for a Sustainable Society. ISEE meeting, Cost Rica, The Natural Step Environmental Institute Ltd.

Roth, H.P. and A.F. Borthick (1991). "Are You Distorting Costs By Violating ABC Assumptions." Management Accounting(November): pp. 112 - 115.

Rubinstein, R.Y. (1981). Simulation and the Monte Carlo Method. New York, John Wiley & Sons.

Ruhl, J.M. (1998). Applying the Theory of Constraints to Enhance Profitability. Handbook of Cost Management. Edwards, J.A. Boston, MA, Warren, Gorham & Lamont: pp. F5-1-F5-41.

Schneiderman, A.M. (1996). "Metrics for the Order Fulfillment Process (Part 1)." Journal of Cost Management for the Manufacturing Industry(Summer): pp. 30 - 42.

Scow, K.M. and J. Hutson (1992). "Effect of Diffusion and Sorption on the Kinetics of Biodegradation: Theoretical Considerations." Soil Science Society of America Journal **56**(January - February): pp. 119 - 127.

Seehusen, J. (1998). "Ozonforvirring (Ozone Confusion)." Teknisk Ukeblad **145**(23): pp. 6.

Seidelmann, P.K., Ed. (1992). Explanatory Supplement to the Astronomical Almanac. Mill Valley, CA, University Science Books.

Seki, M. and R. Christ (1995). African Regional Workshop on Greenhouse Gas Emissions Inventories and Emission Mitigation Options: Forestry, Land-use Change and Agriculture, UNEP.

Shapiro, A. (1992). We're Number One: Where America Stands and Falls in the New World Order. New York, Vintage Books.

Sharman, P.A. (1991). "Activity-Based Costing: A Practitioner's Update." CMA Magazine(July/August): pp. 22 - 25.

Shields, M.D. (1995). "An Empirical Analysis of Firms' Implementation Experiences with Activity-Based Costing." Journal of Management Accounting Research **7**(Fall): pp. 148 - 166.

Shields, M.D. and S.M. Young (1989). "A Behavioral Model for Implementing Cost Management Systems." Journal of Cost Management for the Manufacturing Industry(Winter): pp. 17 - 27.

Shields, M.D. and S.M. Young (1991). "Managing Product Life Cycle Costs: An Organizational Model." Journal of Cost Management for the Manufacturing Industry **5**(No. 3 Fall): pp. 39-51.

Shine, K.P., R.G. Derwent, D.J. Wuebbles and J.J. Morcrette (1993). Radiative Forcing of Climate. Climate Change; The Scientific Assessment. Houghton, J.T., G.J. Jenkins and J.J. Ephraums. Cambridge, UK, Cambridge University Press: pp. 45 - 67.

Sivard, R. (1993). World Military and Social Expenditures 1993. Washington, DC, World Priorities.

Skinner, W. (1986). "The Productivity Paradox." Harvard Business Review(July-August): pp. 55 - 59.

Sollenberger, H.M. and A. Schneider (1996). Managerial Accounting. Cincinnati, OH, South-Western College Publishing.

Sprigge, T.L.S. (1995). Verification Principle. The Oxford Companion to Philosophy. Honderich, T. New York, Oxford University Press: pp. 898.

Tatikonda, L.U. and R.J. Tatikonda (1991). "Overhead Cost Control - Through Allocation or Elimination?" Production and Inventory Management Journal(1st Quarter): pp. 37 - 41.

Thabit, S.S. and J. Stark (1984). Combined Heat and Power Plant: A Thermoeconomic Analysis. Energy Economics and Management in Industry, Proceedings of the European Congress, Algarve, Portugal, 2-5 April, Pergamon Press. pp. 165 - 170.

The Economist (2000a). Change management - An inside job. The Economist. **356**:(8179), pp. 65.

The Economist (2000b). Changing the climate of opinion. The Economist. **356**:(8183), pp. 67.

The Encyclopædia Britannica (1911). Cambridge, England, Cambridge University Press.

The National Advisory Council for Environmental Policy and Technology (1993). Transforming Environmental Permitting and Compliance Policies to Promote Pollution Prevention: Report and Recommendations of the Technology Innovation and Economics Committee, U.S. Environmental Protection Agency, Washington, DC.

Thingstad, P.G. (1997). Birds as Indicators of Natural and Human Induced Variations in the Environment, with Special Focus on the Suitability of the Pied Flycatcher. Department of Natural History. Trondheim, Norway, Norwegian University of Science and Technology.

Tilley, D.R. (1998). Emergy Basis for Ecosystem Management: Valuing the Work of Nature and Humanity. Proceedings of the 42nd Annual Conference of The International Society for the Systems Sciences, Atlanta, Georgia, International Society for the Systems Sciences.

Tsatsaronis, G., L. Lin and J. Pisa (1993). "Exergy Costing in Exergoeconomics." Journal of Energy Resources Technology **115**(March): pp. 9 - 16.

TU Delft (1996). IDEMAT. Delft, The Netherlands, Technische Universiteit Delft.

Turney, P.B.B. (1990). "Ten Myths that Create Barriers to the Implementation of Activity-Based Cost Systems." Journal of Cost Management for the Manufacturing Industry(Spring): pp. 24 - 32.

Turney, P.B.B. (1991a). Common Cents: The ABC Performance Breakthrough - How To Succeed With Activity-Based Costing. Hillboro, OR, Cost Technology.

Turney, P.B.B. (1991b). "How Activity-Based Costing Helps Reduce Cost." Journal of Cost Management for the Manufacturing Industry **4**(No. 4 Winter): pp. 29-35.

Turney, P.B.B. (1992). "What an Activity-Based Cost Model Looks Like." Journal of Cost Management for the Manufacturing Industry(Winter): pp. 54 - 60.

Turney, P.B.B. and A.J. Stratton (1992). "Using ABC to Support Continuous Improvement." Management Accounting(September): pp. 46 - 50.

Ullring, S. (1996). Praise or Absolution - How Will We Be Judged On Our Environmental Stewardship. The 1996 Annual Congress of the International Union of Marine Insurance - Environment and Insurance, Oslo, Norway, Det Norske Veritas.

UN 'Brundtland' Commission (1987). Our Common Future, Report of the World Commission on Environment and Development. Oxford, Oxford University Press.

Union of Concerned Scientists (1992). World Scientist's Warning to Humanity. Cambridge, MA.

United Nations Environment Programme (1992). Environmental Data Report 1991 - 1992. Cambridge, MA, Basil Blackwell.

United Nations Environment Programme (1993). Human Development Report 1993. New York, Oxford University Press.

Vigon, B. (1997). SETAC Foundation Life-Cycle Assessment Newsletter, Society of Environmental Toxicology and Chemistry (SETAC). **17**:(6).

Waldock, M. (1995). Recovery of Oyster Fisheries Following TBT Regulations. Proceedings of the Malacological Society of London Nov. 1995, London, The Malacological Society of London.

Watson, R.T., H. Rodhe, H. Oeschger and U. Siegenthaler (1993). Greenhouse Gases and Aerosols. Climate Change; The Scientific Assessment. Houghton, J.T., G.J. Jenkins and J.J. Ephraums. Cambridge, UK, Cambridge University Press.

Webster (1983). Webster's New Universal Unabridged Dictionary. New York, New World Dictionaries/Simon and Schuster.

Westnofa Industrier (1997). Environmental Report 1997. Åndalsnes, Norway, Westnofa Industrier AS.

Wigley, T.M.L. and T.P. Barnett (1993). Detection of the Greenhouse Effect in the Observations. Climate Change; The Scientific Assessment. Houghton, J.T., G.J. Jenkins and J.J. Ephraums. Cambridge, UK, Cambridge University Press.

Wilson, E.O. (1988). BioDiversity. Washington DC, National Academy Press.

Winner, R.I., J.P. Pennell, H.E. Bertrand and M.M.G. Slusarczuk (1988). The Role of Concurrent Engineering in Weapons System Acquisition. Alexandra, VA, Institute for Defense Analyses.

Wood, J.C. (1998). Environmental Impacts on Life Cycle Costs. Handbook of Cost Management. Edwards, J.A. Boston, MA, Warren, Gorham & Lamont: pp. D6-1 - D6-30.

World Resources Institute and International Institute for Environment and Development (1992). World Resources 1992 - 93. New York, Basic Books.

WTEC (2000). WTEC Panel Report on Environmentally Benign Manufacturing. Baltimore, MD, Loyola College.

Xia, L. (1994). "A Two-Axis Adjusted Vegetation Index (TWVI)." International Journal of Remote Sensing **15**(No. 7): pp. 1447 - 1458.

Yang, D., J.C.Y. Wang, S. Lin and D. Zhang (1992). Methodology for Exergy-Anergy Costing. ECOS '92 International Symposium on Efficiency, Cost, Optimization and Simulation, Zaragoza, Spain, American Society of Mechanical Engineers, New York. pp. 167 - 172.

Yoshikawa, T., J. Innes and F. Mitchell (1994). "Functional Analysis of Activity-Based Cost Information." Journal of Cost Management for the Manufacturing Industry(Spring): pp. 40 - 48.

Youde, R.K. (1992). "Cost-of-Quality Reporting: How We See It." Management Accounting(Jan): pp. 34 - 38.

Appendix A

WASTE INDEX DATA

I have learned to use the word 'impossible' with the greatest caution.

Wernher von Braun

The main WI formula used in this book is (see Chapter 2):

$$WI = \sum_{i=1}^{M} R^i \cdot \left(\frac{V^i \cdot T_N^i}{2 \cdot A_N^i} \right) \tag{1}$$

Our primary sources for Waste Index data were (Mackay, Shiu *et al.* 1992a; IPCC 1993). The numbers concerning 'Natural Amount' (A_N) and 'Balance Time' (T_N) are found in (IPCC 1993). The three major control volumes of the Earth in Table 1 are obtained from (Lang 1992; Seidelmann 1992), (COESA 1976) and the U.S. 1967 Geological Survey. The control volume (V) values are used to compute the $V.T_N/2A_N$ ratio (in 10^{-12} years/kg, or pico-years/kg) in Table 2 with the given T_N and A_N numbers.

Table 1 - Control Volume Data

Major Control Volume	Size [m^3]	V [-]
Atmosphere	4.1927 E+18	0.758928
Oceans	1.3318 E+18	0.241077
Top Soil	2.9800 E+13	0.000005
Total System Control Volume	**5.5245 E+18**	

Table 2 - Waste Index Data and $V.T_N/2A_N$ Ratio Calculation.

Chemical Compound	T_N [years]	Control Volume	V [-]	A_N [kg]	$(V\,T_N/2A_N)$Ratio [10^{-12} years/kg]
1,2,3-Trimethylbenzene	10.230	Atmosphere	0.76	2.200E+12	1.76E+00
CO_2	120.000	Atmosphere	0.76	2.113E+15	2.15E-02
CO	0.210	Atmosphere	0.76	3.602E+11	2.21E-01
CH_4	10.000	Atmosphere	0.76	2.200E+12	1.72E+00
C_5H_{12}	10.136	Atmosphere	0.76	2.200E+12	1.75E+00
C_7H_{16}	10.190	Atmosphere	0.76	2.200E+12	1.76E+00
N_20	150.000	Atmosphere	0.76	1.482E+12	3.85E+01
NO_x	0.075	Atmosphere	0.76	1.578E+09	1.80E+01
p-Xylene	10.537	Atmosphere	0.76	2.200E+12	1.82E+00
SO_2	0.100	Atmosphere	0.76	6.042E+10	6.28E-01
Tributyltin (TBT)	6.000	Ocean	0.24	4.900E+19	1.48E-08

The data in Table 2 can be (re)used for any resource that releases these chemical compounds into the environment, because the data is based on basic properties of the compounds (degradation time and natural amount).

Note that we have used V and A_N on a global scale, but you can adjust both to local values by taking, say, the volumes and natural amounts in your community as measures for V and A_N. Even more precise values for V can be found if actual dispersion data or models are available. The Waste Index (WI) data used in this book, however, has been gathered with a low budget and hence some simplifications have been made, hence a 'global' V value. For heptane and n-pentane we were also unable to establish the entire chain of reactions. In our opinion, this is work for chemists and not for us. In this book, the Waste Index is primarily used to show that the approach is feasible.

Using the data in Table 2, Waste Index calculations of the waste <u>content</u> of a number of materials can be made as shown in Table 3. All data for the releases R^i were obtained from the IDEMAT 96/98 software/databases. The releases include all releases from raw material extraction through production of the material. As a simplification, only atmospheric releases were included in the waste content calculations, and we suspect that some of the data from IDEMAT 96/98 may be flawed. However, we assumed that as a first cut, this set of data was good enough to explain the principles of our approach.

If you have your own (better) release data, then simply multiply your release (R^i) amount times the data in Table 2 ($V^iT^i/2A_N$) to get the Waste Index value according to the Waste Index equation given in Equation 1. For example, if you know that your LPG releases 0.75 grams of SO_2 instead of 1.10 grams (as per Table 3), then simply multiply 0.75×10^{-3} times 0.628 (from Table 2 - note that Table 2 data is in *kilo*grams) and the Waste Index will be 0.471×10^{-3} pWU (down from 0.6908×10^{-3}). With the new SO_2 release data, the waste content for LPG will now be 0.066675 pWU/kg.

Table 3 - Waste Content of Some Common Materials and Electricity

Production of 1 kg (1000 g)	**Generates Releases R^i of [g]**							**Waste Content [picoWU/kg]**
	CO_2	**CO**	**NO_x**	**N_2O**	**SO_2**	**Methane**	**n-Pentane**	
10SPb20			1.23E+00	9.71E-04	3.87E-01	1.61E+00	1.75E-03	0.025168
28NiCrMo4			1.82E+00	9.62E-04	1.55E+00	1.64E+00	1.71E-03	0.036642
CuZn30	7.01E+03	4.70E+00	2.73E+01		2.84E+01	8.65E-02	6.47E-02	0.661251
Fe360			1.16E+00	9.26E-04	3.39E-01	1.53E+00	1.65E-03	0.023686
Zinc			1.75E+00	1.59E-03	5.62E+00	2.26E+00	7.97E-03	0.039037
X10CrNiS 18 9			7.80E+00	5.56E-04	1.29E+01	1.54E+00	1.69E-03	0.151127
Electricity	1.76E+02	9.24E-03	3.35E-01		2.22E-01	2.08E-02		0.009993
E-glass Fibre		8.34E-02	3.02E+00		2.28E+00	1.67E-02		0.055913
Glare 1 3/2-0.3			1.52E+01	2.50E-03		1.12E+01	2.22E-03	0.293229
Hylite 2/1-0.2		1.33E-07	1.24E-03	1.74E-01		2.23E+01		0.045156
Concrete, Plain			1.04E-01	1.22E-04	6.43E-02	3.52E-02	9.89E-05	0.001981
Concrete, Steel			1.50E-01	1.46E-04	7.77E-02	9.55E-02	1.65E-04	0.002923
Coal	5.79E+02	1.39E-02	4.95E-01		1.05E+00	4.90E-01		0.022863
Crude Oil	1.80E+02	7.00E-02	2.20E+00		9.00E-02	1.70E+00		0.046466
Diesel	2.84E+02	8.00E-02	2.90E+00		1.80E+00	2.90E+00		0.064442
LPG (Propane)	2.70E+02	8.60E-02	3.07E+00		1.10E+00	3.02E+00		0.066969
Natural Gas	2.09E+00	5.04E-03	2.01E-04			1.32E-03		0.000052
Petrol	2.70E+02	8.60E-02	3.07E+00		1.10E+00	3.02E+00		0.066969
Leather	2.38E+03	2.47E-01	3.56E+00		1.79E+00	1.21E+02		0.324018
Nitrile Rubber			1.08E+01		9.14E+00	1.50E+01		0.226590
HDPE	9.40E+02	6.00E-01	1.00E+01		6.00E+00	2.10E+01		0.240231
PE	1.14E+03	9.00E-01	1.22E+01		9.01E+00	1.41E+02		0.493269
PVC, hard	1.75E+03	2.50E+00	1.50E+01		1.30E+01	1.90E+01		0.348957
PVdC	3.55E+03	8.60E+00	3.30E+01		4.90E+01	3.30E+01		0.759758
Cotton			7.39E-01	8.15E-01	7.02E-02	1.03E+00		0.046491
Polyester	1.63E+04	4.59E+01	5.36E+01		7.06E+01	5.67E+01		1.467814
Pitch Pine				3.20E-03	5.98E-02	9.77E-01	8.23E-03	0.001856

INDEX

As a rule, men worry more about what they can't see than about what they can.

Julius Caesar